Mathematical Lectures from Peking University

Mathematical Lectures from Peking University includes monographs, lecture notes, and proceedings based on research pursued and events held at Beijing International Center for Mathematical Research (BICMR). BICMR is a mathematical research institute sponsored by the national government of China. The center was created in 2005 by national government decree. Its goal is to build a world-class mathematical center for research and training young talents; to foster a new generation of leading world-class mathematicians in China; to support the application of mathematics in sciences and practical fields; to promote research and improve the level of mathematics education at Peking University, as well as all over China.

More information about this series at http://www.springer.com/series/11574

Ying Jiao
Editor

From Probability to Finance

Lecture Notes of BICMR Summer School on Financial Mathematics

Springer

Editor
Ying Jiao
Institut de Science Financière et
d'Assurances
Université Claude Bernard Lyon 1
Lyon, France

ISSN 2197-4209 ISSN 2197-4217 (electronic)
Mathematical Lectures from Peking University
ISBN 978-981-15-1575-0 ISBN 978-981-15-1576-7 (eBook)
https://doi.org/10.1007/978-981-15-1576-7

Mathematics Subject Classification (2010): 91Gxx, 60Gxx, 60Hxx

This Springer imprint is published by the registered company Springer Nature Singapore Pte Ltd.
The registered company address is: 152 Beach Road, #21-01/04 Gateway East, Singapore 189721, Singapore

Preface

Since the last 70s, financial mathematics has experienced increasingly rapid and deep development where various theories and ideas from different fields of mathematics confront and innovate. The probability theory plays in particular a fundamental role. The BICMR Summer School on Financial Mathematics was held at Beijing International Center for Mathematical Research, Peking University during 29 May–9 June 2017. The summer school has welcomed young students and researchers from different backgrounds, and its aim was to cover both theoretical probability themes and the financial applications in order to provide their interactive perspective to the audience. Based on this principle, this volume of lecture notes contains five chapters, corresponding, respectively, to five mini-courses, which include two lectures on fundamental probability topics, one on continuous-state branching processes and the other on enlargement of filtrations. Both theories have been initially introduced during the 70s but have witnessed rich development and interesting financial applications in diverse directions recently. The following three chapters are motivated by more concrete and practical problems from financial markets and some of them address relevant applications related to the theoretical chapters.

In the first chapter, Zenghu Li provides a comprehensive introduction to the continuous-state branching processes. In particular, the two cases without (CB process) and with immigration (CBI process) are distinguished and detailed. Different aspects of this family of processes are explained such as the construction, the martingale problems and the related stochastic differential equations.

The second chapter contains a discrete-time-based enlargement of filtration theory by Christophette Blanchet-Scalliet and Monique Jeanblanc. Both initial and progressive enlargements are presented. In the progressive enlargement, special attentions are accorded to situations before and after a random time, respectively. The chapter is concluded by a brief application to credit risk.

Guillaume Bernis and Simone Scotti study the Hawkes processes for the phenomenon of jump clusters in finance. They revisit standard Hawkes processes and further generalizations, before focusing on financial applications such as modelling and calibration problems related to the self-exciting property. The last part consists of a natural extension of Hawkes processes in the framework of CBI processes, in link with the first chapter.

Jingping Yang, Fang Wang and Zongkai Xie review the Bernstein and composite Bernstein copulas, which can be applied to approximate an arbitrary copula function. Probability properties, examples and simulation methods are presented with a numerical study by using market data to complete the chapter.

In the framework of valuation adjustment metrics (XVA), Claudio Albanese, Marc Chataigner and Stéphane Crépey propose a balance sheet approach from the perspective of wealth conservation and transfer. They introduce the pricing principle and explain how XVAs aim to represent a switch of paradigm in derivative management, from hedging to balance sheet optimization in view of credit and counterparty risks.

We are grateful to Beijing International Center for Mathematical Research for financial and administrative support and in particular to its staff for their constant help during the summer school. The summer school has also benefited from funding from National Natural Science Foundation of China and Institut Europlace de Finance of France. We thank warmly and sincerely all lecturers of the summer school and authors of this lecture note for their precious contributions, and also the referees for their careful reading and their constructive comments and suggestions. We also thank the Springer editors Ramon Peng and Daniel Wang who have accompanied this volume during its preparation.

Lyon, France Ying Jiao

Contents

Continuous-State Branching Processes with Immigration

Zenghu Li

Abstract This work provides a brief introduction to continuous-state branching processes with or without immigration. The processes are constructed by taking rescaling limits of classical discrete-state branching models. We give quick developments of the martingale problems and stochastic equations of the continuous-state processes. The proofs here are more elementary than those appearing in the literature before. We have made them readable without requiring too much preliminary knowledge on branching processes and stochastic analysis. Using the stochastic equations, we give characterizations of the local and global maximal jumps of the processes. Under suitable conditions, their strong Feller property and exponential ergodicity are studied by a coupling method based on one of the stochastic equations.

Keywords Continuous-state branching process · Immigration · Rescaling limit · Martingale problem · Stochastic equation · Strong Feller property · Exponential ergodicity

Mathematics Subject Classification (2010) 60J80 · 60J85 · 60H10 · 60H20

Introduction

Continuous-state branching processes (CB-processes) and continuous-state branching processes with immigration (CBI-processes) constitute important classes of Markov processes taking values in the positive (=nonnegative) half line. They were introduced as probabilistic models describing the evolution of large populations with small individuals. The study of CB-processes was initiated by Feller [18], who noticed that a diffusion process may arise in a limit theorem of Galton–Watson discrete branching processes; see also Aliev and Shchurenkov [2], Grimvall [22], and Lamperti [34]. A characterization of CB-processes by random time changes of

Z. Li (✉)
School of Mathematical Sciences, Beijing Normal University, Beijing 100875, China
e-mail: lizh@bnu.edu.cn
URL: http://math0.bnu.edu.cn/~lizh/

© Springer Nature Singapore Pte Ltd. 2020
Y. Jiao (ed.), *From Probability to Finance*, Mathematical Lectures
from Peking University, https://doi.org/10.1007/978-981-15-1576-7_1

Lévy processes was given by Lamperti [35]. The convergence of rescaled discrete branching processes with immigration to CBI-processes was studied in Aliev [1], Kawazu and Watanabe [29] and Li [39]. From a mathematical point of view, the continuous-state processes are usually easier to deal with because both their time and state spaces are smooth, and the distributions that appear are infinitely divisible. For general treatments and backgrounds of CB- and CBI-processes, the reader may refer to Kyprianou [31] and Li [40]. In the recent work of Pardoux [46], more complicated probabilistic population models involving competition were studied, which extend the stochastic logistic growth model of Lambert [32].

A continuous CBI-process with subcritical branching mechanism was used by Cox et al. [8] to describe the evolution of interest rates and has been known in mathematical finance as the *Cox–Ingersoll–Ross model* (CIR-model). Compared with other financial models introduced before, the CIR-model is more appealing as it is positive and mean-reverting. The asymptotic behavior of the estimators of the parameters in this model was studied by Overbeck and Rydén [45]; see also Li and Ma [43]. Applications of stochastic calculus to finance including those of the CIR-model were discussed systematically in Lamberton and Lapeyre [33]. A natural generalization of the CBI-process is the so-called affine Markov process, which has also been used a lot in mathematical finance; see, e.g., Duffie et al. [13] and the references therein.

A strong stochastic equation for general CBI-processes was first established in Dawson and Li [10]. A flow of discontinuous CB-processes was constructed in Bertoin and Le Gall [5] by weak solutions to a stochastic equation. Their results were extended to flows of CBI-processes in Dawson and Li [11] using strong solutions; see also Li [41] and Li and Ma [42]. For the stable branching CBI-process, a strong stochastic differential equation driven by Lévy processes was established in Fu and Li [19]. The approach of stochastic equations has played an important role in recent developments of the theory and applications of CB- and CBI-processes.

The purpose of these notes is to provide a brief introduction to CB- and CBI-processes accessible to graduate students with reasonable background in probability theory and stochastic processes. In particular, we give a quick development of the stochastic equations of the processes and some immediate applications. The proofs given here are more elementary than those appearing in the literature before. We have made them readable without requiring too much preliminary knowledge on branching processes and stochastic analysis.

In Sect. 1, we review some properties of Laplace transforms of finite measures on the positive half line. In Sect. 2, a construction of CB-processes is given as rescaling limits of Galton–Watson branching processes. This approach also gives the physical interpretation of the CB-processes. Some basic properties of the processes are developed in Sect. 3. The Laplace transforms of some positive integral functionals are calculated explicitly in Sect. 4. In Sect. 5, the CBI-processes are constructed as rescaling limits of Galton–Watson branching processes with immigration. In Sect. 6, we present reconstructions of the CB- and CBI-processes by Poisson random measures determined by entrance laws, which reveal the structures of the trajectories of the processes. Several equivalent formulations of martingale problems for CBI-processes are given in Sect. 7. From those we derive the stochastic equations of the processes in Sect. 8. Using the stochastic equations, some characterizations of local and global maximal jumps of the CB- and CBI-processes are given in Sect. 9. In Sect. 10, we prove the strong Feller property and the exponential ergodicity of the CBI-process under suitable conditions using a coupling based on one of the stochastic equations.

These lecture notes originated from graduate courses I gave at Beijing Normal University in the past years. They were also used for mini courses at Peking University in 2017 and at the University of Verona in 2018. I would like to thank Professors Ying Jiao and Simone Scotti, who invited me to give the mini courses. I am grateful to the participants of all those courses for their helpful comments. I would also like to thank NSFC for the financial supports. I am indebted to the Laboratory of Mathematics and Complex Systems (Ministry of Education) for providing me the research facilities.

1 Laplace Transforms of Measures

Let $\mathscr{B}[0,\infty)$ be the Borel σ-algebra on the positive half line $[0,\infty)$. Let $B[0,\infty) = \mathrm{b}\mathscr{B}[0,\infty)$ be the set of bounded Borel functions on $[0,\infty)$. Given a finite measure μ on $[0,\infty)$, we define the *Laplace transform* L_μ of μ by

$$L_\mu(\lambda) = \int_{[0,\infty)} \mathrm{e}^{-\lambda x}\mu(\mathrm{d}x), \qquad \lambda \ge 0. \tag{1}$$

Theorem 1.1 *A finite measure on* $[0,\infty)$ *is uniquely determined by its Laplace transform.*

Proof Suppose that μ_1 and μ_2 are finite measures on $[0,\infty)$ and $L_{\mu_1}(\lambda) = L_{\mu_2}(\lambda)$ for all $\lambda \ge 0$. Let $\mathscr{K} = \{x \mapsto \mathrm{e}^{-\lambda x} : \lambda \ge 0\}$ and let $\mathscr{L}$ be the class of functions $F \in B[0,\infty)$ so that

$$\int_{[0,\infty)} F(x)\mu_1(\mathrm{d}x) = \int_{[0,\infty)} F(x)\mu_2(\mathrm{d}x).$$

Then $\mathscr{K}$ is closed under multiplication and $\mathscr{L}$ is a monotone vector space containing $\mathscr{K}$. It is easy to see $\sigma(\mathscr{K}) = \mathscr{B}[0,\infty)$. Then the monotone class theorem implies $\mathscr{L} \supset \mathrm{b}\sigma(\mathscr{K}) = B[0,\infty)$. That proves the desired result.

Theorem 1.2 *Let* $\{\mu_n\}$ *be a sequence of finite measures on* $[0,\infty)$ *and* $\lambda \mapsto L(\lambda)$ *a continuous function on* $[0,\infty)$. *If* $\lim_{n\to\infty} L_{\mu_n}(\lambda) = L(\lambda)$ *for every* $\lambda \ge 0$, *then there is a finite measure* μ *on* $[0,\infty)$ *such that* $L_\mu = L$ *and* $\lim_{n\to\infty}\mu_n = \mu$ *by weak convergence.*

Proof We can regard each μ_n as a finite measure on $[0, \infty]$, the one-point compactification of $[0, \infty)$. Let F_n denote the distribution function of μ_n. By Helly's theorem we infer that $\{F_n\}$ contains a subsequence $\{F_{n_k}\}$ that converges weakly on $[0, \infty]$ to some distribution function F. Then the corresponding subsequence $\{\mu_{n_k}\}$ converges weakly on $[0, \infty]$ to the finite measure μ determined by F. It follows that

$$\mu[0, \infty] = \lim_{k\to\infty} \mu_{n_k}[0, \infty] = \lim_{k\to\infty} \mu_{n_k}[0, \infty) = \lim_{k\to\infty} L_{\mu_{n_k}}(0) = L(0). \tag{2}$$

Moreover, for $\lambda > 0$ we have

$$\begin{aligned}\int_{[0,\infty]} e^{-\lambda x}\mu(dx) &= \lim_{k\to\infty}\int_{[0,\infty]} e^{-\lambda x}\mu_{n_k}(dx)\\ &= \lim_{k\to\infty}\int_{[0,\infty)} e^{-\lambda x}\mu_{n_k}(dx) = L(\lambda),\end{aligned} \tag{3}$$

where $e^{-\lambda\cdot\infty} = 0$ by convention. By letting $\lambda \to 0+$ in (3) and using the continuity of L at $\lambda = 0$ we find $\mu[0, \infty) = L(0)$. From this and (2) we see μ is supported by $[0, \infty)$. By Theorem 1.7 of Li [40, p. 4] we have $\lim_{n\to\infty} \mu_{n_k} = \mu$ weakly on $[0, \infty)$. It follows that, for $\lambda \geq 0$,

$$\int_{[0,\infty)} e^{-\lambda x}\mu(dx) = \lim_{k\to\infty}\int_{[0,\infty)} e^{-\lambda x}\mu_{n_k}(dx) = L(\lambda).$$

Then $L_\mu = L$. If μ_n does not converge weakly to μ, then F_n does not converge weakly to F, so there is a subsequence $\{F_{n'_k}\} \subset \{F_n\}$ that converges weakly to a limit $G \neq F$. The above arguments show that G corresponds to a finite measure on $[0, \infty)$ with Laplace transform $L = L_\mu$, yielding a contradiction. Then $\lim_{n\to\infty} \mu_n = \mu$ weakly on $[0, \infty)$.

Corollary 1.3 *Let $\mu_1, \mu_2, \ldots$ and μ be finite measures on $[0, \infty)$. Then $\mu_n \to \mu$ weakly if and only if $L_{\mu_n}(\lambda) \to L_\mu(\lambda)$ for every $\lambda \geq 0$.*

Proof If $\mu_n \to \mu$ weakly, we have $\lim_{n\to\infty} L_{\mu_n}(\lambda) = L_\mu(\lambda)$ for every $\lambda \geq 0$. The converse assertion is a consequence of Theorem 1.2.

Given two probability measures μ_1 and μ_2 on $[0, \infty)$, we denote by $\mu_1 \times \mu_2$ their product measure on $[0, \infty)^2$. The image of $\mu_1 \times \mu_2$ under the mapping $(x_1, x_2) \mapsto x_1 + x_2$ is called the *convolution* of μ_1 and μ_2 and is denoted by $\mu_1 * \mu_2$, which is a probability measure on $[0, \infty)$. According to the definition, for any $F \in B[0, \infty)$ we have

$$\int_{[0,\infty)} F(x)(\mu_1 * \mu_2)(dx) = \int_{[0,\infty)} \mu_1(dx_1)\int_{[0,\infty)} F(x_1 + x_2)\mu_2(dx_2). \tag{4}$$

Clearly, if ξ_1 and ξ_2 are independent random variables with distributions μ_1 and μ_2 on $[0, \infty)$, respectively, then the random variable $\xi_1 + \xi_2$ has distribution $\mu_1 * \mu_2$.

It is easy to show that

$$L_{\mu_1 * \mu_2}(\lambda) = L_{\mu_1}(\lambda) L_{\mu_2}(\lambda), \qquad \lambda \geq 0. \tag{5}$$

Let $\mu^{*0} = \delta_0$ and define $\mu^{*n} = \mu^{*(n-1)} * \mu$ inductively for integers $n \geq 1$.

We say a probability distribution μ on $[0, \infty)$ is *infinitely divisible* if for each integer $n \geq 1$, there is a probability μ_n such that $\mu = \mu_n^{*n}$. In this case, we call μ_n the *n-th root* of μ. A positive random variable ξ is said to be *infinitely divisible* if it has infinitely divisible distribution on $[0, \infty)$. Write $\psi \in \mathscr{I}$ if $\lambda \mapsto \psi(\lambda)$ is a positive function on $[0, \infty)$ with the *Lévy–Khintchine representation*:

$$\psi(\lambda) = h\lambda + \int_{(0,\infty)} (1 - \mathrm{e}^{-\lambda u}) l(\mathrm{d}u), \tag{6}$$

where $h \geq 0$ and $l(\mathrm{d}u)$ is a σ-finite measure on $(0, \infty)$ satisfying

$$\int_{(0,\infty)} (1 \wedge u) l(\mathrm{d}u) < \infty.$$

The relation $\psi = -\log L_\mu$ establishes a one-to-one correspondence between the functions $\psi \in \mathscr{I}$ and infinitely divisible probability measures μ on $[0, \infty)$; see, e.g., Theorem 1.39 in Li [40, p. 20].

2 Construction of CB-Processes

Let $\{p(j) : j \in \mathbb{N}\}$ be a probability distribution on the space of positive integers $\mathbb{N} := \{0, 1, 2, \ldots\}$. It is well known that $\{p(j) : j \in \mathbb{N}\}$ is uniquely determined by its generating function g defined by

$$g(z) = \sum_{j=0}^{\infty} p(j) z^j, \qquad |z| \leq 1.$$

Suppose that $\{\xi_{n,i} : n, i = 1, 2, \ldots\}$ is a family of $\mathbb{N}$-valued i.i.d. random variables with distribution $\{p(j) : j \in \mathbb{N}\}$. Given an $\mathbb{N}$-valued random variable $x(0)$ independent of $\{\xi_{n,i}\}$, we define inductively

$$x(n) = \sum_{i=1}^{x(n-1)} \xi_{n,i}, \qquad n = 1, 2, \ldots. \tag{7}$$

Here we understand $\sum_{i=1}^{0} = 0$. For $i \in \mathbb{N}$ let $\{Q(i, j) : j \in \mathbb{N}\}$ denote the i-fold convolution of $\{p(j) : j \in \mathbb{N}\}$, that is, $Q(i, j) = p^{*i}(j)$ for $i, j \in \mathbb{N}$. For any $n \geq 1$

and $\{i_0, i_1, \cdots, i_{n-1} = i, j\} \subset \mathbb{N}$ it is easy to see that

$$\begin{aligned}\mathbf{P}\Big(x(n) = j \Big| x(0) = i_0, x(1) = i_1, \cdots, x(n-1) = i_{n-1}\Big) \\ = \mathbf{P}\Big(\sum_{i=1}^{x(n-1)} \xi_{n,i} = j \Big| x(n-1) = i_{n-1}\Big) \\ = \mathbf{P}\Big(\sum_{k=1}^{i} \xi_{n,k} = j\Big) = Q(i, j).\end{aligned}$$

Then $\{x(n) : n \geq 0\}$ is an $\mathbb{N}$-valued Markov chain with one-step transition matrix $Q = (Q(i, j) : i, j \in \mathbb{N})$. The random variable $x(n)$ can be thought of as the number of individuals in generation n of an evolving population system. After one unit time, each individual in the population splits independently of others into a random number of offspring according to the distribution $\{p(j) : j \in \mathbb{N}\}$. Clearly, we have, for $i \in \mathbb{N}$ and $|z| \leq 1$,

$$\sum_{j=0}^{\infty} Q(i, j)z^j = \sum_{j=0}^{\infty} p^{*i}(j)z^j = g(z)^i. \tag{8}$$

Clearly, the transition matrix Q satisfies the *branching property*:

$$Q(i_1 + i_2, \cdot) = Q(i_1, \cdot) * Q(i_2, \cdot), \qquad i_1, i_2 \in \mathbb{N}. \tag{9}$$

This means that different individuals in the population propagate independently each other.

A Markov chain in $\mathbb{N}$ with one-step transition matrix defined by (8) is called a *Galton–Watson branching process* (GW-process) or a *Bienaymé–Galton–Watson branching process* (BGW-process) with *branching distribution* given by g; see, e.g., Athreya and Ney [3] and Harris [23]. The study of the model goes back to Bienaymé [6] and Galton and Watson [20].

By a general result in the theory of Markov chains, for any $n \geq 1$ the n-step transition matrix of the GW-process is just the n-fold product matrix $Q^n = (Q^n(i, j) : i, j \in \mathbb{N})$.

Proposition 2.1 *For any $n \geq 1$ and $i \in \mathbb{N}$ we have*

$$\sum_{j=0}^{\infty} Q^n(i, j)z^j = g^{\circ n}(z)^i, \qquad |z| \leq 1, \tag{10}$$

where $g^{\circ n}(z)$ is defined by $g^{\circ n}(z) = g \circ g^{\circ(n-1)}(z) = g(g^{\circ(n-1)}(z))$ successively with $g^{\circ 0}(z) = z$ by convention.

Proof From (8) we know (10) holds for $n = 1$. Now suppose that (10) holds for some $n \geq 1$. We have

$$\sum_{j=0}^{\infty} Q^{n+1}(i, j)z^j = \sum_{j=0}^{\infty}\sum_{k=0}^{\infty} Q(i, k)Q^n(k, j)z^j$$
$$= \sum_{k=0}^{\infty} Q(i, k)g^{\circ n}(z)^k = g^{\circ(n+1)}(z)^i.$$

Then (10) also holds when n is replaced by $n + 1$. That proves the result by induction.

It is easy to see that zero is a trap for the GW-process. If $g'(1-) < \infty$, by differentiating both sides of (10) we see the first moment of the distribution $\{Q^n(i, j) : j \in \mathbb{N}\}$ is given by

$$\sum_{j=1}^{\infty} jQ^n(i, j) = ig'(1-)^n. \tag{11}$$

Example 2.2 Given a GW-process $\{x(n) : n \geq 0\}$, we can define its *extinction time* $\tau_0 = \inf\{n \geq 0 : x(n) = 0\}$. In view of (7), we have $x(n) = 0$ on the event $\{n \geq \tau_0\}$. Let $q = \mathbf{P}(\tau_0 < \infty|x(0) = 1)$ be the *extinction probability*. By the independence of the propagation of different individuals we have $\mathbf{P}(\tau_0 < \infty|x(0) = i) = q^i$ for any $i = 0, 1, 2, \ldots$. By the total probability formula,

$$q = \sum_{j=0}^{\infty} \mathbf{P}(x(1) = j|x(0) = 1)\mathbf{P}(\tau_0 < \infty|x(0) = 1, x(1) = j)$$
$$= \sum_{j=0}^{\infty} \mathbf{P}(\xi_{1,1} = j)\mathbf{P}(\tau_0 < \infty|x(1) = j) = \sum_{j=0}^{\infty} p(j)q^j = g(q).$$

Then the extinction probability q is a solution to the equation $z = g(z)$ on $[0, 1]$. Clearly, in the case of $p(1) < 1$ we have $q = 1$ if and only if $\sum_{j=1}^{\infty} jp(j) \leq 1$.

Now suppose we have a sequence of GW-processes $\{x_k(n) : n \geq 0\}$, $k = 1, 2, \ldots$ with branching distributions given by the probability generating functions g_k, $k = 1, 2, \ldots$. Let $z_k(n) = k^{-1}x_k(n)$. Then $\{z_k(n) : n \geq 0\}$ is a Markov chain with state space $E_k := \{0, k^{-1}, 2k^{-1}, \ldots\}$ and n-step transition probability $Q_k^n(x, \mathrm{d}y)$ determined by

$$\int_{E_k} \mathrm{e}^{-\lambda y} Q_k^n(x, \mathrm{d}y) = g_k^{\circ n}(\mathrm{e}^{-\lambda/k})^{kx}, \qquad \lambda \geq 0. \tag{12}$$

Suppose that $\{\gamma_k\}$ is a positive sequence so that $\gamma_k \to \infty$ increasingly as $k \to \infty$. Let $\lfloor \gamma_k t \rfloor$ denote the integer part of $\gamma_k t$. Clearly, given $z_k(0) = x \in E_k$, for any $t \geq 0$ the random variable $z_k(\lfloor \gamma_k t \rfloor) = k^{-1}x_k(\lfloor \gamma_k t \rfloor)$ has distribution $Q_k^{\lfloor \gamma_k t \rfloor}(x, \cdot)$ on E_k determined by

$$\int_{E_k} \mathrm{e}^{-\lambda y} Q_k^{\lfloor \gamma_k t \rfloor}(x, \mathrm{d}y) = \exp\{-x v_k(t, \lambda)\}, \tag{13}$$

where

$$v_k(t, \lambda) = -k \log g_k^{\circ \lfloor \gamma_k t \rfloor}(\mathrm{e}^{-\lambda/k}). \tag{14}$$

We are interested in the asymptotic behavior of the sequence of continuous time processes $\{z_k(\lfloor \gamma_k t \rfloor) : t \geq 0\}$ as $k \to \infty$. By (14), for $\gamma_k^{-1}(i-1) \leq t < \gamma_k^{-1} i$ we have

$$v_k(t, \lambda) = v_k(\gamma_k^{-1} \lfloor \gamma_k t \rfloor, \lambda) = v_k(\gamma_k^{-1}(i-1), \lambda).$$

It follows that

$$\begin{aligned}
v_k(t, \lambda) =& v_k(0, \lambda) + \sum_{j=1}^{\lfloor \gamma_k t \rfloor} [v_k(\gamma_k^{-1} j, \lambda) - v_k(\gamma_k^{-1}(j-1), \lambda)] \\
=& \lambda - k \sum_{j=1}^{\lfloor \gamma_k t \rfloor} [\log g_k^{\circ j}(\mathrm{e}^{-\lambda/k}) - \log g_k^{\circ (j-1)}(\mathrm{e}^{-\lambda/k})] \\
=& \lambda - k \sum_{j=1}^{\lfloor \gamma_k t \rfloor} \log \big[g_k(g_k^{\circ (j-1)}(\mathrm{e}^{-\lambda/k})) g_k^{\circ (j-1)}(\mathrm{e}^{-\lambda/k})^{-1} \big] \\
=& \lambda - \gamma_k^{-1} \sum_{j=1}^{\lfloor \gamma_k t \rfloor} \bar{\phi}_k(-k \log g_k^{\circ (j-1)}(\mathrm{e}^{-\lambda/k})) \\
=& \lambda - \gamma_k^{-1} \sum_{j=1}^{\lfloor \gamma_k t \rfloor} \bar{\phi}_k(v_k(\gamma_k^{-1}(j-1), \lambda)) \\
=& \lambda - \int_0^{\gamma_k^{-1} \lfloor \gamma_k t \rfloor} \bar{\phi}_k(v_k(s, \lambda)) \mathrm{d}s,
\end{aligned} \tag{15}$$

where

$$\bar{\phi}_k(z) = k\gamma_k \log \big[g_k(\mathrm{e}^{-z/k}) \mathrm{e}^{z/k} \big], \qquad z \geq 0. \tag{16}$$

It is easy to see that

$$\bar{\phi}_k(z) = k\gamma_k \log \big[1 + (k\gamma_k)^{-1} \tilde{\phi}_k(z) \mathrm{e}^{z/k} \big], \tag{17}$$

where

$$\tilde{\phi}_k(z) = k\gamma_k [g_k(\mathrm{e}^{-z/k}) - \mathrm{e}^{-z/k}]. \tag{18}$$

The sequence $\{\tilde{\phi}_k\}$ is sometimes easier to handle than the original sequence $\{\bar{\phi}_k\}$. The following lemma shows that the two sequences are really not very different.

Lemma 2.3 *Suppose that either* $\{\bar{\phi}_k\}$ *or* $\{\tilde{\phi}_k\}$ *is uniformly bounded on each bounded interval. Then, we have* (i) $\lim_{k\to\infty} |\bar{\phi}_k(z) - \tilde{\phi}_k(z)| = 0$ *uniformly on each bounded interval;* (ii) $\{\bar{\phi}_k\}$ *is uniformly Lipschitz on each bounded interval if and only if so is* $\{\tilde{\phi}_k\}$.

Proof The first assertion follows immediately from (17). By the same relation, we have

$$\bar{\phi}_k'(z) = \frac{[\tilde{\phi}_k'(z) + k^{-1}\tilde{\phi}_k(z)]\mathrm{e}^{z/k}}{1 + (k\gamma_k)^{-1}\tilde{\phi}_k(z)\mathrm{e}^{z/k}}, \qquad z \geq 0.$$

Then $\{\bar{\phi}_k'\}$ is uniformly bounded on each bounded interval if and only if so is $\{\tilde{\phi}_k'\}$. That gives the second assertion.

By the above lemma, if either $\{\tilde{\phi}_k\}$ or $\{\bar{\phi}_k\}$ is uniformly Lipschitz on each bounded interval, then they converge or diverge simultaneously and in the convergent case they have the same limit. For the convenience of statement of the results, we formulate the following condition.

Condition 2.4 The sequence $\{\tilde{\phi}_k\}$ is uniformly Lipschitz on $[0, a]$ for every $a \geq 0$ and there is a function ϕ on $[0, \infty)$ so that $\tilde{\phi}_k(z) \to \phi(z)$ uniformly on $[0, a]$ for every $a \geq 0$ as $k \to \infty$.

Proposition 2.5 *Suppose that Condition 2.4 is satisfied. Then the limit function* ϕ *has representation*

$$\phi(z) = bz + cz^2 + \int_{(0,\infty)} \big(\mathrm{e}^{-zu} - 1 + zu\big)m(\mathrm{d}u), \quad z \geq 0, \tag{19}$$

where $c \geq 0$ *and* b *are constants and* $m(\mathrm{d}u)$ *is a* σ*-finite measure on* $(0, \infty)$ *satisfying*

$$\int_{(0,\infty)} (u \wedge u^2)m(\mathrm{d}u) < \infty.$$

Proof For each $k \geq 1$ let us define the function ϕ_k on $[0, k]$ by

$$\phi_k(z) = k\gamma_k[g_k(1 - z/k) - (1 - z/k)]. \tag{20}$$

From (18) and (20) we have

$$\tilde{\phi}_k'(z) = \gamma_k \mathrm{e}^{-z/k}[1 - g_k'(\mathrm{e}^{-z/k})], \qquad z \geq 0,$$

and

$$\phi_k'(z) = \gamma_k[1 - g_k'(1 - z/k)], \qquad 0 \le z \le k.$$

Since $\{\tilde{\phi}_k\}$ is uniformly Lipschitz on each bounded interval, the sequence $\{\tilde{\phi}_k'\}$ is uniformly bounded on each bounded interval. Then $\{\phi_k'\}$ is also uniformly bounded on each bounded interval, and so the sequence $\{\phi_k\}$ is uniformly Lipschitz on each bounded interval. Let $a \ge 0$. By the mean-value theorem, for $k \ge a$ and $0 \le z \le a$ we have

$$\begin{aligned}\tilde{\phi}_k(z) - \phi_k(z) &= k\gamma_k\big[g_k(\mathrm{e}^{-z/k}) - g_k(1 - z/k) - \mathrm{e}^{-z/k} + (1 - z/k)\big] \\ &= k\gamma_k[g_k'(\eta_k) - 1](\mathrm{e}^{-z/k} - 1 + z/k),\end{aligned}$$

where

$$1 - a/k \le 1 - z/k \le \eta_k \le \mathrm{e}^{-z/k} \le 1.$$

Choose $k_0 \ge a$ so that $\mathrm{e}^{-2a/k_0} \le 1 - a/k_0$. Then $\mathrm{e}^{-2a/k} \le 1 - a/k$ for $k \ge k_0$ and hence

$$\gamma_k|g_k'(\eta_k) - 1| \le \sup_{0\le z\le 2a} \gamma_k|g_k'(\mathrm{e}^{-z/k}) - 1| = \sup_{0\le z\le 2a} \mathrm{e}^{z/k}|\tilde{\phi}_k'(z)|.$$

Since $\{\tilde{\phi}_k'\}$ is uniformly bounded on $[0, 2a]$, the sequence $\{\gamma_k|g_k'(\eta_k) - 1| : k \ge k_0\}$ is bounded. Then $\lim_{k\to\infty} |\phi_k(z) - \tilde{\phi}_k(z)| = 0$ uniformly on each bounded interval. It follows that $\lim_{k\to\infty} \phi_k(z) = \phi(z)$ uniformly on each bounded interval. Then the result follows by Corollary 1.46 in Li [40, p. 26].

Proposition 2.6 *For any function ϕ with representation* (19) *there is a sequence $\{\tilde{\phi}_k\}$ in the form of* (18) *satisfying Condition 2.4.*

Proof By the proof of Proposition 2.5 it suffices to construct a sequence $\{\phi_k\}$ with the expression (20) that is uniformly Lipschitz on $[0, a]$ and $\phi_k(z) \to \phi(z)$ uniformly on $[0, a]$ for every $a \ge 0$. To simplify the formulations we decompose the function ϕ into two parts. Let $\phi_0(z) = \phi(z) - bz$. We first define

$$\gamma_{0,k} = (1 + 2c)k + \int_{(0,\infty)} u(1 - \mathrm{e}^{-ku})m(\mathrm{d}u)$$

and

$$g_{0,k}(z) = z + k^{-1}\gamma_{0,k}^{-1}\phi_0(k(1 - z)), \qquad |z| \le 1.$$

It is easy to see that $z \mapsto g_{0,k}(z)$ is an analytic function satisfying $g_{0,k}(1) = 1$ and

$$\frac{\mathrm{d}^n}{\mathrm{d}z^n} g_{0,k}(0) \geq 0, \qquad n \geq 0.$$

Therefore $g_{0,k}(\cdot)$ is a probability generating function. Let $\phi_{0,k}$ be defined by (20) with (γ_k, g_k) replaced by $(\gamma_{0,k}, g_{0,k})$. Then $\phi_{0,k}(z) = \phi_0(z)$ for $0 \leq z \leq k$. That completes the proof if $b = 0$. In the case $b \neq 0$, we set

$$g_{1,k}(z) = \frac{1}{2}\left(1 + \frac{b}{|b|}\right) + \frac{1}{2}\left(1 - \frac{b}{|b|}\right)z^2.$$

Let $\gamma_{1,k} = |b|$ and let $\phi_{1,k}(z)$ be defined by (20) with (γ_k, g_k) replaced by $(\gamma_{1,k}, g_{1,k})$. Then

$$\phi_{1,k}(z) = bz + \frac{1}{2k}(|b| - b)z^2.$$

Finally, let $\gamma_k = \gamma_{0,k} + \gamma_{1,k}$ and $g_k = \gamma_k^{-1}(\gamma_{0,k} g_{0,k} + \gamma_{1,k} g_{1,k})$. Then the sequence $\phi_k(z)$ defined by (20) is equal to $\phi_{0,k}(z) + \phi_{1,k}(z)$ which satisfies the required condition.

Lemma 2.7 *Suppose that the sequence* $\{\tilde{\phi}_k\}$ *defined by* (18) *is uniformly Lipschitz on* $[0, 1]$. *Then there are constants* $B, N \geq 0$ *such that* $v_k(t, \lambda) \leq \lambda \mathrm{e}^{Bt}$ *for every* $t, \lambda \geq 0$ *and* $k \geq N$.

Proof Let $b_k := \tilde{\phi}_k'(0+)$ for $k \geq 1$. Since $\{\tilde{\phi}_k\}$ is uniformly Lipschitz on $[0, 1]$, the sequence $\{b_k\}$ is bounded. From (18) we have $b_k = \gamma_k[1 - g_k'(1-)]$. By (11) and (18) it is not hard to obtain

$$\int_{E_k} y Q_k^{\lfloor \gamma_k t \rfloor}(x, \mathrm{d}y) = x g_k'(1-)^{\lfloor \gamma_k t \rfloor} = x\left(1 - \frac{b_k}{\gamma_k}\right)^{\lfloor \gamma_k t \rfloor}.$$

Let $B \geq 0$ be a constant such that $2|b_k| \leq B$ for all $k \geq 1$. Since $\gamma_k \to \infty$ as $k \to \infty$, there is $N \geq 1$ so that

$$0 \leq \left(1 - \frac{b_k}{\gamma_k}\right)^{\gamma_k/B} \leq \left(1 + \frac{B}{2\gamma_k}\right)^{\gamma_k/B} \leq \mathrm{e}, \qquad k \geq N.$$

It follows that, for $t \geq 0$ and $k \geq N$,

$$\int_{E_k} y Q_k^{\lfloor \gamma_k t \rfloor}(x, \mathrm{d}y) \leq x \exp\left\{B\lfloor \gamma_k t \rfloor / \gamma_k\right\} \leq x\mathrm{e}^{Bt}.$$

Then the desired estimate follows from (11) and Jensen's inequality.

Theorem 2.8 *Suppose that Condition 2.4 holds. Then for every $a \geq 0$ we have $v_k(t, \lambda) \to$ some $v_t(\lambda)$ uniformly on $[0, a]^2$ as $k \to \infty$ and the limit function solves the integral equation*

$$v_t(\lambda) = \lambda - \int_0^t \phi(v_s(\lambda))\mathrm{d}s, \qquad \lambda, t \geq 0. \tag{21}$$

Proof The following argument is a modification of that of Aliev and Shchurenkov [2] and Aliev [1]. In view of (15), we can write

$$v_k(t, \lambda) = \lambda + \varepsilon_k(t, \lambda) - \int_0^t \bar{\phi}_k(v_k(s, \lambda))\mathrm{d}s, \tag{22}$$

where

$$\varepsilon_k(t, \lambda) = \big(t - \gamma_k^{-1}\lfloor \gamma_k t \rfloor\big)\bar{\phi}_k\big(v_k(\gamma_k^{-1}\lfloor \gamma_k t \rfloor, \lambda)\big).$$

By Lemma 2.3 and Condition 2.4, for any $0 < \varepsilon \leq 1$ we can choose $N \geq 1$ so that $|\bar{\phi}_k(z) - \phi(z)| \leq \varepsilon$ for $k \geq N$ and $0 \leq z \leq a\mathrm{e}^{Ba}$. It follows that, for $0 \leq t \leq a$ and $0 \leq \lambda \leq a$,

$$|\varepsilon_k(t, \lambda)| \leq \gamma_k^{-1}\big|\bar{\phi}_k\big(v_k(\gamma_k^{-1}\lfloor \gamma_k t \rfloor, \lambda)\big)\big| \leq \gamma_k^{-1}M, \tag{23}$$

where

$$M = 1 + \sup_{0 \leq z \leq a\mathrm{e}^{Ba}} |\phi(z)|.$$

For $n \geq k \geq N$ let

$$K_{k,n}(t, \lambda) = \sup_{0 \leq s \leq t} |v_n(s, \lambda) - v_k(s, \lambda)|.$$

By (22) and (23) we obtain, for $0 \leq t \leq a$ and $0 \leq \lambda \leq a$,

$$\begin{aligned} K_{k,n}(t, \lambda) &\leq 2\gamma_k^{-1}M + \int_0^t |\bar{\phi}_k(v_k(s, \lambda)) - \bar{\phi}_n(v_n(s, \lambda))|\mathrm{d}s \\ &\leq 2(\gamma_k^{-1}M + \varepsilon a) + \int_0^t |\phi_k(v_k(s, \lambda)) - \phi_n(v_n(s, \lambda))|\mathrm{d}s \\ &\leq 2(\gamma_k^{-1}M + \varepsilon a) + L\int_0^t K_{k,n}(s, \lambda)\mathrm{d}s, \end{aligned}$$

where $L = \sup_{0 \leq z \leq a\mathrm{e}^{Ba}} |\phi'(z)|$. By Gronwall's inequality,

$$K_{k,n}(t, \lambda) \leq 2(\gamma_k^{-1}M + \varepsilon a)\exp\{Lt\}, \qquad 0 \leq t, \lambda \leq a.$$

Then $v_k(t, \lambda) \to$ some $v_t(\lambda)$ uniformly on $[0, a]^2$ as $k \to \infty$ for every $a \geq 0$. From (22) we get (21).

Theorem 2.9 *Suppose that ϕ is a function given by* (19). *Then for any $\lambda \geq 0$ there is a unique positive solution $t \mapsto v_t(\lambda)$ to* (21). *Moreover, the solution satisfies the semigroup property:*

$$v_{r+t}(\lambda) = v_r \circ v_t(\lambda) = v_r(v_t(\lambda)), \qquad r, t, \lambda \geq 0. \tag{24}$$

Proof By Proposition 2.6 there is a sequence $\{\tilde{\phi}_k\}$ in form (18) satisfying Condition 2.4. Let $v_k(t, \lambda)$ be given by (13) and (14). By Theorem 2.8 the limit $v_t(\lambda) = \lim_{k\to\infty} v_k(t, \lambda)$ exists and solves (21). Clearly, any positive solution $t \mapsto v_t(\lambda)$ to (21) is locally bounded. The uniqueness of the solution follows by Gronwall's inequality. The relation (24) is a consequence of the uniqueness of the solution.

Theorem 2.10 *Suppose that ϕ is a function given by* (19). *For any $\lambda \geq 0$ let $t \mapsto v_t(\lambda)$ be the unique positive solution to* (21). *Then we can define a transition semigroup $(Q_t)_{t\geq 0}$ on $[0, \infty)$ by*

$$\int_{[0,\infty)} \mathrm{e}^{-\lambda y} Q_t(x, \mathrm{d}y) = \mathrm{e}^{-x v_t(\lambda)}, \qquad \lambda \geq 0, x \geq 0. \tag{25}$$

Proof By Proposition 2.6, there is a sequence $\{\tilde{\phi}_k\}$ in form (18) satisfying Condition 2.4. By Theorem 2.8 we have $v_k(t, \lambda) \to v_t(\lambda)$ uniformly on $[0, a]^2$ as $k \to \infty$ for every $a \geq 0$. Taking $x_k \in E_k$ satisfying $x_k \to x$ as $k \to \infty$, we see by Theorem 1.2 that (25) defines a probability measure $Q_t(x, \mathrm{d}y)$ on $[0, \infty)$ and $\lim_{k\to\infty} Q_k^{\lfloor \gamma_k t \rfloor}(x_k, \cdot) = Q_t(x, \cdot)$ by weak convergence. By a monotone class argument one can see that $Q_t(x, \mathrm{d}y)$ is a kernel on $[0, \infty)$. The semigroup property of the family of kernels $(Q_t)_{t\geq 0}$ follows from (24) and (25).

Proposition 2.11 *For every $t \geq 0$ the function $\lambda \mapsto v_t(\lambda)$ is strictly increasing on $[0, \infty)$.*

Proof By the continuity of $t \mapsto v_t(\lambda)$, for any $\lambda_0 > 0$ there is $t_0 > 0$ so that $v_t(\lambda_0) > 0$ for $0 \leq t \leq t_0$. Then (25) implies $Q_t(x, \{0\}) < 1$ for $x > 0$ and $0 \leq t \leq t_0$, and so $\lambda \mapsto v_t(\lambda)$ is strictly increasing for $0 \leq t \leq t_0$. By the semigroup property (24) we infer $\lambda \mapsto v_t(\lambda)$ is strictly increasing for all $t \geq 0$.

Theorem 2.12 *The transition semigroup $(Q_t)_{t\geq 0}$ defined by* (25) *is a Feller semigroup.*

Proof For $\lambda > 0$ and $x \geq 0$ set $e_\lambda(x) = \mathrm{e}^{-\lambda x}$. We denote by $\mathscr{D}_0$ the linear span of $\{e_\lambda : \lambda > 0\}$. By Proposition 2.11, the operator Q_t preserves $\mathscr{D}_0$ for every $t \geq 0$. By the continuity of $t \mapsto v_t(\lambda)$ it is easy to show that $t \mapsto Q_t e_\lambda(x)$ is continuous for $\lambda > 0$ and $x \geq 0$. Then $t \mapsto Q_t f(x)$ is continuous for every $f \in \mathscr{D}_0$ and $x \geq 0$. Let $C_0[0, \infty)$ be the space of continuous functions on $[0, \infty)$ vanishing at infinity. By the Stone–Weierstrass theorem, the set $\mathscr{D}_0$ is uniformly dense in $C_0[0, \infty)$; see, e.g.,

Hewitt and Stromberg [25, pp. 98–99]. Then each operator Q_t preserves $C_0[0,\infty)$ and $t \mapsto Q_t f(x)$ is continuous for $x \geq 0$ and $f \in C_0[0,\infty)$. That gives the Feller property of the semigroup $(Q_t)_{t\geq 0}$.

A Markov process in $[0,\infty)$ is called a *continuous-state branching process* (CB-process) with *branching mechanism* ϕ if it has transition semigroup $(Q_t)_{t\geq 0}$ defined by (25). It is simple to see that $(Q_t)_{t\geq 0}$ satisfies the *branching property*:

$$Q_t(x_1 + x_2, \cdot) = Q_t(x_1, \cdot) * Q_t(x_2, \cdot), \qquad t, x_1, x_2 \geq 0. \tag{26}$$

The family of functions $(v_t)_{t\geq 0}$ is called the *cumulant semigroup* of the CB-process. Since $(Q_t)_{t\geq 0}$ is a Feller semigroup, the process has a Hunt realization; see, e.g., Chung [7, p. 75]. Clearly, zero is a trap for the CB-process.

Proposition 2.13 *Suppose that* $\{(x_1(t), \mathscr{F}_t^1) : t \geq 0\}$ *and* $\{(x_2(t), \mathscr{F}_t^2) : t \geq 0\}$ *are two independent CB-processes with branching mechanism* ϕ. *Let* $x(t) = x_1(t) + x_2(t)$ *and* $\mathscr{F}_t = \sigma(\mathscr{F}_t^1 \cup \mathscr{F}_t^2)$. *Then* $\{(x(t), \mathscr{F}_t) : t \geq 0\}$ *is also a CB-processes with branching mechanism* ϕ.

Proof Let $t \geq r \geq 0$ and for $i = 1, 2$ let F_i be a bounded $\mathscr{F}_r^i$-measurable random variable. For any $\lambda \geq 0$ we have

$$\begin{aligned}\mathbf{P}\big[F_1 F_2 \mathrm{e}^{-\lambda x(t)}\big] &= \mathbf{P}\big[F_1 \mathrm{e}^{-\lambda x_1(t)}\big]\mathbf{P}\big[F_2 \mathrm{e}^{-\lambda x_2(t)}\big] \\ &= \mathbf{P}\big[F_1 \mathrm{e}^{-x_1(r)v_{t-r}(\lambda)}\big]\mathbf{P}\big[F_2 \mathrm{e}^{-x_2(r)v_{t-r}(\lambda)}\big] \\ &= \mathbf{P}\big[F_1 F_2 \mathrm{e}^{-x(r)v_{t-r}(\lambda)}\big].\end{aligned}$$

A monotone class argument shows that

$$\mathbf{P}\big[F \mathrm{e}^{-\lambda x(t)}\big] = \mathbf{P}\big[F \mathrm{e}^{-x(r)v_{t-r}(\lambda)}\big]$$

for any bounded $\mathscr{F}_r$-measurable random variable F. Then $\{(x(t), \mathscr{F}_t) : t \geq 0\}$ is a Markov processes with transition semigroup $(Q_t)_{t\geq 0}$.

Let $D[0,\infty)$ denote the space of positive càdlàg paths on $[0,\infty)$ furnished with the Skorokhod topology. The following theorem is a slight modification of Theorem 2.1 of Li [39], which gives a physical interpretation of the CB-process as an approximation of the GW-process with small individuals.

Theorem 2.14 *Suppose that Condition 2.4 holds. Let* $\{x(t) : t \geq 0\}$ *be a càdlàg CB-process with transition semigroup* $(Q_t)_{t\geq 0}$ *defined by* (25). *For* $k \geq 1$ *let* $\{z_k(n) : n \geq 0\}$ *be a Markov chain with state space* $E_k := \{0, k^{-1}, 2k^{-1}, \ldots\}$ *and* n*-step transition probability* $Q_k^n(x, \mathrm{d}y)$ *determined by* (12). *If* $z_k(0)$ *converges to* $x(0)$ *in distribution, then* $\{z_k(\lfloor \gamma_k t \rfloor) : t \geq 0\}$ *converges as* $k \to \infty$ *to* $\{x(t) : t \geq 0\}$ *in distribution on* $D[0,\infty)$.

Proof For $\lambda > 0$ and $x \geq 0$ set $e_\lambda(x) = \mathrm{e}^{-\lambda x}$. Let $C_0[0, \infty)$ be the space of continuous functions on $[0, \infty)$ vanishing at infinity. By (13), (25) and Theorem 2.8 it is easy to show

$$\lim_{k\to\infty} \sup_{x\in E_k} \left|Q_k^{\lfloor\gamma_k t\rfloor} e_\lambda(x) - Q_t e_\lambda(x)\right| = 0, \qquad \lambda > 0.$$

Then the Stone–Weierstrass theorem implies

$$\lim_{k\to\infty} \sup_{x\in E_k} \left|Q_k^{\lfloor\gamma_k t\rfloor} f(x) - Q_t f(x)\right| = 0, \qquad f \in C_0[0, \infty).$$

By Ethier and Kurtz [17, p. 226 and pp. 233–234] we conclude that $\{z_k(\lfloor\gamma_k t\rfloor) : t \geq 0\}$ converges to the CB-process $\{x(t) : t \geq 0\}$ in distribution on $D[0, \infty)$.

For any $w \in D[0, \infty)$ let $\tau_0(w) = \inf\{s > 0 : w(s) \text{ or } w(s-) = 0\}$. Let $D_0[0, \infty)$ be the set of paths $w \in D[0, \infty)$ such that $w(t) = 0$ for $t \geq \tau_0(w)$. Then $D_0[0, \infty)$ is a Borel subset of $D[0, \infty)$. It is not hard to show that the distributions of the processes $\{z_k(\lfloor\gamma_k t\rfloor) : t \geq 0\}$ and $\{x(t) : t \geq 0\}$ are all supported by $D_0[0, \infty)$. By Theorem 1.7 of Li [40, p. 4] we have the following.

Corollary 2.15 *Under the conditions of Theorem 2.14, the sequence* $\{z_k(\lfloor\gamma_k t\rfloor) : t \geq 0\}$ *converges as* $k \to \infty$ *to* $\{x(t) : t \geq 0\}$ *in distribution on* $D_0[0, \infty)$.

The convergence of rescaled GW-processes to diffusion processes was first studied by Feller [18]. Lamperti [34] showed that all CB-processes are weak limits of rescaled GW-processes. A characterization of CB-processes by random time changes of Lévy processes was given by Lamperti [35]; see also Kyprianou [31]. We have followed Aliev and Shchurenkov [2] and Li [39, 40] in some of the above calculations.

Example 2.16 For any $0 \leq \alpha \leq 1$ the function $\phi(\lambda) = \lambda^{1+\alpha}$ can be represented in the form of (19). In particular, for $0 < \alpha < 1$ we can use integration by parts to see

$$\begin{aligned}
&\int_{(0,\infty)} (\mathrm{e}^{-\lambda u} - 1 + \lambda u)\frac{\mathrm{d}u}{u^{2+\alpha}} \\
&\quad = \lambda^{1+\alpha} \int_{(0,\infty)} (\mathrm{e}^{-v} - 1 + v)\frac{\mathrm{d}v}{v^{2+\alpha}} \\
&\quad = \lambda^{1+\alpha}\left[-\frac{\mathrm{e}^{-v} - 1 + v}{(1+\alpha)v^{1+\alpha}}\bigg|_0^\infty + \int_{(0,\infty)} \frac{(1-\mathrm{e}^{-v})\mathrm{d}v}{(1+\alpha)v^{1+\alpha}}\right] \\
&\quad = \frac{\lambda^{1+\alpha}}{1+\alpha}\left[-(1-\mathrm{e}^{-v})\frac{1}{\alpha v^\alpha}\bigg|_0^\infty + \int_{(0,\infty)} \mathrm{e}^{-v}\frac{\mathrm{d}v}{\alpha v^\alpha}\right] \\
&\quad = \frac{\Gamma(1-\alpha)}{\alpha(1+\alpha)}\lambda^{1+\alpha}.
\end{aligned}$$

Thus we have

$$\lambda^{1+\alpha} = \frac{\alpha(1+\alpha)}{\Gamma(1-\alpha)} \int_{(0,\infty)} (\mathrm{e}^{-\lambda u} - 1 + \lambda u)\frac{\mathrm{d}u}{u^{2+\alpha}}, \qquad \lambda \geq 0. \tag{27}$$

Example 2.17 Suppose that there are constants $c > 0$, $0 < \alpha \le 1$ and b so that $\phi(z) = cz^{1+\alpha} + bz$. Let $q_\alpha^0(t) = \alpha t$ and $q_\alpha^b(t) = b^{-1}(1 - \mathrm{e}^{-\alpha bt})$ for $b \ne 0$. By solving the equation

$$\frac{\partial}{\partial t} v_t(\lambda) = -cv_t(\lambda)^{1+\alpha} - bv_t(\lambda), \qquad v_0(\lambda) = \lambda$$

we get

$$v_t(\lambda) = \frac{\mathrm{e}^{-bt}\lambda}{\left[1 + cq_\alpha^b(t)\lambda^\alpha\right]^{1/\alpha}}, \qquad t \ge 0, \lambda \ge 0. \tag{28}$$

3 Some Basic Properties

In this section we prove some basic properties of CB-processes. Most of the results presented here can be found in Grey [21] and Li [37]. We here use the treatments in Li [40]. Suppose that ϕ is a branching mechanism defined by (19). This is a convex function on $[0, \infty)$. In fact, it is easy to see that

$$\phi'(z) = b + 2cz + \int_{(0,\infty)} u\big(1 - \mathrm{e}^{-zu}\big)m(\mathrm{d}u), \qquad z \ge 0, \tag{29}$$

which is an increasing function. In particular, we have $\phi'(0) = b$. The limit $\phi(\infty) := \lim_{z\to\infty} \phi(z)$ exists in $[-\infty, \infty]$ and $\phi'(\infty) := \lim_{z\to\infty} \phi'(z)$ exists in $(-\infty, \infty]$. In particular, we have

$$\phi'(\infty) := b + 2c \cdot \infty + \int_{(0,\infty)} u\, m(\mathrm{d}u) \tag{30}$$

with $0 \cdot \infty = 0$ by convention. Observe also that $-\infty \le \phi(\infty) \le 0$ if and only if $\phi'(\infty) \le 0$, and $\phi(\infty) = \infty$ if and only if $\phi'(\infty) > 0$.

The transition semigroup $(Q_t)_{t\ge 0}$ of the CB-process is defined by (21) and (25). From the branching property (26), we see that the probability measure $Q_t(x, \cdot)$ is infinitely divisible. Then $(v_t)_{t\ge 0}$ has the *canonical representation*:

$$v_t(\lambda) = h_t\lambda + \int_{(0,\infty)} (1 - \mathrm{e}^{-\lambda u})l_t(\mathrm{d}u), \qquad t \ge 0, \lambda \ge 0, \tag{31}$$

where $h_t \ge 0$ and $l_t(\mathrm{d}u)$ is a σ-finite measure on $(0, \infty)$ satisfying

$$\int_{(0,\infty)} (1 \wedge u)l_t(\mathrm{d}u) < \infty.$$

The pair (h_t, l_t) is uniquely determined by (31); see, e.g., Proposition 1.30 in Li [40, p. 16]. By differentiating both sides of the equation and using (21), it is easy to find

$$h_t + \int_{(0,\infty)} ul_t(\mathrm{d}u) = \frac{\partial}{\partial \lambda} v_t(0+) = \mathrm{e}^{-bt}, \qquad t \geq 0. \tag{32}$$

Then we infer that $l_t(\mathrm{d}u)$ satisfies

$$\int_{(0,\infty)} ul_t(\mathrm{d}u) < \infty.$$

From (25) and (32) we get

$$\int_{[0,\infty)} y Q_t(x, \mathrm{d}y) = x\mathrm{e}^{-bt}, \qquad t \geq 0, x \geq 0. \tag{33}$$

We say the branching mechanism ϕ is *critical*, *subcritical*, or *supercritical* accordingly as $b = 0$, $b \geq 0$, or $b \leq 0$, respectively.

From (21) we see that $t \mapsto v_t(\lambda)$ is first continuous and then continuously differentiable. Moreover, it is easy to show that

$$\frac{\partial}{\partial t} v_t(\lambda)\Big|_{t=0} = -\phi(\lambda), \qquad \lambda \geq 0.$$

By the semigroup property $v_{t+s} = v_s \circ v_t = v_t \circ v_s$, we get the backward differential equation

$$\frac{\partial}{\partial t} v_t(\lambda) = -\phi(v_t(\lambda)), \qquad v_0(\lambda) = \lambda, \tag{34}$$

and forward differential equation

$$\frac{\partial}{\partial t} v_t(\lambda) = -\phi(\lambda) \frac{\partial}{\partial \lambda} v_t(\lambda), \quad v_0(\lambda) = \lambda. \tag{35}$$

The corresponding equations for a branching process with continuous time and discrete state were given in Athreya and Ney [3, p. 106].

Proposition 3.1 *Suppose that* $\lambda > 0$ *and* $\phi(\lambda) \neq 0$. *Then the equation* $\phi(z) = 0$ *has no root between* λ *and* $v_t(\lambda)$ *for every* $t \geq 0$. *Moreover, we have*

$$\int_{v_t(\lambda)}^{\lambda} \phi(z)^{-1} \mathrm{d}z = t, \qquad t \geq 0. \tag{36}$$

Proof By (19) we see $\phi(0) = 0$ and $z \mapsto \phi(z)$ is a convex function. Since $\phi(\lambda) \neq 0$ for some $\lambda > 0$ according to the assumption, the equation $\phi(z) = 0$ has at most one root in $(0, \infty)$. Suppose that $\lambda_0 \geq 0$ is a root of $\phi(z) = 0$. Then (35) implies $v_t(\lambda_0) = \lambda_0$ for all $t \geq 0$. By Proposition 2.11 we have $v_t(\lambda) > \lambda_0$ for $\lambda > \lambda_0$ and $0 < v_t(\lambda) < \lambda_0$ for $0 < \lambda < \lambda_0$. Then $\lambda > 0$ and $\phi(\lambda) \neq 0$ imply there is no root of $\phi(z) = 0$ between λ and $v_t(\lambda)$. From (34) we get (36).

Corollary 3.2 *Suppose that* $\phi(z_0) \neq 0$ *for some* $z_0 > 0$. *Let* $\theta_0 = \inf\{z > 0 : \phi(z) \geq 0\}$ *with the convention* $\inf \emptyset = \infty$. *Then* $\lim_{t\to\infty} v_t(\lambda) = \theta_0$ *increasingly for* $0 < \lambda < \theta_0$ *and decreasingly* $\lambda > \theta_0$.

Proof In the case $\theta_0 = \infty$, we have $\phi(z) < 0$ for all $z > 0$. From (34) we see $\lambda \mapsto v_t(\lambda)$ is increasing. Then (34) implies $\lim_{t\to\infty} v_t(\lambda) = \infty$ for every $\lambda > 0$. In the case $\theta_0 < \infty$, we have clearly $\phi(\theta_0) = 0$. Furthermore, $\phi(z) < 0$ for $0 < z < \theta_0$ and $\phi(z) > 0$ for $z > \theta_0$. From (35) we see $v_t(\theta_0) = \theta_0$ for all $t \geq 0$. Then (36) implies that $\lim_{t\to\infty} v_t(\lambda) = \theta_0$ increasingly for $0 < \lambda < \theta_0$ and decreasingly $\lambda > \theta_0$.

Corollary 3.3 *Suppose that* $\phi(z_0) \neq 0$ *for some* $z_0 > 0$. *Then for any* $x > 0$ *we have*

$$\lim_{t\to\infty} Q_t(x, \cdot) = \mathrm{e}^{-x\theta_0}\delta_0 + (1 - \mathrm{e}^{-x\theta_0})\delta_\infty$$

by weak convergence of probability measures on $[0, \infty]$.

Proof The space of probability measures on $[0, \infty]$ endowed the topology of weak convergence is compact and metrizable; see, e.g., Parthasarathy [47, p. 45]. Let $\{t_n\}$ be any positive sequence so that $t_n \to \infty$ and $Q_{t_n}(x, \cdot) \to$ some $Q_\infty(x, \cdot)$ weakly as $n \to \infty$. By (25) and Corollary 3.2, for every $\lambda > 0$ we have

$$\begin{aligned}\int_{[0,\infty]} \mathrm{e}^{-\lambda y} Q_\infty(x, \mathrm{d}y) &= \lim_{n\to\infty} \int_{[0,\infty]} \mathrm{e}^{-\lambda y} Q_{t_n}(x, \mathrm{d}y) \\ &= \lim_{n\to\infty} \mathrm{e}^{-x v_{t_n}(\lambda)} = \mathrm{e}^{-x\theta_0}.\end{aligned}$$

It follows that

$$Q_\infty(x, \{0\}) = \lim_{\lambda\to\infty} \int_{[0,\infty]} \mathrm{e}^{-\lambda y} Q_\infty(x, \mathrm{d}y) = \mathrm{e}^{-x\theta_0}$$

and

$$Q_\infty(x, \{\infty\}) = \lim_{\lambda\to 0} \int_{[0,\infty]} (1 - \mathrm{e}^{-\lambda y}) Q_\infty(x, \mathrm{d}y) = 1 - \mathrm{e}^{-x\theta_0}.$$

That shows $Q_\infty(x, \cdot) = \mathrm{e}^{-x\theta_0}\delta_0 + (1 - \mathrm{e}^{-x\theta_0})\delta_\infty$, which is independent of the particular choice of the sequence $\{t_n\}$. Then we have $Q_t(x, \cdot) \to Q_\infty(x, \cdot)$ weakly as $t \to \infty$.

A simple asymptotic behavior of the CB-process is described in Corollary 3.3. Clearly, we have (i) $\theta_0 > 0$ if and only if $b < 0$; (ii) $\theta_0 = \infty$ if and only if $\phi'(\infty) \leq 0$.

The reader can refer to Grey [21] and Li [40, Sect. 3.2] for more asymptotic results on the CB-process.

Since $(Q_t)_{t\geq 0}$ is a Feller transition semigroup, the CB-process has a Hunt process realization $X = (\Omega, \mathscr{F}, \mathscr{F}_t, x(t), \mathbf{Q}_x)$; see, e.g., Chung [7, p. 75]. Let $\tau_0 := \inf\{s \geq 0 : x(s) = 0\}$ denote the *extinction time* of the CB-process.

Theorem 3.4 *For every $t \geq 0$ the limit $\bar{v}_t = \uparrow\lim_{\lambda\to\infty} v_t(\lambda)$ exists in $(0, \infty]$. Moreover, the mapping $t \mapsto \bar{v}_t$ is decreasing and for any $t \geq 0$ and $x > 0$ we have*

$$\mathbf{Q}_x\{\tau_0 \leq t\} = \mathbf{Q}_x\{x(t) = 0\} = \exp\{-x\bar{v}_t\}. \tag{37}$$

Proof By Proposition 2.11 the limit $\bar{v}_t = \uparrow\lim_{\lambda\to\infty} v_t(\lambda)$ exists in $(0, \infty]$ for every $t \geq 0$. For $t \geq r \geq 0$ we have

$$\bar{v}_t = \uparrow\lim_{\lambda\to\infty} v_r(v_{t-r}(\lambda)) = v_r(\bar{v}_{t-r}) \leq \bar{v}_r. \tag{38}$$

Since zero is a trap for the CB-process, we get (37) by letting $\lambda \to \infty$ in (25).

For the convenience of statement of the results in the sequel, we formulate the following condition on the branching mechanism, which is known as *Grey's condition*.

Condition 3.5 There is some constant $\theta > 0$ so that

$$\phi(z) > 0 \text{ for } z \geq \theta \text{ and } \int_\theta^\infty \phi(z)^{-1}\mathrm{d}z < \infty.$$

Theorem 3.6 *We have $\bar{v}_t < \infty$ for some and hence all $t > 0$ if and only if Condition 3.5 holds.*

Proof By (38) it is simple to see that $\bar{v}_t = \uparrow\lim_{\lambda\to\infty} v_t(\lambda) < \infty$ for all $t > 0$ if and only if this holds for some $t > 0$. If Condition 3.5 holds, we can let $\lambda \to \infty$ in (36) to obtain

$$\int_{\bar{v}_t}^\infty \phi(z)^{-1}\mathrm{d}z = t \tag{39}$$

and hence $\bar{v}_t < \infty$ for $t > 0$. For the converse, suppose that $\bar{v}_t < \infty$ for some $t > 0$. By (34) there exists some $\theta > 0$ so that $\phi(\theta) > 0$, for otherwise we would have $\bar{v}_t \geq v_t(\lambda) \geq \lambda$ for all $\lambda \geq 0$, yielding a contradiction. Then $\phi(z) > 0$ for all $z \geq \theta$ by the convexity of the branching mechanism. As in the above we see that (39) still holds, so Condition 3.5 is satisfied.

Theorem 3.7 *Let $\bar{v} = \downarrow\lim_{t\to\infty} \bar{v}_t \in [0, \infty]$. Then for any $x > 0$ we have*

$$\mathbf{Q}_x\{\tau_0 < \infty\} = \exp\{-x\bar{v}\}. \tag{40}$$

Moreover, we have $\bar{v} < \infty$ if and only if Condition 3.5 holds, and in this case $\bar{v}$ is the largest root of $\phi(z) = 0$.

Proof The first assertion follows immediately from Theorem 3.4. By Theorem 3.6 we have $\bar{v}_t < \infty$ for some and hence all $t > 0$ if and only if Condition 3.5 holds. This is clearly equivalent to $\bar{v} < \infty$. From (39) we see $\bar{v}$ is the largest root of $\phi(z) = 0$.

Corollary 3.8 *Suppose that Condition 3.5 holds. Then for any $x > 0$ we have $\mathbf{Q}_x\{\tau_0 < \infty\} = 1$ if and only if $b \geq 0$.*

By Corollary 3.2 and Theorem 3.7 we see $0 \leq \theta_0 \leq \bar{v} \leq \infty$. In fact, we have $0 \leq \theta_0 = \bar{v} < \infty$ if Condition 3.5 holds and $0 \leq \theta_0 < \bar{v} = \infty$ if there is $\theta > 0$ so that

$$\phi(z) > 0 \text{ for } z \geq \theta \text{ and } \int_\theta^\infty \phi(z)^{-1}\mathrm{d}z = \infty.$$

Proposition 3.9 *For any $t \geq 0$ and $\lambda \geq 0$ let $v_t'(\lambda) = (\partial/\partial\lambda)v_t(\lambda)$. Then we have*

$$v_t'(\lambda) = \exp\bigg\{-\int_0^t \phi'(v_s(\lambda))\mathrm{d}s\bigg\}, \tag{41}$$

where ϕ' is given by (29).

Proof Based on (21) and (34) it is elementary to see that

$$\frac{\partial}{\partial t}v_t'(\lambda) = \frac{\partial}{\partial\lambda}\frac{\partial}{\partial t}v_t(\lambda) = -\phi'(v_t(\lambda))v_t'(\lambda).$$

It follows that

$$\frac{\partial}{\partial t}\big[\log v_t'(\lambda)\big] = v_t'(\lambda)^{-1}\frac{\partial}{\partial t}v_t'(\lambda) = -\phi'(v_t(\lambda)).$$

Since $v_0'(\lambda) = 1$, we get (41).

Theorem 3.10 *Let $\phi_0'(z) = \phi'(z) - b$ for $z \geq 0$, where ϕ' is given by (29). We can define a Feller transition semigroup $(Q_t^b)_{t\geq 0}$ on $[0, \infty)$ by*

$$\int_{[0,\infty)} \mathrm{e}^{-\lambda y} Q_t^b(x, \mathrm{d}y) = \exp\bigg\{-xv_t(\lambda) - \int_0^t \phi_0'(v_s(\lambda))\mathrm{d}s\bigg\}. \tag{42}$$

Moreover, we have $Q_t^b(x, \mathrm{d}y) = \mathrm{e}^{bt}x^{-1}yQ_t(x, \mathrm{d}y)$ for $x > 0$ and

$$Q_t^b(0, \mathrm{d}y) = \mathrm{e}^{bt}[h_t\delta_0(\mathrm{d}y) + yl_t(\mathrm{d}y)], \qquad t, y \geq 0. \tag{43}$$

Proof In view of (33), it is simple to check that $Q_t^b(x, \mathrm{d}y) := \mathrm{e}^{bt} x^{-1} y Q_t(x, \mathrm{d}y)$ defines a Markov transition semigroup $(Q_t^b)_{t \geq 0}$ on $(0, \infty)$. Let $q_t(\lambda) = \mathrm{e}^{bt} v_t(\lambda)$ and let $q_t'(\lambda) = (\partial/\partial\lambda) q_t(\lambda)$. By differentiating both sides of (25) we see

$$\int_{(0,\infty)} \mathrm{e}^{-\lambda y} Q_t^b(x, \mathrm{d}y) = \exp\{-x v_t(\lambda)\} q_t'(\lambda), \qquad x > 0, \lambda \geq 0.$$

From (31) and (41) we have

$$q_t'(\lambda) = \mathrm{e}^{bt}\Big[h_t + \int_{(0,\infty)} \mathrm{e}^{-\lambda u} u l_t(\mathrm{d}u)\Big] = \exp\Big\{-\int_0^t \phi_0'(v_s(\lambda))\mathrm{d}s\Big\}.$$

Then we can define $Q_t^b(0, \mathrm{d}y)$ by (43) and extend $(Q_t^b)_{t \geq 0}$ to a Markov transition semigroup on $[0, \infty)$. The Feller property of the semigroup is immediate by (42).

Corollary 3.11 *Let $(Q_t^b)_{t \geq 0}$ be the transition semigroup defined by* (42). *Then we have $Q_t^b(0, \{0\}) = \mathrm{e}^{bt} h_t$ and $Q_t^b(x, \{0\}) = 0$ for $t \geq 0$ and $x > 0$.*

Theorem 3.12 *Suppose that $T > 0$ and $x > 0$. Then $\mathbf{P}_x^{b,T}(\mathrm{d}\omega) = x^{-1}\mathrm{e}^{bT} x(\omega, T)$ $\mathbf{Q}_x(\mathrm{d}\omega)$ defines a probability measure on $(\Omega, \mathscr{F}_T)$. Moreover, the process $\{(x(t), \mathscr{F}_t): 0 \leq t \leq T\}$ under this measure is a Markov process with transition semigroup $(Q_t^b)_{t \geq 0}$ given by* (42).

Proof Clearly, the probability measure $\mathbf{P}_x^{b,T}$ is carried by $\{x(T) > 0\} \in \mathscr{F}_T$. Then we have $\mathbf{P}_x^{b,T}\{x(t) > 0\} = 1$ for every $0 \leq t \leq T$. Let $0 \leq r \leq t \leq T$. Let F be a bounded $\mathscr{F}_r$-measurable random variable and f a bounded Borel function on $[0, \infty)$. By (33) and the Markov property under $\mathbf{Q}_x$,

$$\begin{aligned}
\mathbf{P}_x^{b,T}[Ff(x(t))] &= x^{-1}\mathrm{e}^{bT}\mathbf{Q}_x\big[Ff(x(t))x(T)\big] \\
&= x^{-1}\mathrm{e}^{bt}\mathbf{Q}_x\big[Ff(x(t))x(t)\big] \\
&= x^{-1}\mathrm{e}^{br}\mathbf{Q}_x\big[Fx(r)Q_{t-r}^b f(x(r))\big] \\
&= \mathbf{P}_x^{b,T}\big[FQ_{t-r}^b f(x(r))\big],
\end{aligned}$$

where we have used the relation $Q_{t-r}^b(x, \mathrm{d}y) = \mathrm{e}^{bt} x^{-1} y Q_{t-r}(x, \mathrm{d}y)$ for the third equality. Then $\{(x(t), \mathscr{F}_t) : 0 \leq t \leq T\}$ under $\mathbf{P}_x^{b,T}$ is a Markov process with transition semigroup $(Q_t^b)_{t \geq 0}$.

Recall that zero is a trap for the CB-process. Let $(Q_t^\circ)_{t \geq 0}$ denote the restriction of its transition semigroup $(Q_t)_{t \geq 0}$ to $(0, \infty)$. For a σ-finite measure μ on $(0, \infty)$ write

$$\mu Q_t^\circ(\mathrm{d}y) = \int_{(0,\infty)} \mu(\mathrm{d}x) Q_t^\circ(x, \mathrm{d}y), \qquad t \geq 0, y > 0.$$

A family of σ-finite measures $(\kappa_t)_{t>0}$ on $(0, \infty)$ is called an *entrance rule* for $(Q_t^\circ)_{t \geq 0}$ if $\kappa_r Q_{t-r}^\circ \leq \kappa_t$ for all $t > r > 0$ and $\kappa_r Q_{t-r}^\circ \to \kappa_t$ as $r \to t$. We call $(\kappa_t)_{t>0}$ an *entrance law* if $\kappa_r Q_{t-r}^\circ = \kappa_t$ for all $t > r > 0$.

The special case of the canonical representation (31) with $h_t = 0$ for all $t > 0$ is particularly interesting. In this case, we have

$$v_t(\lambda) = \int_{(0,\infty)} (1 - \mathrm{e}^{-\lambda u}) l_t(\mathrm{d}u), \qquad t > 0, \lambda \geq 0. \tag{44}$$

From this and (25) we have, for $t > 0$ and $\lambda \geq 0$,

$$\int_{(0,\infty)} (1 - \mathrm{e}^{-y\lambda}) l_t(\mathrm{d}y) = \lim_{x \to 0} x^{-1} \int_{(0,\infty)} (1 - \mathrm{e}^{-y\lambda}) Q_t^\circ(x, \mathrm{d}y).$$

Then, formally,

$$l_t = \lim_{x \to 0} x^{-1} Q_t(x, \cdot). \tag{45}$$

Theorem 3.13 *The cumulant semigroup $(v_t)_{t \geq 0}$ admits representation* (44) *if and only if $\phi'(\infty) = \infty$. In this case, the family $(l_t)_{t>0}$ is an entrance law for $(Q_t^\circ)_{t \geq 0}$.*

Proof By differentiating both sides of the general representation (31) we get

$$v_t'(\lambda) = h_t + \int_{(0,\infty)} u \mathrm{e}^{-\lambda u} l_t(\mathrm{d}u), \qquad t \geq 0, \lambda \geq 0. \tag{46}$$

From this and (41) it follows that

$$h_t = v_t'(\infty) = \exp\left\{ - \int_0^t \phi'(\bar{v}_s)\underset{\cdot}{s} \right\}.$$

Then we have $\phi'(\infty) = \infty$ if $h_t = 0$ for any $t > 0$. For the converse, assume that $\phi'(\infty) = \infty$. If Condition 3.5 holds, we have $\bar{v}_t < \infty$ for $t > 0$ by Theorem 3.6, so $h_t = 0$ by (31). If Condition 3.5 does not hold, we have $\bar{v}_t = \infty$ by Theorem 3.6. Since $\phi'(\infty) = \infty$, by (46) and (41) we see $h_t = v_t'(\infty) = 0$ for $t > 0$. If $(v_t)_{t \geq 0}$ admits the representation (44), we can use (24) to see, for $t > r > 0$ and $\lambda \geq 0$,

$$\begin{aligned} \int_{(0,\infty)} (1 - \mathrm{e}^{-\lambda u}) l_t(\mathrm{d}u) &= \int_{(0,\infty)} (1 - \mathrm{e}^{-u v_{t-r}(\lambda)}) l_r(\mathrm{d}u) \\ &= \int_{(0,\infty)} l_r(\mathrm{d}x) \int_{(0,\infty)} (1 - \mathrm{e}^{-\lambda u}) Q_{t-r}^\circ(x, \underset{\cdot}{u}). \end{aligned}$$

Then $(l_t)_{t>0}$ is an entrance law for $(Q_t^\circ)_{t \geq 0}$.

Corollary 3.14 *If Condition 3.5 holds, the cumulant semigroup admits the representation* (44) *and $t \mapsto \bar{v}_t = l_t(0, \infty)$ is the unique solution to the differential equation*

$$\frac{\mathrm{d}}{\mathrm{d}t} \bar{v}_t = -\phi(\bar{v}_t), \qquad t > 0 \tag{47}$$

with singular initial condition $\bar{v}_{0+} = \infty$.

Proof Under Condition 3.5, for every $t > 0$ we have $\bar{v}_t < \infty$ by Theorem 3.6. Moreover, the condition and the convexity of $z \mapsto \phi(z)$ imply $\phi'(\infty) = \infty$. Then we have the representation (44) by Theorem 3.13. The semigroup property of $(v_t)_{t\geq 0}$ implies $\bar{v}_{s+t} = v_s(\bar{v}_t)$ for $s > 0$ and $t > 0$. Then $t \mapsto \bar{v}_t$ satisfies (47). From (39) it is easy to see $\bar{v}_{0+} = \infty$. Suppose that $t \mapsto u_t$ and $t \mapsto v_t$ are two solutions to (47) with $u_{0+} = v_{0+} = \infty$. For any $\varepsilon > 0$ there exits $\delta > 0$ so that $u_s \geq v_\varepsilon$ for every $0 < s \leq \delta$. Since both $t \mapsto u_{s+t}$ and $t \mapsto v_{\varepsilon+t}$ are solutions to (47), we have $u_{s+t} \geq v_{\varepsilon+t}$ for $t \geq 0$ and $0 < s \leq \delta$ by Proposition 2.11. Then we can let $s \to 0$ and $\varepsilon \to 0$ to see $u_t \geq v_t$ for $t > 0$. By symmetry we get the uniqueness of the solution.

Theorem 3.15 *If* $\delta := \phi'(\infty) < \infty$, *then we have, for* $t \geq 0$ *and* $\lambda \geq 0$,

$$v_t(\lambda) = \mathrm{e}^{-\delta t}\lambda + \int_0^t \mathrm{e}^{-\delta s}\mathrm{d}s \int_{(0,\infty)} (1 - \mathrm{e}^{-uv_{t-s}(\lambda)})m(\mathrm{d}u), \tag{48}$$

that is, we have (31) *with*

$$h_t = \mathrm{e}^{-\delta t}, \quad l_t = \int_0^t \mathrm{e}^{-\delta s} m Q^\circ_{t-s}\mathrm{d}s, \quad t \geq 0. \tag{49}$$

In this case, the family $(l_t)_{t>0}$ *is an entrance rule for* $(Q^\circ_t)_{t\geq 0}$.

Proof If $\delta := \phi'(\infty) < \infty$, by (30) we must have $c = 0$. In this case, we can write the branching mechanism into

$$\phi(\lambda) = \delta\lambda + \int_{(0,\infty)} (\mathrm{e}^{-\lambda z} - 1)m(\mathrm{d}z), \qquad \lambda \geq 0. \tag{50}$$

By (21) and integration by parts,

$$\begin{aligned} v_t(\lambda)\mathrm{e}^{\delta t} =& \lambda + \int_0^t \delta v_s(\lambda)\mathrm{e}^{\delta s}\mathrm{d}s - \int_0^t \phi(v_s(\lambda))\mathrm{e}^{\delta s}\mathrm{d}s \\ =& \lambda + \int_0^t \mathrm{e}^{\delta s}\mathrm{d}s \int_{(0,\infty)} (1 - \mathrm{e}^{-uv_s(\lambda)})m(\mathrm{d}u). \end{aligned}$$

That gives (48) and (49). It is easy to see that $(l_t)_{t>0}$ is an entrance rule for $(Q^\circ_t)_{t\geq 0}$.

Example 3.16 Suppose that there are constants $c > 0$, $0 < \alpha \leq 1$ and b so that $\phi(z) = cz^{1+\alpha} + bz$. Then Condition 3.5 is satisfied. Let $q^0_\alpha(t)$ be defined as in Example 2.17. By letting $\lambda \to \infty$ in (28) we get $\bar{v}_t = c^{-1/\alpha}\mathrm{e}^{-bt}q^b_\alpha(t)^{-1/\alpha}$ for $t > 0$. In particular, if $\alpha = 1$, then (44) holds with

$$l_t(\mathrm{d}u) = \frac{\mathrm{e}^{-bt}}{c^2 q^b_1(t)^2}\exp\left\{-\frac{u}{cq^b_1(t)}\right\}\mathrm{d}u, \qquad t > 0, u > 0.$$

4 Positive Integral Functionals

In this section, we give characterizations of a class of positive integral functionals of the CB-process in terms of Laplace transforms. The corresponding results in the measure-valued setting can be found in Li [40]. For our purpose, it is more convenient to start the process from an arbitrary initial time $r \geq 0$. Let $X = (\Omega, \mathscr{F}, \mathscr{F}_{r,t}, x(t), \mathbf{Q}_{r,x})$ a càdlàg realization of the CB-process with transition semigroup $(Q_t)_{t\geq 0}$ defined by (21) and (25). For any $t \geq r \geq 0$ and $\lambda \geq 0$ we have

$$\mathbf{Q}_{r,x} \exp\{-\lambda x(t)\} = \exp\{-x u_r(\lambda)\}, \tag{51}$$

where $r \mapsto u_r(\lambda) := v_{t-r}(\lambda)$ is the unique bounded positive solution to

$$u_r(\lambda) + \int_r^t \phi(u_s(\lambda))\mathrm{d}s = \lambda, \qquad 0 \leq r \leq t. \tag{52}$$

Proposition 4.1 *For $\{t_1 < \cdots < t_n\} \subset [0, \infty)$ and $\{\lambda_1, \ldots, \lambda_n\} \subset [0, \infty)$ we have*

$$\mathbf{Q}_{r,x} \exp\left\{ -\sum_{j=1}^n \lambda_j x(t_j) 1_{\{r \leq t_j\}} \right\} = \exp\{-x u(r)\}, \quad 0 \leq r \leq t_n, \tag{53}$$

where $r \mapsto u(r)$ is a bounded positive function on $[0, t_n]$ solving

$$u(r) + \int_r^{t_n} \phi(u(s))\mathrm{d}s = \sum_{j=1}^n \lambda_j 1_{\{r \leq t_j\}}. \tag{54}$$

Proof We shall give the proof by induction in $n \geq 1$. For $n = 1$ the result follows from (51) and (52). Now supposing (53) and (54) are satisfied when n is replaced by $n - 1$, we prove they are also true for n. It is clearly sufficient to consider the case with $0 \leq r \leq t_1 < \cdots < t_n$. By the Markov property,

$$\begin{aligned}
&\mathbf{Q}_{r,x} \exp\left\{ -\sum_{j=1}^n \lambda_j x(t_j) \right\} \\
&= \mathbf{Q}_{r,x}\left[\mathbf{Q}_{r,x}\left(\exp\left\{ -\sum_{j=1}^n \lambda_j x(t_j) \right\} \Big| \mathscr{F}_{r,t_1} \right) \right] \\
&= \mathbf{Q}_{r,x}\left[\mathrm{e}^{-x(t_1)\lambda_1} \mathbf{Q}_{r,x}\left(\exp\left\{ -\sum_{j=2}^n \lambda_j x(t_j) \right\} \Big| \mathscr{F}_{r,t_1} \right) \right] \\
&= \mathbf{Q}_{r,x}\left[\mathrm{e}^{-x(t_1)\lambda_1} \mathbf{Q}_{t_1,x(t_1)}\left(\exp\left\{ -\sum_{j=2}^n \lambda_j x(t_j) \right\} \right) \right]
\end{aligned}$$

$$= \mathbf{Q}_{r,x} \exp\Big\{ - x(t_1)\lambda_1 - x(t_1)w(t_1)\Big\},$$

where $r \mapsto w(r)$ is a bounded positive Borel function on $[0, t_n]$ satisfying

$$w(r) + \int_r^{t_n} \phi(w(s))\mathrm{d}s = \sum_{j=2}^{n} \lambda_j 1_{\{r\le t_j\}}. \tag{55}$$

Then the result for $n = 1$ implies that

$$\mathbf{Q}_{r,x} \exp\Big\{ - \sum_{j=1}^{n} \lambda_j x(t_j)\Big\} = \exp\{-xu(r)\}$$

with $r \mapsto u(r)$ being a bounded positive Borel function on $[0, t_1]$ satisfying

$$u(r) + \int_r^{t_1} \phi(u(s))\mathrm{d}s = \lambda_1 + w(t_1). \tag{56}$$

Setting $u(r) = w(r)$ for $t_1 < r \le t_n$, from (55) and (56) one checks that $r \mapsto u(r)$ is a bounded positive solution to (54) on $[0, t_n]$.

Theorem 4.2 *Suppose that $t \ge 0$ and μ is a finite measure supported by $[0, t]$. Let $s \mapsto \lambda(s)$ be a bounded positive Borel function on $[0, t]$. Then we have*

$$\mathbf{Q}_{r,x} \exp\Big\{ - \int_{[r,t]} \lambda(s)x(s)\mu(\mathrm{d}s)\Big\} = \exp\{-xu(r)\}, \quad 0 \le r \le t, \tag{57}$$

where $r \mapsto u(r)$ is the unique bounded positive solution on $[0, t]$ to

$$u(r) + \int_r^{t} \phi(u(s))\mathrm{d}s = \int_{[r,t]} \lambda(s)\mu(\mathrm{d}s). \tag{58}$$

Proof *Step 1.* We first consider a bounded positive continuous function $s \mapsto \lambda(s)$ on $[0, t]$. To avoid triviality we assume $t > 0$. For any integer $n \ge 1$ define the finite measure μ_n on $[0, t]$ by

$$\mu_n(\mathrm{d}s) = \sum_{k=1}^{2^n} \mu[(k-1)t/2^n, kt/2^n)\delta_{kt/2^n}(\mathrm{d}s) + \mu(\{t\})\delta_t(\mathrm{d}s).$$

By Proposition 4.1 we see that

$$\mathbf{Q}_{r,x} \exp\Big\{ - \int_{[r,t]} \lambda(s)x(s)\mu_n(\mathrm{d}s)\Big\} = \exp\{-xu_n(r)\}, \tag{59}$$

where $r \mapsto u_n(r)$ is a bounded positive solution on $[0, t]$ to

$$u_n(r) + \int_r^t \phi(u_n(s))\mathrm{d}s = \int_{[r,t]} \lambda(s)\mu_n(\mathrm{d}s). \tag{60}$$

Let $v_n(r) = u_n(t-r)$ for $0 \leq r \leq t$. Observe that $\phi(z) \geq bz \geq -|b|z$ for every $z \geq 0$. From (60) we have

$$\begin{aligned} v_n(r) =& \int_{[t-r,t]} \lambda(s)\mu_n(\mathrm{d}s) - \int_0^r \phi(v_n(s))\mathrm{d}s \\ &\leq \sup_{0 \leq s \leq t} \lambda(s)\mu[0,t] + |b| \int_0^r v_n(s)\mathrm{d}s. \end{aligned}$$

By Gronwall's inequality it is easy to show that $\{v_n\}$ and hence $\{u_n\}$ is uniformly bounded on $[0, t]$. Let $q_n(t) = t$. For any $0 \leq s < t$ let $q_n(s) = (\lfloor 2^n s/t \rfloor + 1)t/2^n$, where $\lfloor 2^n s/t \rfloor$ denotes the integer part of $2^n s/t$. Then $s \leq q_n(s) \leq s + t/2^n$. It is easy to see that

$$\int_{[r,t]} f(s)\mu_n(\mathrm{d}s) = \int_{[r,t]} f(q_n(s))\mu(\mathrm{d}s)$$

for any bounded Borel function f on $[0, t]$. By the right continuity of $s \mapsto \lambda(s)$ and $s \mapsto x(s)$ we have

$$\lim_{n\to\infty} \int_{[r,t]} \lambda(s)\mu_n(\mathrm{d}s) = \int_{[r,t]} \lambda(s)\mu(\mathrm{d}s)$$

and

$$\lim_{n\to\infty} \int_{[r,t]} \lambda(s)x(s)\mu_n(\mathrm{d}s) = \int_{[r,t]} \lambda(s)x(s)\mu(\mathrm{d}s).$$

From (59) we see the limit $u(r) = \lim_{n\to\infty} u_n(r)$ exists and (57) holds for $0 \leq r \leq t$. Then we get (58) by letting $n \to \infty$ in (60).

Step 2. Let $B_0[0, \infty)$ be the set of bounded Borel functions $s \mapsto \lambda(s)$ for which there exist bounded positive solutions $r \mapsto u(r)$ of (58) such that (57) holds. Then $B_0[0, \infty)$ is closed under bounded pointwise convergence. The result of the first step shows that $B_0[0, \infty)$ contains all positive continuous functions on $[0, t]$. By Proposition 1.3 in Li [40, p. 3] we infer that $B_0[0, \infty)$ contains all bounded positive Borel functions on $[0, t]$.

Step 3. To show the uniqueness of the solution to (58), suppose that $r \mapsto v(r)$ is another bounded positive Borel function on $[0, t]$ satisfying this equation. Since $z \mapsto \phi(z)$ is locally Lipschitz, it is easy to find a constant $K \geq 0$ such that

$$\begin{aligned}|u(r)-v(r)| &\leq \int_r^t |\phi(u(s))-\phi(v(s))|\mathrm{d}s\\ &\leq K\int_r^t |u(s)-v(s)|\mathrm{d}s.\end{aligned}$$

Let $U(r) = |u(t-r) - v(t-r)|$ for $0 \leq r \leq t$. We have

$$U(r) \leq K\int_0^r U(s)\mathrm{d}s, \quad 0 \leq r \leq t.$$

Then Gronwall's inequality implies $U(r) = 0$ for every $0 \leq r \leq t$.

Suppose that $\mu(\mathrm{d}s)$ is a locally bounded Borel measure on $[0, \infty)$ and $s \mapsto \lambda(s)$ is a locally bounded positive Borel function on $[0, \infty)$. We define the *positive integral functional*:

$$A[r, t] := \int_{[r,t]} \lambda(s)x(s)\mu(\mathrm{d}s), \qquad t \geq r \geq 0.$$

By replacing $\lambda(s)$ with $\theta\lambda(s)$ in Theorem 4.2 for $\theta \geq 0$ we get a characterization of the Laplace transform of the random variable $A[r, t]$.

Theorem 4.3 *Let $t \geq 0$ be given. Let $\lambda \geq 0$ and let $s \mapsto \theta(s)$ be a bounded positive Borel function on $[0, t]$. Then for $0 \leq r \leq t$ we have*

$$\mathbf{Q}_{r,x}\exp\left\{-\lambda x(t) - \int_r^t \theta(s)x(s)\mathrm{d}s\right\} = \exp\{-xu(r)\}, \tag{61}$$

where $r \mapsto u(r)$ is the unique bounded positive solution on $[0, t]$ to

$$u(r) + \int_r^t \phi(u(s))\mathrm{d}s = \lambda + \int_r^t \theta(s)\mathrm{d}s. \tag{62}$$

Proof This follows by an application of Theorem 4.2 to the measure $\mu(\mathrm{d}s) = \mathrm{d}s + \delta_t(\mathrm{d}s)$ and the function $\lambda(s) = 1_{\{s<t\}}\theta(s) + 1_{\{s=t\}}\lambda$.

Corollary 4.4 *Let $X = (\Omega, \mathscr{F}, \mathscr{F}_t, x(t), \mathbf{Q}_x)$ be a Hunt realization of the CB-process started from time zero. Then we have, for $t, \lambda, \theta \geq 0$,*

$$\mathbf{Q}_x\exp\left\{-\lambda x(t) - \theta\int_0^t x(s)\mathrm{d}s\right\} = \exp\{-xv(t)\}, \tag{63}$$

where $t \mapsto v(t) = v(t, \lambda, \theta)$ is the unique positive solution to

$$\frac{\partial}{\partial t}v(t) = \theta - \phi(v(t)), \quad v(0) = \lambda. \tag{64}$$

Proof By Theorem 4.3 we have (63) with $v(t) = u_t(0)$, where $r \mapsto u_t(r)$ is the unique bounded positive solution on $[0, t]$ to

$$u(r) + \int_r^t \phi(u(s))\mathrm{d}s = \lambda + (t - r)\theta.$$

Then $r \mapsto v(r) := u_t(t - r)$ is the unique bounded positive solution on $[0, t]$ of

$$v(r) + \int_0^r \phi(v(s))\mathrm{d}s = \lambda + r\theta. \tag{65}$$

Clearly, we can extend (65) to all $r \geq 0$ and the extended equation is equivalent with the differential equation (64). The uniqueness of the solution follows by Gronwall's inequality.

Corollary 4.5 *Let $X = (\Omega, \mathscr{F}, \mathscr{F}_t, x(t), \mathbf{Q}_x)$ be a Hunt realization of the CB-process started from time zero. Then we have, for $t, \theta \geq 0$,*

$$\mathbf{Q}_x \exp\left\{ -\theta \int_0^t x(s)\mathrm{d}s \right\} = \exp\{-xv(t)\}. \tag{66}$$

where $t \mapsto v(t) = v(t, \theta)$ is the unique positive solution to

$$\frac{\partial}{\partial t} v(t) = \theta - \phi(v(t)), \quad v(0) = 0. \tag{67}$$

Corollary 4.6 *Let $t \geq 0$ be given. Let ϕ_1 and ϕ_2 be two branching mechanisms in form* (19) *satisfying $\phi_1(z) \geq \phi_2(z)$ for all $z \geq 0$. Let $t \mapsto v_i(t)$ be the solution to* (64) *or* (65) *with $\phi = \phi_i$. Then $v_1(t) \leq v_2(t)$ for all $t \geq 0$.*

Proof Fix $t \geq 0$ and let $u_i(r) = v_i(t - r)$ for $0 \leq r \leq t$. Then $r \mapsto u_1(r)$ is the unique bounded positive solution on $[0, t]$ of

$$u(r) + \int_r^t \phi_1(u(s))\mathrm{d}s = \lambda + (t - r)\theta$$

and $r \mapsto u_2(r)$ is the unique bounded positive solution on $[0, t]$ of

$$u(r) + \int_r^t \phi_1(u(s))\mathrm{d}s = \lambda + \int_r^t [\theta + g(s)]\mathrm{d}s,$$

where $g(s) = \phi_1(u_2(s)) - \phi_2(u_2(s)) \geq 0$. By Theorem 4.3 one can see $u_1(r) \leq u_2(r)$ for all $0 \leq r \leq t$.

Recall that $\phi'(\infty)$ is given by (30). Under the condition $\phi'(\infty) > 0$, we have $\phi(z) \to \infty$ as $z \to \infty$, so the inverse $\phi^{-1}(\theta) := \inf\{z \geq 0 : \phi(z) > \theta\}$ is well defined for $\theta \geq 0$.

Proposition 4.7 *For $\theta > 0$ let $t \mapsto v(t, \theta)$ be the unique positive solution to* (67). *Then* $\lim_{t\to\infty} v(t, \theta) = \infty$ *if* $\phi'(\infty) \leq 0$, *and* $\lim_{t\to\infty} v(t, \theta) = \phi^{-1}(\theta)$ *if* $\phi'(\infty) > 0$.

Proof By Proposition 2.11 we have $\mathbf{Q}_x\{x(t) > 0\} > 0$ for every $x > 0$ and $t \geq 0$. From (66) we see $t \mapsto v(t, \theta)$ is strictly increasing, so $(\partial/\partial t)v(t, \theta) > 0$ for all $\theta > 0$. Let $v(\infty, \theta) = \lim_{t\to\infty} v(t, \theta) \in (0, \infty]$. In the case $\phi'(\infty) \leq 0$, we clearly have $\phi(z) \leq 0$ for all $z \geq 0$. Then $(\partial/\partial t)v(t, \theta) \geq \theta > 0$ and $v(\infty, \theta) = \infty$. In the case $\phi'(\infty) > 0$, we note

$$\phi(v(t, \theta)) = \theta - \frac{\partial}{\partial t} v(t, \theta) < \theta,$$

and hence $v(t, \theta) < \phi^{-1}(\theta)$, implying $v(\infty, \theta) \leq \phi^{-1}(\theta) < \infty$. It follows that

$$0 = \lim_{t\to\infty} \frac{\partial}{\partial t} v(t, \theta) = \theta - \lim_{t\to\infty} \phi(v(t, \theta)) = \theta - \phi(v(\infty, \theta)).$$

Then we have $v(\infty, \theta) = \phi^{-1}(\theta)$.

Theorem 4.8 *Let $X = (\Omega, \mathscr{F}, \mathscr{F}_t, x(t), \mathbf{Q}_x)$ be a Hunt realization of the CB-process started from time zero. If $\phi'(\infty) > 0$, then for $x > 0$ and $\theta > 0$ we have*

$$\mathbf{Q}_x \exp\left\{ -\theta \int_0^\infty x(s)\mathrm{d}s \right\} = \exp\{-x\phi^{-1}(\theta)\}$$

and

$$\mathbf{Q}_x\left\{ \int_0^\infty x(s)\mathrm{d}s < \infty \right\} = \exp\{-x\phi^{-1}(0)\},$$

where $\phi^{-1}(0) = \inf\{z > 0 : \phi(z) \geq 0\}$. If $\phi'(\infty) \leq 0$, then for any $x > 0$ we have

$$\mathbf{Q}_x\left\{ \int_0^\infty x(s)\mathrm{d}s < \infty \right\} = 0.$$

Proof In view of (66), we have

$$\mathbf{Q}_x \exp\left\{ -\theta \int_0^\infty x(s)\mathrm{d}s \right\} = \lim_{t\to\infty} \exp\{-xv(t, \theta)\}.$$

Then the result follows from Proposition 4.7.

5 Construction of CBI-Processes

Let $\{p(j) : j \in \mathbb{N}\}$ and $\{q(j) : j \in \mathbb{N}\}$ be probability distributions on $\mathbb{N} := \{0, 1, 2, \ldots\}$ with generating functions g and h, respectively. Suppose that $\{\xi_{n,i} : n, i = 1, 2, \ldots\}$ is a family of $\mathbb{N}$-valued i.i.d. random variables with distribution $\{p(j) : j \in \mathbb{N}\}$ and $\{\eta_n : n = 1, 2, \ldots\}$ is a family of $\mathbb{N}$-valued i.i.d. random variables with distribution $\{q(j) : j \in \mathbb{N}\}$. We assume the two families are independent of each other. Given an $\mathbb{N}$-valued random variable $y(0)$ independent of $\{\xi_{n,i}\}$ and $\{\eta_n\}$, we define inductively

$$y(n) = \sum_{i=1}^{y(n-1)} \xi_{n,i} + \eta_n, \qquad n = 1, 2, \ldots. \tag{68}$$

This is clearly a generalization of (7). For $i \in \mathbb{N}$ let $\{Q(i, j) : j \in \mathbb{N}\}$ denote the i-fold convolution of $\{p(j) : j \in \mathbb{N}\}$. Let

$$P(i, j) = (Q(i, \cdot) * q)(j) = (p^{*i} * q)(j), \qquad i, j \in \mathbb{N}.$$

For any $n \geq 1$ and $\{i_0, \cdots, i_{n-1} = i, j\} \subset \mathbb{N}$ we have

$$\begin{aligned}
&\mathbf{P}\Big(y(n) = j \Big| y(0) = i_0, y(1) = i_1, \cdots, y(n-1) = i_{n-1}\Big) \\
&\quad = \mathbf{P}\bigg(\sum_{k=1}^{y(n-1)} \xi_{n,k} + \eta_n = j \bigg| y(n-1) = i\bigg) \\
&\quad = \mathbf{P}\bigg(\sum_{k=1}^{i} \xi_{n,k} + \eta_n = j\bigg) = P(i, j).
\end{aligned}$$

Then $\{y(n) : n \geq 0\}$ is a Markov chain with one-step transition matrix $P = (P(i, j) : i, j \in \mathbb{N})$. The random variable $y(n)$ can be thought of as the number of individuals in generation n of a population system with immigration. After one unit time, each of the $y(n)$ individuals splits independently of others into a random number of offsprings according to the distribution $\{p(j) : j \in \mathbb{N}\}$ and a random number of immigrants are added to the system according to the distribution $\{q(j) : j \in \mathbb{N}\}$. It is easy to see that

$$\sum_{j=0}^{\infty} P(i, j) z^j = g(z)^i h(z), \qquad |z| \leq 1. \tag{69}$$

A Markov chain in $\mathbb{N}$ with one-step transition matrix defined by (69) is called a *Galton–Watson branching process with immigration* (GWI-process) or a *Bienaymé–Galton–Watson branching process with immigration* (BGWI-process) with *branching distribution* given by g and *immigration distribution* given by h. When $h \equiv 1$, this reduces to the GW-process defined before. For any $n \geq 1$ the n-step transition matrix of the GWI-process is just the n-fold product $P^n = (P^n(i,j) : i, j \in \mathbb{N})$.

Proposition 5.1 *For any $n \geq 1$ and $i \in \mathbb{N}$ we have*

$$\sum_{j=0}^{\infty} P^n(i,j)z^j = g^{\circ n}(z)^i \prod_{j=1}^{n} h(g^{\circ(j-1)}(z)), \qquad |z| \leq 1. \tag{70}$$

Proof From (69) we see (70) holds for $n = 1$. Now suppose that (70) holds for some $n \geq 1$. We have

$$\begin{aligned}\sum_{j=0}^{\infty} P^{n+1}(i,j)z^j &= \sum_{j=0}^{\infty}\sum_{k=0}^{\infty} P(i,k)P^n(k,j)z^j \\ &= \sum_{k=0}^{\infty} P(i,k)g^{\circ n}(z)^k \prod_{j=1}^{n} h(g^{\circ j-1}(z)) \\ &= g(g^{\circ n}(z))^i h(g^{\circ n}(z)) \prod_{j=1}^{n} h(g^{\circ j-1}(z)) \\ &= g^{\circ(n+1)}(z)^i \prod_{j=1}^{n+1} h(g^{\circ j-1}(z)).\end{aligned}$$

Then (70) also holds when n is replaced by $n+1$. That gives the result by induction.

Suppose that for each integer $k \geq 1$ we have a GWI-process $\{y_k(n) : n \geq 0\}$ with branching distribution given by the probability generating function g_k and immigration distribution given by the probability generating function h_k. Let $z_k(n) = y_k(n)/k$. Then $\{z_k(n) : n \geq 0\}$ is a Markov chain with state space $E_k := \{0, 1/k, 2/k, \ldots\}$ and n-step transition probability $P_k^n(x, dy)$ determined by

$$\int_{E_k} e^{-\lambda y} P_k^n(x, dy) = g_k^{\circ n}(e^{-\lambda/k})^{kx} \prod_{j=1}^{n} h_k(g_k^{\circ(j-1)}(e^{-\lambda/k})). \tag{71}$$

Suppose that $\{\gamma_k\}$ is a positive real sequence so that $\gamma_k \to \infty$ increasingly as $k \to \infty$. Let $\lfloor \gamma_k t \rfloor$ denote the integer part of $\gamma_k t$. In view of (71), given $z_k(0) = x \in E_k$, the random variable $z_k(\lfloor \gamma_k t \rfloor) = k^{-1} y_k(\lfloor \gamma_k t \rfloor)$ has distribution $P_k^{\lfloor \gamma_k t \rfloor}(x, \cdot)$ on E_k determined by

$$\begin{aligned}&\int_{E_k} e^{-\lambda y} P_k^{\lfloor \gamma_k t \rfloor}(x, dy) \\ &\quad = g_k^{\circ \lfloor \gamma_k t \rfloor}(e^{-\lambda/k})^{kx} \prod_{j=1}^{\lfloor \gamma_k t \rfloor} h_k(g_k^{j-1}(e^{-\lambda/k}))\end{aligned}$$

$$= \exp\left\{xk \log g_k^{\circ\lfloor\gamma_k t\rfloor}(\mathrm{e}^{-\lambda/k})\right\} \exp\left\{\sum_{j=1}^{\lfloor\gamma_k t\rfloor} \log h_k(g_k^{j-1}(\mathrm{e}^{-\lambda/k}))\right\}$$

$$= \exp\left\{-xv_k(t,\lambda) - \int_0^{\gamma_k^{-1}\lfloor\gamma_k t\rfloor} \bar{\psi}_k(v_k(s,\lambda))\mathrm{d}s\right\}, \tag{72}$$

where $v_k(t,\lambda)$ is given by (14) and

$$\bar{\psi}_k(z) = -\gamma_k \log h_k(\mathrm{e}^{-z/k}). \tag{73}$$

For any $z \geq 0$ we have

$$\bar{\psi}_k(z) = -\gamma_k \log\left[1 - \gamma_k^{-1}\tilde{\psi}_k(z)\right], \tag{74}$$

where

$$\tilde{\psi}_k(z) = \gamma_k[1 - h_k(\mathrm{e}^{-z/k})]. \tag{75}$$

Lemma 5.2 *Suppose that the sequence $\{\tilde{\psi}_k\}$ is uniformly bounded on each bounded interval. Then we have $\lim_{k\to\infty} |\bar{\psi}_k(z) - \tilde{\psi}_k(z)| = 0$ uniformly on each bounded interval.*

Proof This is immediate by the relation (74).

Condition 5.3 There is a function ψ on $[0,\infty)$ such that $\tilde{\psi}_k(z) \to \psi(z)$ uniformly on $[0,a]$ for every $a \geq 0$ as $k \to \infty$.

Proposition 5.4 *Suppose that Condition 5.3 is satisfied. Then the limit function ψ has representation*

$$\psi(z) = \beta z + \int_{(0,\infty)} \left(1 - \mathrm{e}^{-zu}\right)\nu(\mathrm{d}u), \qquad z \geq 0, \tag{76}$$

where $\beta \geq 0$ is a constant and $\nu(\mathrm{d}u)$ is a σ-finite measure on $(0,\infty)$ satisfying

$$\int_{(0,\infty)} (1 \wedge u)\nu(\mathrm{d}u) < \infty.$$

Proof It is well known that ψ has representation (76) if and only if $\mathrm{e}^{-\psi} = L_\mu$ is the Laplace transform of an infinitely divisible probability distribution μ on $[0,\infty)$; see, e.g., Theorem 1.39 in Li [40, p. 20]. In view of (75), the function $\tilde{\psi}_k$ can be represented by a special form (76), so $\mathrm{e}^{-\tilde{\psi}_k} = L_{\mu_k}$ is the Laplace transform of an infinitely divisible distribution μ_k on $[0,\infty)$. By Lemma 5.2 and Condition 5.3 we have $\bar{\psi}_k(z) \to \psi(z)$ uniformly on $[0,a]$ for every $a \geq 0$ as $k \to \infty$. By Theorem 1.2 there is a probability distribution μ on $[0,\infty)$ so that $\mu = \lim_{k\to\infty} \mu_k$ weakly and $\mathrm{e}^{-\psi} = L_\mu$. Clearly μ is also infinitely divisible, so ψ has representation (76).

Proposition 5.5 *For any function* ψ *with representation* (76) *there is a sequence* $\{\tilde{\psi}_k\}$ *in the form of* (73) *satisfying Condition 5.3.*

Proof This is similar to the proof of Proposition 2.6 and is left to the reader as an exercise.

Theorem 5.6 *Suppose that* ϕ *and* ψ *are given by* (19) *and* (76), *respectively. For any* $\lambda \geq 0$ *let* $t \mapsto v_t(\lambda)$ *be the unique positive solution to* (21). *Then there is a Feller transition semigroup* $(P_t)_{t\geq 0}$ *on* $[0, \infty)$ *defined by*

$$\int_{[0,\infty)} \mathrm{e}^{-\lambda y} P_t(x, \mathrm{d}y) = \exp\bigg\{ - x v_t(\lambda) - \int_0^t \psi(v_s(\lambda))\mathrm{d}s \bigg\}. \tag{77}$$

Proof This follows by arguments similar to those in Sect. 2.

If a Markov process in $[0, \infty)$ has transition semigroup $(P_t)_{t\geq 0}$ defined by (77), we call it a *continuous-state branching process with immigration* (CBI-process) with *branching mechanism* ϕ and *immigration mechanism* ψ. In particular, if

$$\int_{(0,\infty)} u\nu(\mathrm{d}u) < \infty, \tag{78}$$

one can differentiate both sides of (77) and use (32) to see

$$\int_{[0,\infty)} y P_t(x, \mathrm{d}y) = x\mathrm{e}^{-bt} + \psi'(0) \int_0^t \mathrm{e}^{-bs}\mathrm{d}s, \tag{79}$$

where

$$\psi'(0) = \beta + \int_{(0,\infty)} u\nu(\mathrm{d}u). \tag{80}$$

Proposition 5.7 *Suppose that* $\{(y_1(t), \mathscr{G}_t^1) : t \geq 0\}$ *and* $\{(y_2(t), \mathscr{G}_t^2) : t \geq 0\}$ *are two independent CBI-processes with branching mechanism* ϕ *and immigration mechanisms* ψ_1 *and* ψ_2, *respectively. Let* $y(t) = y_1(t) + y_2(t)$ *and* $\mathscr{G}_t = \sigma(\mathscr{G}_t^1 \cup \mathscr{G}_t^2)$. *Then* $\{(y(t), \mathscr{G}_t) : t \geq 0\}$ *is a CBI-processes with branching mechanism* ϕ *and immigration mechanism* $\psi = \psi_1 + \psi_2$.

Proof Let $t \geq r \geq 0$ and for $i = 1, 2$ let F_i be a bounded positive $\mathscr{G}_r^i$-measurable random variable. For any $\lambda \geq 0$ we have

$$\begin{aligned}
\mathbf{P}\big[F_1 F_2 \mathrm{e}^{-\lambda y(t)}\big] &= \mathbf{P}\big[F_1 \mathrm{e}^{-\lambda y_1(t)}\big]\mathbf{P}\big[F_2 \mathrm{e}^{-\lambda y_2(t)}\big] \\
&= \mathbf{P}\bigg[F_1 \exp\bigg\{ - y_1(r)v_{t-r}(\lambda) - \int_0^{t-r} \psi_1(v_s(\lambda))\mathrm{d}s\bigg\}\bigg] \\
&\quad \cdot \mathbf{P}\bigg[F_2 \exp\bigg\{ - y_2(r)v_{t-r}(\lambda) - \int_0^{t-r} \psi_2(v_s(\lambda))\mathrm{d}s\bigg\}\bigg]
\end{aligned}$$

$$=\mathbf{P}\Big[F_1 F_2 \exp\Big\{ - y(r)v_{t-r}(\lambda) - \int_0^{t-r} \psi(v_s(\lambda))\mathrm{d}s\Big\}\Big].$$

As in the proof of Proposition 2.13, one can see $\{(y(t), \mathscr{G}_t) : t \geq 0\}$ is a CBI-processes with branching mechanism ϕ and immigration mechanism ψ.

The next theorem follows by a modification of the proof of Theorem 2.14.

Theorem 5.8 *Suppose that Conditions 2.4 and 5.3 are satisfied. Let $\{y(t) : t \geq 0\}$ be a CBI-process with transition semigroup $(P_t)_{t\geq 0}$ defined by (77). For $k \geq 1$ let $\{z_k(n) : n \geq 0\}$ be a Markov chain with state space $E_k := \{0, k^{-1}, 2k^{-1}, \ldots\}$ and n-step transition probability $P_k^n(x, \mathrm{d}y)$ determined by (71). If $z_k(0)$ converges to $y(0)$ in distribution, then $\{z_k(\lfloor \gamma_k t\rfloor) : t \geq 0\}$ converges to $\{y(t) : t \geq 0\}$ in distribution on $D[0, \infty)$.*

The convergence of rescaled GWI-processes to CBI-processes have been studied by many authors; see, e.g., Aliev [1], Kawazu and Watanabe [29] and Li [39, 40].

Example 5.9 The transition semigroup $(Q_t^b)_{t\geq 0}$ defined by (42) corresponds to a CBI-process with branching mechanism ϕ and immigration mechanism ϕ_0'.

Example 5.10 Suppose that $c > 0$, $0 < \alpha \leq 1$ and b are constants and let $\phi(z) = cz^{1+\alpha} + bz$. Let $v_t(\lambda)$ and $q_\alpha^0(t)$ be defined as in Example 2.17. Let $\beta \geq 0$ and let $\psi(z) = \beta z^\alpha$. We can use (77) to define the transition semigroup $(P_t)_{t\geq 0}$. It is easy to show that

$$\int_{[0,\infty)} \mathrm{e}^{-\lambda y} P_t(x, \mathrm{d}y) = \frac{1}{\big[1 + cq_\alpha^b(t)\lambda^\alpha\big]^{\beta/c\alpha}} \mathrm{e}^{-xv_t(\lambda)}, \qquad \lambda \geq 0.$$

6 Structures of Sample Paths

In this section, we give some reconstructions of the type of Pitman and Yor [48] for the CB- and CBI-processes, which reveal the structures of their sample paths. Let $(Q_t)_{t\geq 0}$ be the transition semigroup of the CB-process with branching mechanism ϕ given by (19). Let $(Q_t^\circ)_{t\geq 0}$ be the restriction of $(Q_t)_{t\geq 0}$ on $(0, \infty)$. Let $D[0, \infty)$ denote the space of positive càdlàg paths on $[0, \infty)$. On this space, we define the σ-algebras $\mathscr{A} = \sigma(\{w(s) : 0 \leq s < \infty\})$ and $\mathscr{A}_t = \sigma(\{w(s) : 0 \leq s \leq t\})$ for $t \geq 0$. For any $w \in D[0, \infty)$ let $\tau_0(w) = \inf\{s > 0 : w(s) \text{ or } w(s-) = 0\}$. Let $D_0[0, \infty)$ be the set of paths $w \in D[0, \infty)$ such that $w(t) = 0$ for $t \geq \tau_0(w)$. Let $D_1[0, \infty)$ be the set of paths $w \in D_0[0, \infty)$ satisfying $w(0) = 0$. Then both $D_0[0, \infty)$ and $D_1[0, \infty)$ are $\mathscr{A}$-measurable subsets of $D[0, \infty)$.

Theorem 6.1 *Suppose that $\phi'(\infty) = \infty$ and let $(l_t)_{t>0}$ be the entrance law for $(Q_t^\circ)_{t\geq 0}$ determined by (44). Then there is a unique σ-finite measure $\mathbf{N}_0$ on $(D[0, \infty)$,*

$\mathscr{A}$) supported by $D_1[0,\infty)$ such that, for $0 < t_1 < t_2 < \cdots < t_n$ and $x_1, x_2, \ldots, x_n \in (0,\infty)$,

$$\begin{aligned}&\mathbf{N}_0(w(t_1) \in \mathrm{d}x_1, w(t_2) \in \mathrm{d}x_2, \ldots, w(t_n) \in \mathrm{d}x_n)\\&\quad = l_{t_1}(\mathrm{d}x_1) Q^\circ_{t_2-t_1}(x_1, \mathrm{d}x_2) \cdots Q^\circ_{t_n-t_{n-1}}(x_{n-1}, \mathrm{d}x_n).\end{aligned} \tag{81}$$

Proof Recall that $(Q^b_t)_{t\geq 0}$ is the transition semigroup on $[0,\infty)$ given by (42). Let $X = (D[0,\infty), \mathscr{A}, \mathscr{A}_t, w(t), \mathbf{Q}_x)$ be the canonical realization of $(Q_t)_{t\geq 0}$ and $Y = (D[0,\infty), \mathscr{A}, \mathscr{A}_t, w(t), \mathbf{Q}^b_x)$ the canonical realization of $(Q^b_t)_{t\geq 0}$. For any $T > 0$ there is a probability measure $\mathbf{P}^{b,T}_0$ on $(D[0,\infty), \mathscr{A})$ supported by $D_1[0,\infty)$ so that

$$\begin{aligned}&\mathbf{P}^{b,T}_0[F(\{w(s) : s \geq 0\})G(\{w(T+s) : s \geq 0\})]\\&\quad = \mathbf{Q}^b_0[F(\{w(s) : s \geq 0\})\mathbf{Q}_{w(T)}G(\{w(s) : s \geq 0\})],\end{aligned}$$

where F is a bounded $\mathscr{A}_T$-measurable function and G is a bounded $\mathscr{A}$-measurable function. The formula above means that under $\mathbf{P}^{b,T}_0$ the random path $\{w(s) : s \in [0,T]\}$ is a Markov process with initial state $w(0) = 0$ and transition semigroup $(Q^b_t)_{t\geq 0}$ and $\{w(s) : s \in [T,\infty)\}$ is a Markov process with transition semigroup $(Q_t)_{t\geq 0}$. Then Corollary 3.11 implies $\mathbf{P}^{b,T}_0(w(s) = 0) = Q^b_s(0,\{0\}) = 0$ for every $0 < s \leq T$. Let $\mathbf{N}^{b,T}_0(\mathrm{d}w) = \mathrm{e}^{-bT} w(T)^{-1} 1_{\{w(T)>0\}} \mathbf{P}^{b,T}_0(\mathrm{d}w)$. We have

$$\begin{aligned}&\mathbf{N}^{b,T}_0(w(s_1) \in \mathrm{d}x_1, w(s_2) \in \mathrm{d}x_2, \ldots, w(s_m) \in \mathrm{d}x_m, w(T) \in \mathrm{d}z,\\&\qquad\qquad w(t_1) \in \mathrm{d}y_1, w(t_2) \in \mathrm{d}y_2, \ldots, w(t_n) \in \mathrm{d}y_n)\\&\quad = Q^b_{s_1}(0, \mathrm{d}x_1) Q^b_{s_2-s_1}(x_1, \mathrm{d}x_2) \cdots Q^b_{s_m-s_{m-1}}(x_{m-1}, \mathrm{d}x_m) Q^b_{T-s_m}(x_m, \mathrm{d}z)\\&\qquad \mathrm{e}^{-bT} z^{-1} Q_{t_1-T}(z, \mathrm{d}y_1) Q_{t_2-t_1}(y_1, \mathrm{d}y_2) \cdots Q_{t_n-t_{n-1}}(y_{n-1}, \mathrm{d}y_n)\\&\quad = Q^b_{s_1}(0, \mathrm{d}x_1) x_1^{-1} x_1 \cdots Q^b_{s_m-s_{m-1}}(x_{m-1}, \mathrm{d}x_m) x_m^{-1} x_m Q^b_{T-s_m}(x_m, \mathrm{d}z)\\&\qquad \mathrm{e}^{-bT} z^{-1} Q_{t_1-T}(z, \mathrm{d}y_1) Q_{t_2-t_1}(y_1, \mathrm{d}y_2) \cdots Q_{t_n-t_{n-1}}(y_{n-1}, \mathrm{d}y_n)\\&\quad = l_{s_1}(\mathrm{d}x_1) Q^\circ_{s_2-s_1}(x_1, \mathrm{d}x_2) \cdots Q^\circ_{s_m-s_{m-1}}(x_{m-1}, \mathrm{d}x_m) Q^\circ_{T-s_m}(x_m, \mathrm{d}z)\\&\qquad Q_{t_1-T}(z, \mathrm{d}y_1) Q_{t_2-t_1}(y_1, \mathrm{d}y_2) \cdots Q_{t_n-t_{n-1}}(y_{n-1}, \mathrm{d}y_n),\end{aligned} \tag{82}$$

where $0 < s_1 < \cdots < s_m < T < t_1 < \cdots < t_n$ and $x_1, \ldots, x_m, z, y_1, \ldots, y_n \in (0,\infty)$. Then for any $T_1 \geq T_2 > 0$ the two measures $\mathbf{N}^{b,T_1}_0$ and $\mathbf{N}^{b,T_2}_0$ coincide on $\{w \in D_1[0,\infty) : \tau_0(w) > T_1\}$, so the increasing limit $\mathbf{N}_0 := \lim_{T\to 0} \mathbf{N}^{b,T}_0$ exists and defines a σ-finite measure supported by $D_1[0,\infty)$. From (82) we get (81). The uniqueness of the measure $\mathbf{N}_0$ satisfying (81) follows by the measure extension theorem.

The elements of $D_1[0,\infty)$ are called *excursions* and the measure $\mathbf{N}_0$ is referred to as the *excursion law* for the CB-process. In view of (45) and (81), we have formally, for $0 < t_1 < t_2 < \cdots < t_n$ and $x_1, x_2, \ldots, x_n \in (0,\infty)$,

$$\begin{aligned}
&\mathbf{N}_0(w(t_1) \in \mathrm{d}x_1, w(t_2) \in \mathrm{d}x_2, \ldots, w(t_n) \in \mathrm{d}x_n) \\
&\quad = \lim_{x \to 0} x^{-1} Q^\circ_{t_1}(x, \mathrm{d}x_1) Q^\circ_{t_2 - t_1}(x_1, \mathrm{d}x_2) \cdots Q^\circ_{t_n - t_{n-1}}(x_{n-1}, \mathrm{d}x_n) \\
&\quad = \lim_{x \to 0} x^{-1} \mathbf{Q}_x(w(t_1) \in \mathrm{d}x_1, w(t_2) \in \mathrm{d}x_2, \ldots, w(t_n) \in \mathrm{d}x_n),
\end{aligned} \tag{83}$$

which explains why $\mathbf{N}_0$ is supported by $D_1[0, \infty)$.

From (81) we see that the excursion law is Markovian, namely, the path $\{w(t) : t > 0\}$ behaves under this law as a Markov process with transition semigroup $(Q^\circ_t)_{t \geq 0}$. Based on the excursion law, a reconstruction of the CB-process is given in the following theorem.

Theorem 6.2 *Suppose that $\phi'(\infty) = \infty$. Let $z \geq 0$ and let $N_z = \sum_{i=1}^{\infty} \delta_{w_i}$ be a Poisson random measure on $D[0, \infty)$ with intensity $z\mathbf{N}_0(\mathrm{d}w)$. Let $X_0 = z$ and for $t > 0$ let*

$$X^z_t = \int_{D[0,\infty)} w(t) N_z(\mathrm{d}w) = \sum_{i=1}^{\infty} w_i(t). \tag{84}$$

For $t \geq 0$ let $\mathscr{G}^z_t$ be the σ-algebra generated by the collection of random variables $\{N_z(A) : A \in \mathscr{A}_t\}$. Then $\{(X^z_t, \mathscr{G}^z_t) : t \geq 0\}$ is a CB-process with branching mechanism ϕ.

Proof It is easy to see that $\{X^z_t : t \geq 0\}$ is adapted relative to the filtration $\{\mathscr{G}^z_t : t \geq 0\}$. We claim that the random variable X^z_t has distribution $Q_t(z, \cdot)$ on $[0, \infty)$. Indeed, for $t = 0$ this is immediate. For any $t > 0$ and $\lambda \geq 0$ we have

$$\begin{aligned}
\mathbf{P}\big[\exp\{-\lambda X^z_t\}\big] &= \exp\bigg\{-z \int_{D[0,\infty)} (1 - \mathrm{e}^{-\lambda w(t)}) \mathbf{N}_0(\mathrm{d}w)\bigg\} \\
&= \exp\bigg\{-z \int_{(0,\infty)} (1 - \mathrm{e}^{-\lambda u}) l_t(\mathrm{d}u)\bigg\} = \exp\{-z v_t(\lambda)\}.
\end{aligned}$$

By the Markov property (81), for any $t \geq r > 0$ and any bounded $\mathscr{A}_r$-measurable function H on $D[0, \infty)$ we have

$$\begin{aligned}
&\int_{D[0,\infty)} H(w)(1 - \mathrm{e}^{-\lambda w(t)}) \mathbf{N}_0(\mathrm{d}w) \\
&\qquad = \int_{D[0,\infty)} H(w)(1 - \mathrm{e}^{-v_{t-r}(\lambda) w(r)}) \mathbf{N}_0(\mathrm{d}w).
\end{aligned}$$

It follows that, for any bounded positive $\mathscr{A}_r$-measurable function F on $D[0, \infty)$,

$$\begin{aligned}
&\mathbf{P}\bigg[\exp\bigg\{-\int_{D[0,\infty)} F(w) N_z(\mathrm{d}w)\bigg\} \cdot \exp\Big\{-\lambda X^z_t\Big\}\bigg] \\
&\qquad = \mathbf{P}\bigg[\exp\bigg\{-\int_{D[0,\infty)} \big[F(w) + \lambda w(t)\big] N_z(\mathrm{d}w)\bigg\}\bigg]
\end{aligned}$$

$$
\begin{aligned}
&= \exp\bigg\{-z\int_{D[0,\infty)}\big(1-\mathrm{e}^{-F(w)-\lambda w(t)}\big)\mathbf{N}_0(\mathrm{d}w)\bigg\} \\
&= \exp\bigg\{-z\int_{D[0,\infty)}\big(1-\mathrm{e}^{-F(w)}\big)\mathbf{N}_0(\mathrm{d}w)\bigg\} \\
&\qquad\cdot\exp\bigg\{-z\int_{D[0,\infty)}\mathrm{e}^{-F(w)}\big(1-\mathrm{e}^{-\lambda w(t)}\big)\mathbf{N}_0(\mathrm{d}w)\bigg\} \\
&= \exp\bigg\{-z\int_{D[0,\infty)}\big(1-\mathrm{e}^{-F(w)}\big)\mathbf{N}_0(\mathrm{d}w)\bigg\} \\
&\qquad\cdot\exp\bigg\{-z\int_{D[0,\infty)}\mathrm{e}^{-F(w)}\big(1-\mathrm{e}^{-w(r)v_{t-r}(\lambda)}\big)\mathbf{N}_0(\mathrm{d}w)\bigg\} \\
&= \exp\bigg\{-z\int_{D[0,\infty)}\big(1-\mathrm{e}^{-F(w)}\mathrm{e}^{-w(r)v_{t-r}(\lambda)}\big)\mathbf{N}_0(\mathrm{d}w)\bigg\} \\
&= \mathbf{P}\bigg[\exp\bigg\{-\int_{D[0,\infty)}\big[F(w)+v_{t-r}(\lambda)w(r)\big]N_z(\mathrm{d}w)\bigg\}\bigg] \\
&= \mathbf{P}\bigg[\exp\bigg\{-\int_{D[0,\infty)}F(w)N_z(\mathrm{d}w)\bigg\}\cdot\exp\big\{-v_{t-r}(\lambda)X_r^z\big\}\bigg].
\end{aligned}
$$

Clearly, the σ-algebra $\mathscr{G}_r^z$ is generated by the collection of random variables

$$
\exp\bigg\{-\int_{D[0,\infty)}F(w)N_z(\mathrm{d}w)\bigg\},
$$

where F runs over all bounded positive $\mathscr{A}_r$-measurable functions on $D[0,\infty)$. Then $\{(X_t^z,\mathscr{G}_t^z): t\ge 0\}$ is a Markov process with transition semigroup $(Q_t)_{t\ge 0}$.

The above theorem gives a description of the structures of the population represented by the CB-process. From (84) we see that the population at any time $t>0$ consists of at most countably many families, which evolve as the excursions $\{w_i: i=1,2,\cdots\}$ selected by the Poisson random measure $N_z(\mathrm{d}w)$. Unfortunately, this reconstruction is only available under the condition $\phi'(\infty)=\infty$. To give reconstructions of the CB- and CBI-processes when this condition is not necessarily satisfied, we need to consider some inhomogeneous immigration structures and more general Markovian measures on the path space. As a consequence of Theorem 6.1 we obtain the following.

Proposition 6.3 *Let $\beta\ge 0$ be a constant and ν a σ-finite measure on $(0,\infty)$ such that $\int_{(0,\infty)}u\nu(\mathrm{d}u)<\infty$. If $\beta>0$, assume in addition $\phi'(\infty)=\infty$. Then we can define a σ-finite measure $\mathbf{N}$ on $(D[0,\infty),\mathscr{A})$ by*

$$
\mathbf{N}(\mathrm{d}w)=\beta\mathbf{N}_0(\mathrm{d}w)+\int_{(0,\infty)}\nu(\mathrm{d}x)\mathbf{Q}_x(\mathrm{d}w),\qquad w\in D[0,\infty). \tag{85}
$$

Moreover, for $0<t_1<t_2<\cdots<t_n$ and $x_1,x_2,\ldots,x_n\in(0,\infty)$, we have

$$\begin{aligned}&\mathbf{N}(w(t_1) \in \mathrm{d}x_1, w(t_2) \in \mathrm{d}x_2, \ldots, w(t_n) \in \mathrm{d}x_n) \\ &\quad = H_{t_1}(\mathrm{d}x_1) Q^\circ_{t_2-t_1}(x_1, \mathrm{d}x_2) \cdots Q^\circ_{t_n-t_{n-1}}(x_{n-1}, \mathrm{d}x_n),\end{aligned} \tag{86}$$

where

$$H_t(\mathrm{d}y) = \beta l_t(\mathrm{d}y) + \int_{(0,\infty)} \nu(\mathrm{d}x) Q^\circ_t(x, \mathrm{d}y), \qquad y > 0.$$

Clearly, the measure $\mathbf{N}$ defined by (85) is actually supported by $D_0[0,\infty)$. Under this law, the path $\{w(t) : t > 0\}$ behaves as a Markov process with transition semigroup $(Q^\circ_t)_{t\geq 0}$ and one-dimensional distributions $(H_t)_{t>0}$.

Theorem 6.4 *Suppose that the conditions of Proposition 6.3 are satisfied and let* $\mathbf{N}$ *be defined by* (85). *Let* ρ *be a Borel measure on* $(0,\infty)$ *such that* $\rho(0,t] < \infty$ *for each* $0 < t < \infty$. *Suppose that* $N = \sum_{i=1}^\infty \delta_{(s_i,w_i)}$ *is a Poisson random measure on* $(0,\infty) \times D[0,\infty)$ *with intensity* $\rho(\mathrm{d}s)\mathbf{N}(\mathrm{d}w)$. *For* $t \geq 0$ *let*

$$Y_t = \int_{(0,t]} \int_{D[0,\infty)} w(t-s) N(\mathrm{d}s, \mathrm{d}w) = \sum_{0<s_i\leq t} w_i(t-s_i) \tag{87}$$

and let $\mathscr{G}_t$ *be the* σ*-algebra generated by the random variables* $\{N((0,u] \times A) : A \in \mathscr{A}_{t-u}, 0 \leq u \leq t\}$. *Then* $\{(Y_t, \mathscr{G}_t) : t \geq 0\}$ *is a Markov process with inhomogeneous transition semigroup* $(P_{r,t})_{t\geq r\geq 0}$ *given by*

$$\int_{[0,\infty)} \mathrm{e}^{-\lambda y} P_{r,t}(x, \mathrm{d}y) = \exp\bigg\{ - x v_{t-r}(\lambda) - \int_{(r,t]} \psi(v_{t-s}(\lambda)) \rho(\mathrm{d}s) \bigg\}, \tag{88}$$

where the function ψ *is given by* (76).

Proof From (87) we see that $\{Y_t : t \geq 0\}$ is adapted to the filtration $\{\mathscr{G}_t : t \geq 0\}$. Let $t \geq r \geq u \geq 0$ and let F be a bounded positive function on $D[0,\infty)$ measurable relative to $\mathscr{A}_{r-u}$. For $\lambda \geq 0$, we can use the Markov property (86) to see

$$\begin{aligned}
&\mathbf{P}\bigg[\exp\bigg\{ - \int_{(0,u]} \int_{D[0,\infty)} F(w) N(\mathrm{d}s, \mathrm{d}w) - \lambda Y_t \bigg\}\bigg] \\
&\quad = \mathbf{P}\bigg[\exp\bigg\{ - \int_{(0,t]} \int_{D[0,\infty)} \big[F(w)1_{\{s\leq u\}} + \lambda w(t-s)\big] N(\mathrm{d}s, \mathrm{d}w) \bigg\}\bigg] \\
&\quad = \exp\bigg\{ - \int_{(0,t]} \rho(\mathrm{d}s) \int_{D[0,\infty)} \big(1 - \mathrm{e}^{-F(w)1_{\{s\leq u\}}} \mathrm{e}^{-\lambda w(t-s)}\big) \mathbf{N}(\mathrm{d}w) \bigg\} \\
&\quad = \exp\bigg\{ - \int_{(0,r]} \rho(\mathrm{d}s) \int_{D[0,\infty)} \big(1 - \mathrm{e}^{-F(w)1_{\{s\leq u\}}} \mathrm{e}^{-w(t-s)}\big) \mathbf{N}(\mathrm{d}w) \bigg\} \\
&\qquad \cdot \exp\bigg\{ - \int_{(r,t]} \rho(\mathrm{d}s) \int_{D[0,\infty)} \big(1 - \mathrm{e}^{-\lambda w(t-s)}\big) \mathbf{N}(\mathrm{d}w) \bigg\} \\
&\quad = \exp\bigg\{ - \int_{(0,r]} \rho(\mathrm{d}s) \int_{D[0,\infty)} \big(1 - \mathrm{e}^{-F(w)1_{\{s\leq u\}}}\big) \mathbf{N}(\mathrm{d}w) \bigg\}
\end{aligned}$$

$$
\begin{aligned}
&\cdot \exp\Big\{-\int_{(0,r]}\rho(\mathrm{d}s)\int_{D[0,\infty)}\mathrm{e}^{-F(w)1_{\{s\le u\}}}\big(1-\mathrm{e}^{-w(t-s)}\big)\mathbf{N}(\mathrm{d}w)\Big\}\\
&\cdot \exp\Big\{-\int_{(r,t]}\rho(\mathrm{d}s)\int_{D[0,\infty)}\big(1-\mathrm{e}^{-\lambda w(t-s)}\big)\mathbf{N}(\mathrm{d}w)\Big\}\\
= &\exp\Big\{-\int_{(0,r]}\rho(\mathrm{d}s)\int_{D[0,\infty)}\big(1-\mathrm{e}^{-F(w)1_{\{s\le u\}}}\big)\mathbf{N}(\mathrm{d}w)\Big\}\\
&\cdot \exp\Big\{-\int_{(0,r]}\rho(\mathrm{d}s)\int_{D[0,\infty)}\mathrm{e}^{-F(w)1_{\{s\le u\}}}\big(1-\mathrm{e}^{-v_{t-r}(\lambda)w(r-s)}\big)\mathbf{N}(\mathrm{d}w)\Big\}\\
&\cdot \exp\Big\{-\int_{(r,t]}\rho(\mathrm{d}s)\int_{(0,\infty)}\big(1-\mathrm{e}^{-\lambda y}\big)H_{t-s}(\mathrm{d}w)\Big\}\\
= &\exp\Big\{-\int_{(0,r]}\rho(\mathrm{d}s)\int_{D[0,\infty)}\big(1-\mathrm{e}^{-F(w)1_{\{s\le u\}}}\mathrm{e}^{-v_{t-r}(\lambda)w(r-s)}\big)\mathbf{N}(\mathrm{d}w)\Big\}\\
&\cdot \exp\Big\{-\beta\int_{(r,t]}\rho(\mathrm{d}s)\int_{(0,\infty)}(1-\mathrm{e}^{-\lambda y})l_{t-s}(\mathrm{d}y)\\
&\qquad\qquad -\int_{(r,t]}\rho(\mathrm{d}s)\int_{(0,\infty)}(1-\mathrm{e}^{-\lambda y})\nu Q^\circ_{t-s}(\mathrm{d}y)\Big\}\\
= &\,\mathbf{P}\Big[\exp\Big\{-\int_{(0,r]}\int_{D[0,\infty)}\big[F(w)1_{\{s\le u\}}+v_{t-r}(\lambda)w(r-s)\big]N(\mathrm{d}s,\mathrm{d}w)\Big\}\Big]\\
&\cdot \exp\Big\{-\int_{(r,t]}\Big[\beta v_{t-s}(\lambda)+\int_{(0,\infty)}(1-\mathrm{e}^{-yv_{t-s}(\lambda)})\nu(\mathrm{d}y)\Big]\rho(\mathrm{d}s)\Big\}\\
= &\,\mathbf{P}\Big[\exp\Big\{-\int_{(0,u]}\int_{D[0,\infty)}F(w)N(\mathrm{d}s,\mathrm{d}w)\Big\}\\
&\qquad\qquad\cdot \exp\Big\{-v_{t-r}(\lambda)Y_r-\int_{(r,t]}\psi(v_{t-s}(\lambda))\rho(\mathrm{d}s)\Big\}\Big].
\end{aligned}
$$

Then $\{(Y_t,\mathscr{G}_t): t\ge 0\}$ is a Markov process in $[0,\infty)$ with inhomogeneous transition semigroup $(P_{r,t})_{t\ge r\ge 0}$ given by (88).

Corollary 6.5 *Suppose that $\phi'(\infty)=\infty$. Let $\beta>0$ and let $N_\beta=\sum_{i=1}^\infty \delta_{(s_i,w_i)}$ be a Poisson random measure on $(0,\infty)\times D[0,\infty)$ with intensity $\beta\mathrm{d}s\mathbf{N}_0(\mathrm{d}w)$. For $t\ge 0$ let*

$$
Y_t^\beta=\int_{(0,t]}\int_{D[0,\infty)}w(t-s)N_\beta(\mathrm{d}s,\mathrm{d}w)=\sum_{0<s_i\le t}w_i(t-s_i)
$$

and let $\mathscr{G}_t^\beta$ be the σ-algebra generated by the random variables $\{N_\beta((0,u]\times A): A\in\mathscr{A}_{t-u}, 0\le u\le t\}$. Then $\{(Y_t^\beta,\mathscr{G}_t^\beta): t\ge 0\}$ is a CBI-process with branching mechanism ϕ and immigration mechanism ψ_β defined by $\psi_\beta(\lambda)=\beta\lambda$, $\lambda\ge 0$.

Corollary 6.6 *Let ν be a σ-finite measure on $(0,\infty)$ such that $\int_{(0,\infty)}u\nu(\mathrm{d}u)<\infty$. Let $N_\nu=\sum_{i=1}^\infty\delta_{(s_i,w_i)}$ be a Poisson random measure on $(0,\infty)\times D[0,\infty)$ with intensity $\mathrm{d}s\mathbf{N}_\nu(\mathrm{d}w)$, where*

$$\mathbf{N}_\nu(\mathrm{d}w) = \int_{(0,\infty)} \nu(\mathrm{d}x)\mathbf{Q}_x(\mathrm{d}w). \tag{89}$$

For $t \geq 0$ let

$$Y_t^\nu = \int_{(0,t]} \int_{D[0,\infty)} w(t-s) N_\nu(\mathrm{d}s, \mathrm{d}w) = \sum_{0<s_i\leq t} w_i(t-s_i)$$

and let $\mathscr{G}_t^\nu$ be the σ-algebra generated by the random variables $\{N_\nu((0,u] \times A) : A \in \mathscr{A}_{t-u}, 0 \leq u \leq t\}$. Then $\{(Y_t^\nu, \mathscr{G}_t^\nu) : t \geq 0\}$ is a CBI-process with branching mechanism ϕ and immigration mechanism ψ_ν defined by

$$\psi_\nu(\lambda) = \int_{(0,\infty)} (1 - \mathrm{e}^{-u\lambda})\nu(\mathrm{d}u), \qquad \lambda \geq 0. \tag{90}$$

The transition semigroup $(P_{r,t})_{t\geq r\geq 0}$ defined by (88) is a generalization of the one given by (77); see also Li [36, 2003] and Li [40, p. 224]. A Markov process with transition semigroup $(P_{r,t})_{t\geq r\geq 0}$ is naturally called a CBI-process with *inhomogeneous immigration rate* ρ. In view of (87), the population $\{Y_t : t \geq 0\}$ consists of a countable families of immigrants, whose immigration times $\{s_i : i = 1, 2, \cdots\}$ and evolution trajectories $\{w_i : i = 1, 2, \cdots\}$ are both selected by the Poisson random measure $N(\mathrm{d}s, \mathrm{d}w)$. The processes constructed in Corollaries 6.5 and 6.6 can be interpreted similarly.

Theorem 6.7 *Suppose that $\delta := \phi'(\infty) < \infty$. Let $z > 0$ and let $N_z = \sum_{i=1}^\infty \delta_{(s_i,w_i)}$ be a Poisson random measure on $(0,\infty) \times D[0,\infty)$ with intensity $z\mathrm{e}^{-\delta s}\mathrm{d}s\mathbf{N}_m(\mathrm{d}w)$, where $\mathbf{N}_m$ is defined by (89) with $\nu = m$. For $t \geq 0$ let*

$$X_t^z = z\mathrm{e}^{-\delta t} + \int_{(0,t]} \int_{D[0,\infty)} w(t-s) N_z(\mathrm{d}s, \mathrm{d}w) \tag{91}$$

and let $\mathscr{G}_t^z$ be the σ-algebra generated by the random variables $\{N_z((0,u] \times A) : A \in \mathscr{A}_{t-u}, 0 \leq u \leq t\}$. Then $\{(X_t^z, \mathscr{G}_t^z) : t \geq 0\}$ is a CB-process with branching mechanism ϕ.

Proof Let $Z_t = X_t^z - z\mathrm{e}^{-\delta t}$ denote the second term on the right-hand side of (91). By Theorem 6.4 we infer that $\{(Z_t, \mathscr{G}_t^z) : t \geq 0\}$ is a Markov process with inhomogeneous transition semigroup $(P_{r,t}^z)_{t\geq r\geq 0}$ given by

$$\int_{[0,\infty)} \mathrm{e}^{-\lambda y} P_{r,t}^z(x, \mathrm{d}y) = \exp\bigg\{ - x v_{t-r}(\lambda) - z \int_r^t \psi_m(v_{t-s}(\lambda))\mathrm{e}^{-\delta s}\mathrm{d}s \bigg\},$$

where ψ_m is defined by (90) with $\nu = m$. Let $t \geq r \geq u \geq 0$ and let F be a bounded positive $\mathscr{A}_{r-u}$-measurable function on $D[0,\infty)$. For $\lambda \geq 0$ we have

$$\begin{aligned}
&\mathbf{P}\Big[\exp\Big\{-\int_{(0,u]}\int_{D[0,\infty)} F(w)N_z(\mathrm{d}s,\mathrm{d}w)-\lambda X_t^z\Big\}\Big]\\
&\quad=\mathbf{P}\Big[\exp\Big\{-\int_{(0,u]}\int_{D[0,\infty)} F(w)N_z(\mathrm{d}s,\mathrm{d}w)-\lambda z\mathrm{e}^{-\delta t}-\lambda Z_t\Big\}\Big]\\
&\quad=\mathbf{P}\Big[\exp\Big\{-\int_{(0,u]}\int_{D[0,\infty)} F(w)N_z(\mathrm{d}s,\mathrm{d}w)-\lambda z\mathrm{e}^{-\delta t}\Big\}\\
&\qquad\cdot\exp\Big\{-v_{t-r}(\lambda)Z_r-z\int_r^t\psi_m(v_{t-s}(\lambda))\mathrm{e}^{-\delta s}\mathrm{d}s\Big\}\Big]\\
&\quad=\mathbf{P}\Big[\exp\Big\{-\int_{(0,u]}\int_{D[0,\infty)} F(w)N_z(\mathrm{d}s,\mathrm{d}w)-\lambda z\mathrm{e}^{-\delta t}\Big\}\\
&\qquad\cdot\exp\Big\{-v_{t-r}(\lambda)Z_r-\mathrm{e}^{-\delta r}z\int_0^{t-r}\psi_m(v_{t-r-s}(\lambda))\mathrm{e}^{-\delta s}\mathrm{d}s\Big\}\Big]\\
&\quad=\mathbf{P}\Big[\exp\Big\{-\int_{(0,u]}\int_{D[0,\infty)} F(w)N_z(\mathrm{d}s,\mathrm{d}w)-v_{t-r}(\lambda)Z_r-z\mathrm{e}^{-\delta r}v_{t-r}(\lambda)\Big\}\Big]\\
&\quad=\mathbf{P}\Big[\exp\Big\{-\int_{(0,u]}\int_{D[0,\infty)} F(w)N_z(\mathrm{d}s,\mathrm{d}w)-v_{t-r}(\lambda)X_r^z\Big\}\Big],
\end{aligned}$$

where we have used (48). Then $\{(X_t^z,\mathscr{G}_t^z): t\geq 0\}$ is a CB-process with transition semigroup $(Q_t)_{t\geq 0}$ defined by (25).

Theorem 6.8 *Suppose that* $\delta := \phi'(\infty) < \infty$. *Let* $\beta > 0$ *and let* $N_\beta = \sum_{i=1}^\infty \delta_{(s_i,w_i)}$ *be a Poisson random measure on* $(0,\infty)\times D[0,\infty)$ *with intensity* $\beta\delta^{-1}(1-\mathrm{e}^{-\delta s})$ $\mathrm{d}s\mathbf{N}_m(\mathrm{d}w)$, *where* $\mathbf{N}_m$ *is defined by* (89) *with* $\nu = m$. *For* $t\geq 0$ *let*

$$Y_t^\beta = \beta\delta^{-1}(1-\mathrm{e}^{-\delta t}) + \int_{(0,t]}\int_{D[0,\infty)} w(t-s)N_\beta(\mathrm{d}s,\mathrm{d}w) \tag{92}$$

and let $\mathscr{G}_t^\beta$ *be the* σ*-algebra generated by the random variables* $\{N_\beta((0,u]\times A): A\in\mathscr{A}_{t-u}, 0\leq u\leq t\}$. *Then* $\{(X_t^\beta,\mathscr{G}_t^\beta): t\geq 0\}$ *is a CBI-process with branching mechanism* ϕ *and immigration mechanism* ψ_β *defined by* $\psi_\beta(\lambda)=\beta\lambda$, $\lambda\geq 0$.

Proof Let Z_t denote the second term on the right-hand side of (92). By Theorem 6.4, the process $\{(Z_t,\mathscr{G}_t^\beta): t\geq 0\}$ is a Markov process with inhomogeneous transition semigroup $(P_{r,t}^\beta)_{t\geq r\geq 0}$ given by

$$\int_{[0,\infty)} \mathrm{e}^{-\lambda y}P_{r,t}^\beta(x,\mathrm{d}y) = \exp\Big\{-xv_{t-r}(\lambda)-\beta\delta^{-1}\int_r^t\psi_m(v_{t-s}(\lambda))(1-\mathrm{e}^{-\delta s})\mathrm{d}s\Big\},$$

where ψ_m is defined by (90) with $\nu = m$. For $t\geq 0$ and $\lambda\geq 0$, we can use Theorem 3.15 to see

$$\begin{aligned}
\int_0^t v_s(\lambda)\mathrm{d}s &= \lambda\int_0^t \mathrm{e}^{-\delta s}\mathrm{d}s + \int_0^t\mathrm{d}s\int_0^s \mathrm{e}^{-\delta u}\psi_m(v_{s-u}(\lambda))\mathrm{d}u\\
&= \lambda\int_0^t \mathrm{e}^{-\delta s}\mathrm{d}s + \int_0^t\mathrm{d}s\int_0^{t-s} \mathrm{e}^{-\delta u}\psi_m(v_{t-s-u}(\lambda))\mathrm{d}u
\end{aligned}$$

$$
\begin{aligned}
&= \lambda\delta^{-1}(1-\mathrm{e}^{-\delta t}) + \int_0^t \mathrm{d}s \int_s^t \mathrm{e}^{-\delta(u-s)} \psi_m(v_{t-u}(\lambda))\mathrm{d}u \\
&= \lambda\delta^{-1}(1-\mathrm{e}^{-\delta t}) + \int_0^t \mathrm{d}u \int_0^u \mathrm{e}^{-\delta(u-s)} \psi_m(v_{t-u}(\lambda))\mathrm{d}s \\
&= \lambda\delta^{-1}(1-\mathrm{e}^{-\delta t}) + \delta^{-1} \int_0^t (1-\mathrm{e}^{-\delta u}) \psi_m(v_{t-u}(\lambda))\mathrm{d}u. \qquad (93)
\end{aligned}
$$

Let $t \geq r \geq u \geq 0$ and let F be a bounded positive $\mathscr{A}_{r-u}$-measurable function on $D[0,\infty)$. Then

$$
\begin{aligned}
&\mathbf{P}\Big[\exp\Big\{-\int_{(0,u]}\int_{D[0,\infty)} F(w)N_\beta(\mathrm{d}s,\mathrm{d}w) - \lambda Y_t^\beta\Big\}\Big] \\
&\quad = \mathbf{P}\Big[\exp\Big\{-\int_{(0,u]}\int_{D[0,\infty)} F(w)N_\beta(\mathrm{d}s,\mathrm{d}w) - \lambda\beta\delta^{-1}(1-\mathrm{e}^{-\delta t}) - \lambda Z_t\Big\}\Big] \\
&\quad = \mathbf{P}\Big[\exp\Big\{-\int_{(0,u]}\int_{D[0,\infty)} F(w)N_\beta(\mathrm{d}s,\mathrm{d}w) - \lambda\beta\delta^{-1}(1-\mathrm{e}^{-\delta t})\Big\}\Big] \\
&\qquad \cdot \exp\Big\{-v_{t-r}(\lambda)Z_r - \beta\delta^{-1}\int_r^t \psi_m(v_{t-s}(\lambda))(1-\mathrm{e}^{-\delta s})\mathrm{d}s\Big\} \\
&\quad = \mathbf{P}\Big[\exp\Big\{-\int_{(0,u]}\int_{D[0,\infty)} F(w)N_\beta(\mathrm{d}s,\mathrm{d}w) - \lambda\beta\delta^{-1}(1-\mathrm{e}^{-\delta t})\Big\}\Big] \\
&\qquad \cdot \exp\Big\{-v_{t-r}(\lambda)Y_r^\beta + v_{t-r}(\lambda)\beta\delta^{-1}(1-\mathrm{e}^{-\delta r})\Big\} \\
&\qquad \cdot \exp\Big\{-\beta\delta^{-1}\int_0^{t-r} \psi_m(v_{t-r-s}(\lambda))(1-\mathrm{e}^{-\delta(r+s)})\mathrm{d}s\Big\} \\
&\quad = \mathbf{P}\Big[\exp\Big\{-\int_{(0,u]}\int_{D[0,\infty)} F(w)N_\beta(\mathrm{d}s,\mathrm{d}w) - \lambda\beta\delta^{-1}(1-\mathrm{e}^{-\delta t})\Big\}\Big] \\
&\qquad \cdot \exp\Big\{-v_{t-r}(\lambda)Y_r^\beta + \lambda\beta\delta^{-1}\mathrm{e}^{-\delta(t-r)}(1-\mathrm{e}^{-\delta r})\Big\} \\
&\qquad \cdot \exp\Big\{\beta\delta^{-1}(1-\mathrm{e}^{-\delta r})\int_0^{t-r} \mathrm{e}^{-\delta s}\psi_m(v_{t-r-s}(\lambda))\mathrm{d}s\Big\} \\
&\qquad \cdot \exp\Big\{-\beta\delta^{-1}\int_0^{t-r} \psi_m(v_{t-r-s}(\lambda))(1-\mathrm{e}^{-\delta(r+s)})\mathrm{d}s\Big\} \\
&\quad = \mathbf{P}\Big[\exp\Big\{-\int_{(0,u]}\int_{D[0,\infty)} F(w)N_\beta(\mathrm{d}s,\mathrm{d}w) - \lambda\beta\delta^{-1}(1-\mathrm{e}^{-\delta(t-r)})\Big\}\Big] \\
&\qquad \cdot \exp\Big\{-v_{t-r}(\lambda)Y_r^\beta - \beta\delta^{-1}\int_0^{t-r} \psi_m(v_{t-r-s}(\lambda))(1-\mathrm{e}^{-\delta s})\mathrm{d}s\Big\} \\
&\quad = \mathbf{P}\Big[\exp\Big\{-\int_{(0,u]}\int_{D[0,\infty)} F(w)N_\beta(\mathrm{d}s,\mathrm{d}w) \\
&\qquad\qquad - v_{t-r}(\lambda)Y_r^\beta - \beta\int_0^{t-r} v_s(\lambda))\mathrm{d}s\Big\}\Big],
\end{aligned}
$$

where we used (93) for the last equality. That gives the desired result.

Since the CB- and CBI-processes have Feller transition semigroups, they have càdlàg realizations. By Proposition A.7 of Li [40], any realization of the processes has a càdlàg modification. In the case of $\phi'(\infty) = \infty$, let $(X_t^z, \mathscr{G}_t^z)$, $(Y_t^\beta, \mathscr{G}_t^\beta)$ and $(Y_t^\nu, \mathscr{G}_t^\nu)$ be defined as in Theorem 6.2, Corollaries 6.5 and 6.6, respectively. In the case of $\phi'(\infty) < \infty$, we define those processes as in Theorems 6.7, 6.8, and Corollary 6.6, respectively. In both cases, let $Y_t = X_t^z + Y_t^\beta + Y_t^\nu$ and $\mathscr{G}_t = \sigma(\mathscr{G}_t^z \cup \mathscr{G}_t^\beta \cup \mathscr{G}_t^\nu)$. It is not hard to show that $\{(Y_t, \mathscr{G}_t) : t \geq 0\}$ is a CBI-process with branching mechanism ϕ given by (19) and immigration mechanism ψ given by (76).

The existence of the excursion law for the branching mechanism $\phi(z) = bz + cz^2$ was first proved by Pitman and Yor [48]. As a special case of the so-called *Kuznetsov measure*, the existence of the law for measure-valued branching processes was derived from a general result on Markov processes in Li [38, 40], where it was also shown that the law only charges sample paths starting with zero. In the setting of measure-valued processes, El Karoui and Roelly [16] used (83) to construct the excursion law; see also Duquesne and Labbé [14]. The construction of the CB- or CBI-process based on a excursion law was first given by Pitman and Yor [48]. This type of constructions has also been used in the measure-valued setting by a number of authors; see, e.g., Dawson and Li [9], El Karoui and Roelly [16], Li [36, 38, 40], and Li and Shiga [44].

7 Martingale Problem Formulations

In this section, we give several formulations of the CBI-process in terms of martingale problems. Let (ϕ, ψ) be given by (19) and (76), respectively. We assume (78) is satisfied and define $\psi'(0)$ by (80). Let $C^2[0, \infty)$ denote the set of bounded continuous real functions on $[0, \infty)$ with bounded continuous derivatives up to the second order. For $f \in C^2[0, \infty)$ define

$$\begin{aligned} Lf(x) =& cxf''(x) + x\int_{(0,\infty)} \big[f(x+z) - f(x) - zf'(x)\big]m(dz) \\ &+ (\beta - bx)f'(x) + \int_{(0,\infty)} \big[f(x+z) - f(x)\big]\nu(z). \end{aligned} \tag{94}$$

We shall identify the operator L as the generator of the CBI-process.

Proposition 7.1 *Let $(P_t)_{t\geq 0}$ be the transition semigroup defined by* (25) *and* (77). *Then for any $t \geq 0$ and $\lambda \geq 0$ we have*

$$\int_{[0,\infty)} e^{-\lambda y} P_t(x, dy) = e^{-x\lambda} + \int_0^t ds \int_{[0,\infty)} [y\phi(\lambda) - \psi(\lambda)]e^{-y\lambda} P_s(x, dy). \tag{95}$$

Proof Recall that $v_t'(\lambda) = (\partial/\partial\lambda)v_t(\lambda)$. By differentiating both sides of (77) we get

$$\int_{[0,\infty)} y\mathrm{e}^{-y\lambda} P_t(x, \mathrm{d}y) = \int_{[0,\infty)} \mathrm{e}^{-y\lambda} P_t(x, \mathrm{d}y) \Big[x v_t'(\lambda) + \int_0^t \psi'(v_s(\lambda)) v_s'(\lambda) \mathrm{d}s \Big].$$

From this and (35) it follows that

$$\begin{aligned}
\frac{\partial}{\partial t} \int_{[0,\infty)} \mathrm{e}^{-y\lambda} P_t(x, \mathrm{d}y) &= -\Big[x\frac{\partial}{\partial t} v_t(\lambda) + \psi(v_t(\lambda)) \Big] \int_{[0,\infty)} \mathrm{e}^{-y\lambda} P_t(x, \mathrm{d}y) \\
&= \Big[x\phi(\lambda) v_t'(\lambda) - \psi(\lambda) \Big] \int_{[0,\infty)} \mathrm{e}^{-y\lambda} P_t(x, \mathrm{d}y) \\
&\quad - \int_0^t \psi'(v_s(\lambda)) \frac{\partial}{\partial s} v_s(\lambda) \mathrm{d}s \int_{[0,\infty)} \mathrm{e}^{-y\lambda} P_t(x, \mathrm{d}y) \\
&= \Big[x\phi(\lambda) v_t'(\lambda) - \psi(\lambda) \Big] \int_{[0,\infty)} \mathrm{e}^{-y\lambda} P_t(x, \mathrm{d}y) \\
&\quad + \phi(\lambda) \int_0^t \psi'(v_s(\lambda)) v_s'(\lambda) \mathrm{d}s \int_{[0,\infty)} \mathrm{e}^{-y\lambda} P_t(x, \mathrm{d}y) \\
&= \int_{[0,\infty)} [y\phi(\lambda) - \psi(\lambda)] \mathrm{e}^{-y\lambda} P_t(x, \mathrm{d}y).
\end{aligned}$$

That gives (95).

Suppose that $(\Omega, \mathscr{G}, \mathscr{G}_t, \mathbf{P})$ is a filtered probability space satisfying the usual hypotheses and $\{y(t) : t \geq 0\}$ is a càdlàg process in $[0, \infty)$ that is adapted to $(\mathscr{G}_t)_{t\geq 0}$ and satisfies $\mathbf{P}[y(0)] < \infty$. Let $C^{1,2}([0, \infty)^2)$ be the set of bounded continuous real functions $(t, x) \mapsto G(t, x)$ on $[0, \infty)^2$ with bounded continuous derivatives up to the first-order relative to $t \geq 0$ and up to the second-order relative to $x \geq 0$. Let us consider the following properties:

(1) For every $T \geq 0$ and $\lambda \geq 0$,

$$\exp\Big\{ - v_{T-t}(\lambda) y(t) - \int_0^{T-t} \psi(v_s(\lambda)) \mathrm{d}s \Big\}, \qquad 0 \leq t \leq T,$$

is a martingale.

(2) For every $\lambda \geq 0$,

$$H_t(\lambda) := \exp\Big\{ - \lambda y(t) + \int_0^t [\psi(\lambda) - y(s)\phi(\lambda)] \mathrm{d}s \Big\}, \quad t \geq 0,$$

is a local martingale.

(3) The process $\{y(t) : t \geq 0\}$ has no negative jumps and the optional random measure

$$N_0(\mathrm{d}s, \mathrm{d}z) := \sum_{s>0} 1_{\{\Delta y(s) \neq 0\}} \delta_{(s, \Delta y(s))}(\mathrm{d}s, \mathrm{d}z),$$

where $\Delta y(s) = y(s) - y(s-)$, has predictable compensator $\hat{N}_0(\mathrm{d}s, \mathrm{d}z) = \mathrm{d}s\nu(\mathrm{d}z) + y(s-)\mathrm{d}sm(\mathrm{d}z)$. Let $\tilde{N}_0(\mathrm{d}s, \mathrm{d}z) = N_0(\mathrm{d}s, \mathrm{d}z) - \hat{N}_0(\mathrm{d}s, \mathrm{d}z)$. We have

$$y(t) = y(0) + M^c(t) + M^d(t) - b\int_0^t y(s-)\mathrm{d}s + \psi'(0)t,$$

where $\{M^c(t) : t \geq 0\}$ is a continuous local martingale with quadratic variation $2cy(t-)\mathrm{d}t$ and

$$M^d(t) = \int_0^t \int_{(0,\infty)} z\tilde{N}_0(\mathrm{d}s, \mathrm{d}z), \qquad t \geq 0,$$

is a purely discontinuous local martingale.

(4) For every $f \in C^2[0, \infty)$ we have

$$f(y(t)) = f(y(0)) + \int_0^t Lf(y(s))\mathrm{d}s + \text{local mart.} \tag{96}$$

(5) For any $G \in C^{1,2}([0, \infty)^2)$ we have

$$G(t, y(t)) = G(0, y(0)) + \int_0^t \big[G_t'(s, y(s)) + LG(s, y(s))\big]\mathrm{d}s + \text{local mart.} \tag{97}$$

where L acts on the function $x \mapsto G(s, x)$.

Theorem 7.2 *The above properties* (1)–(5) *are equivalent to each other. Those properties hold if and only if* $\{(y(t), \mathscr{G}_t) : t \geq 0\}$ *is a CBI-process with branching mechanism* ϕ *and immigration mechanism* ψ.

Proof Clearly, (1) holds if and only if $\{y(t) : t \geq 0\}$ is a Markov process relative to $(\mathscr{G}_t)_{t\geq 0}$ with transition semigroup $(P_t)_{t\geq 0}$ defined by (77). Then we only need to prove the equivalence of the five properties.

(1)⇒(2): Suppose that (1) holds. Then $\{y(t) : t \geq 0\}$ is a CBI-process with transition semigroup $(P_t)_{t\geq 0}$ given by (77). By (95) and the Markov property it is easy to see that

$$Y_t(\lambda) := \mathrm{e}^{-\lambda y(t)} + \int_0^t [\psi(\lambda) - y(s)\phi(\lambda)]\mathrm{e}^{-\lambda y(s)}\mathrm{d}s \tag{98}$$

is a martingale. By integration by parts applied to

$$Z_t(\lambda) := \mathrm{e}^{-\lambda y(t)} \text{ and } W_t(\lambda) := \exp\left\{\int_0^t [\psi(\lambda) - y(s)\phi(\lambda)]\mathrm{s}\right\} \tag{99}$$

we obtain

$$\mathrm{d}H_t(\lambda) = \mathrm{e}^{-\lambda y(t-)}\mathrm{d}W_t(\lambda) + W_t(\lambda)\mathrm{d}\mathrm{e}^{-\lambda y(t)} = W_t(\lambda)\mathrm{d}Y_t(\lambda).$$

Then $\{H_t(\lambda)\}$ is a local martingale.

(2)⇒(3): For any $\lambda \geq 0$ let $Z_t(\lambda)$ and $W_t(\lambda)$ be defined by (99). We have $Z_t(\lambda) = H_t(\lambda)W_t(\lambda)^{-1}$ and so

$$\mathrm{d}Z_t(\lambda) = W_t(\lambda)^{-1}\mathrm{d}H_t(\lambda) - Z_{t-}(\lambda)[\psi(\lambda) - \phi(\lambda)y(t-)]\mathrm{d}t \tag{100}$$

by integration by parts. Then the strictly positive process $t \mapsto Z_t(\lambda)$ is a special semi-martingale; see, e.g., Dellacherie and Meyer [12, p. 213]. By Itô's formula we find $t \mapsto y(t)$ is a semi-martingale. Now define the optional random measure $N_0(\mathrm{d}s, \mathrm{d}z)$ on $[0, \infty) \times \mathbb{R}$ by

$$N_0(\mathrm{d}s, \mathrm{d}z) = \sum_{s>0} 1_{\{\Delta y(s)\neq 0\}}\delta_{(s,\Delta y(s))}(\mathrm{d}s, \mathrm{d}z),$$

where $\Delta y(s) = y(s) - y(s-)$. Let $\hat{N}_0(\mathrm{d}s, \mathrm{d}z)$ denote the predictable compensator of $N_0(\mathrm{d}s, \mathrm{d}z)$ and let $\tilde{N}_0(\mathrm{d}s, \mathrm{d}z)$ denote the compensated random measure; see, e.g., Dellacherie and Meyer [12, p. 375]. It follows that

$$y(t) = y(0) + U(t) + M^c(t) + M^d(t), \tag{101}$$

where $t \mapsto U(t)$ is a right-continuous adapted process with locally bounded variations, $t \mapsto M^c(t)$ is a continuous local martingale and

$$t \mapsto M^d(t) := \int_0^t \int_{\mathbb{R}} z\tilde{N}_0(\mathrm{d}s, \mathrm{d}z)$$

is a purely discontinuous local martingale; see, e.g., Dellacherie and Meyer [12, p. 353 and p. 376] or Jacod and Shiryaev [27, pp. 84–85]. Let $\{C(t)\}$ denote the quadratic variation process of $\{M^c(t)\}$. By Itô's formula,

$$\begin{aligned} Z_t(\lambda) =& Z_0(\lambda) - \lambda\int_0^t Z_{s-}(\lambda)\mathrm{d}y(s) + \frac{1}{2}\lambda^2\int_0^t Z_{s-}(\lambda)\mathrm{d}C(s) \\ &+ \int_0^t\int_{\mathbb{R}} Z_{s-}(\lambda)\big(\mathrm{e}^{-z\lambda} - 1 + z\lambda\big)N_0(\mathrm{d}s, \mathrm{d}z) \\ =& Z_0(\lambda) - \lambda\int_0^t Z_{s-}(\lambda)\mathrm{d}U(s) + \frac{1}{2}\lambda^2\int_0^t Z_{s-}(\lambda)\mathrm{d}C(s) \\ &+ \int_0^t\int_{\mathbb{R}} Z_{s-}(\lambda)\big(\mathrm{e}^{-z\lambda} - 1 + z\lambda\big)\hat{N}_0(\mathrm{d}s, \mathrm{d}z) + \text{local mart.} \end{aligned} \tag{102}$$

In view of (100) and (102) we get

$$[y(s-)\phi(\lambda) - \psi(\lambda)]\mathrm{d}s = -\lambda\mathrm{d}U(s) + \frac{1}{2}\lambda^2\mathrm{d}C(s) + \int_{\mathbb{R}}\big(\mathrm{e}^{-z\lambda} - 1 + z\lambda\big)\hat{N}_0(\mathrm{d}s, \mathrm{d}z)$$

by the uniqueness of canonical decompositions of special semi-martingales; see, e.g., Dellacherie and Meyer [12, p. 213]. By substituting the representations (19) and (76) for ϕ and ψ into the above equation and comparing both sides we find

$$\mathrm{d}C(s) = 2cy(s-)\mathrm{d}s, \ \mathrm{d}U(s) = [\psi'(0) - by(s-)]\mathrm{d}s$$

and

$$\hat{N}_0(\mathrm{d}s, \mathrm{d}z) = \mathrm{d}s\nu(\mathrm{d}z) + y(s-)\mathrm{d}sm(\mathrm{d}z).$$

Then the process $t \mapsto y(t)$ has no negative jumps.

(3)⇒(4): This follows by Itô's formula.

(4)⇒(5): For $t \geq 0$ and $k \geq 1$ we have

$$\begin{aligned} G(t, y(t)) = G(0, y(0)) + \sum_{j=0}^{\infty} \big[& G(t \wedge j/k, y(t \wedge (j+1)/k)) \\ & - G(t \wedge j/k, y(t \wedge j/k))\big] \\ + \sum_{j=0}^{\infty} \big[& G(t \wedge (j+1)/k, y(t \wedge (j+1)/k)) \\ & - G(t \wedge j/k, y(t \wedge (j+1)/k))\big], \end{aligned}$$

where the summations only consist of finitely many nontrivial terms. By applying (4) term by term we obtain

$$\begin{aligned} G(t, y(t)) = G(0, y(0)) + \sum_{j=0}^{\infty} \int_{t\wedge j/k}^{t\wedge(j+1)/k} & \Big\{ [\beta - by(s)]G'_x(t \wedge j/k, y(s)) \\ & + cy(s)G''_{xx}(t \wedge j/k, y(s)) + y(s) \int_{(0,\infty)} \Big[G(t \wedge j/k, y(s) + z) \\ & - G(t \wedge j/k, y(s)) - zG'_x(t \wedge j/k, y(s)) \Big] m(\mathrm{d}z) \\ & + \int_{(0,\infty)} \Big[G(t \wedge j/k, y(s) + z) - G(t \wedge j/k, y(s)) \Big] \nu(\mathrm{d}z) \Big\} \mathrm{d}s \\ & + \sum_{j=0}^{\infty} \int_{t\wedge j/k}^{t\wedge(j+1)/k} G'_t(s, y(t \wedge (j+1)/k))\mathrm{d}s + M_k(t), \end{aligned}$$

where $\{M_k(t)\}$ is a local martingale. Since $\{y(t)\}$ is a càdlàg process, letting $k \to \infty$ in the equation above gives

$$\begin{aligned} G(t, y(t)) = G(0, y(0)) + \int_0^t & \Big\{ G'_t(s, y(s)) + [\beta - by(s)]G'_x(s, y(s)) \\ & + cy(s)G''_{xx}(s, y(s)) + y(s) \int_{(0,\infty)} \Big[G(s, y(s) + z) \end{aligned}$$

$$- G(s, y(s)) - zG'_x(s, y(s))\Big]m(\mathrm{d}z)$$
$$+ \int_{(0,\infty)} \Big[G(s, y(s) + z) - G(s, y(s))\Big]\nu(\mathrm{d}z)\Big\}\mathrm{d}s + M(t),$$

where $\{M(t)\}$ is a local martingale. Then we have (97).

(5)⇒(1): For fixed $T \geq 0$ and $\lambda \geq 0$ we define the function

$$G_T(t, x) = \exp\Big\{ - v_{T-t}(\lambda)x - \int_0^{T-t} \psi(v_s(\lambda))\mathrm{d}s\Big\}, \quad 0 \leq t \leq T, x \geq 0,$$

which can be extended to a function in $C^{1,2}([0, \infty)^2)$. Using (34) we see

$$\frac{\mathrm{d}}{\mathrm{d}t}G_T(t, x) + LG_T(t, x) = 0, \quad 0 \leq t \leq T, x \geq 0,$$

Then (97) implies that $t \mapsto G(t \wedge T, y(t \wedge T))$ is a local martingale, and hence a martingale by the boundedness.

Corollary 7.3 *Let $\{(y(t), \mathscr{G}_t) : t \geq 0\}$ be a càdlàg realization of the CBI-process satisfying $\mathbf{P}[y(0)] < \infty$. Then for every $T \geq 0$ there is a constant $C_T \geq 0$ such that*

$$\mathbf{P}\Big[\sup_{0\leq t\leq T} y(t)\Big] \leq C_T\big\{\mathbf{P}[y(0)] + \psi'(0) + \sqrt{\mathbf{P}[y(0)]} + \sqrt{\psi'(0)}\big\}.$$

Proof By the above property (3) and Doob's martingale inequality we have

$$\begin{aligned}
&\mathbf{P}\Big[\sup_{0\leq t\leq T} |y(t) - y(0)|\Big] \\
&\leq T\psi'(0) + \mathbf{P}\Big[|b|\int_0^T y(s)\mathrm{d}s\Big] + \mathbf{P}\Big[\sup_{0\leq t\leq T} |M_t^c|\Big] \\
&\quad + \mathbf{P}\Big[\int_0^T \int_{(1,\infty)} zN_0(\mathrm{d}s, \mathrm{d}z)\Big] + \mathbf{P}\Big[\int_0^T \int_{(1,\infty)} z\hat{N}_0(\mathrm{d}s, \mathrm{d}z)\Big] \\
&\quad + \mathbf{P}\Big[\sup_{0\leq t\leq T}\Big|\int_{(0,1]}\int_0^1 z\tilde{N}_0(\mathrm{d}s, \mathrm{d}z)\Big|\Big] \\
&\leq T\psi'(0) + \mathbf{P}\Big[|b|\int_0^T y(s)\mathrm{d}s\Big] + 2\Big\{\mathbf{P}\Big[c\int_0^T y(s)\mathrm{d}s\Big]\Big\}^{1/2} \\
&\quad + 2T\int_{(1,\infty)} z\nu(\mathrm{d}z) + 2\mathbf{P}\Big[\int_0^T y(s)\mathrm{d}s \int_{(1,\infty)} zm(\mathrm{d}z)\Big] \\
&\quad + 2\Big\{\mathbf{P}\Big[T\int_{(0,1]} z^2\nu(\mathrm{d}z) + \int_0^T y(s)\mathrm{d}s\int_{(0,1]} z^2 m(\mathrm{d}z)\Big]\Big\}^{1/2}.
\end{aligned}$$

Then the desired inequality follows by simple estimates based on (79).

Corollary 7.4 *Let $\{(y(t), \mathscr{G}_t) : t \geq 0\}$ be a càdlàg realization of the CBI-process satisfying $\mathbf{P}[y(0)] < \infty$. Then the above properties (3)–(5) hold with the local martingales being martingales.*

Proof Since the arguments are similar, we only give those for (4). Let $f \in C^2[0, \infty)$ and let

$$M(t) = f(y(t)) - f(y(0)) - \int_0^t Lf(y(s))\mathrm{d}s, \qquad t \geq 0.$$

By property (4) we know $\{M(t)\}$ is a local martingale. Let $\{\tau_n\}$ be a localization sequence of stopping times for $\{M(t)\}$. For any $t \geq r \geq 0$ and any bounded $\mathscr{G}_r$-measurable random variable F, we have

$$\begin{aligned}\mathbf{P}\bigg\{\bigg[f(y(t\wedge\tau_n)) - f(y(0)) - \int_0^{t\wedge\tau_n} Lf(y(s))\mathrm{d}s\bigg]F\bigg\} \\ = \mathbf{P}\bigg\{\bigg[f(y(r\wedge\tau_n)) - f(y(0)) - \int_0^{r\wedge\tau_n} Lf(y(s))\mathrm{d}s\bigg]F\bigg\}.\end{aligned}$$

In view of (94), there is a constant $C \geq 0$ so that $|Lf(x)| \leq C(1+x)$. By Corollary 7.3 we can let $n \to \infty$ and use dominated convergence in the above equality to see $\{M(t)\}$ is a martingale.

Note that property (4) implies that the generator of the CBI-process is the closure of the operator L in the sense of Ethier and Kurtz [17]. This explicit form of the generator was first given in Kawazu and Watanabe [29]. The results of Theorem 7.2 were presented for measure-valued processes in El Karoui and Roelly [16] and Li [40].

8 Stochastic Equations for CBI-Processes

In this section we establish some stochastic equations for the CBI-processes. Suppose that (ϕ, ψ) are branching and immigration mechanisms given, respectively, by (19) and (76) with $\nu(\mathrm{d}u)$ satisfying condition (78). Let $(P_t)_{t\geq 0}$ be the transition semigroup defined by (25) and (77). In this and the next section, for any $b \geq a \geq 0$ we understand

$$\int_a^b = \int_{(a,b]} \quad \text{and} \quad \int_a^\infty = \int_{(a,\infty)}.$$

Let $\{B(t)\}$ be a standard Brownian motion and $\{M(\mathrm{d}s, \mathrm{d}z, \mathrm{d}u)\}$ a Poisson time-space random measure on $(0, \infty)^3$ with intensity $\mathrm{d}s m(\mathrm{d}z)\mathrm{d}u$. Let $\{\eta(t)\}$ be an increasing Lévy process with $\eta(0) = 0$ and with Laplace exponent $\psi(z) = -\log \mathbf{P}\exp\{-z\eta(1)\}$. We assume that $\{B(t)\}$, $\{M(\mathrm{d}s, \mathrm{d}z, \mathrm{d}u)\}$ and $\{\eta(t)\}$ are defined on a complete probability space and are independent of each other. Consider the stochastic integral equation

$$
\begin{aligned}
y(t)=y(0) &+ \int_0^t \sqrt{2cy(s-)}\mathrm{d}B(s) - b\int_0^t y(s-)\mathrm{d}s \\
&+ \int_0^t \int_0^\infty \int_0^{y(s-)} z\tilde{M}(\mathrm{d}s, \mathrm{d}z, \mathrm{d}u) + \eta(t),
\end{aligned}
\tag{103}
$$

where $\tilde{M}(\mathrm{d}s, \mathrm{d}z, \mathrm{d}u) = M(\mathrm{d}s, \mathrm{d}z, \mathrm{d}u) - \mathrm{d}sm(\mathrm{d}z)\mathrm{d}u$ denotes the compensated measure. We understand the fourth term on the right-hand side of (103) as an integral over the set $\{(s, z, u) : 0 < s \leq t, 0 < z < \infty, 0 < u \leq y(s-)\}$ and give similar interpretations for other stochastic integrals in this section. The reader is referred to Ikeda and Watanabe [26] and Situ [49] for the basic theory of stochastic equations.

Theorem 8.1 *A positive càdlàg process $\{y(t) : t \geq 0\}$ is a CBI-process with branching and immigration mechanisms (ϕ, ψ) given, respectively, by* (19) *and* (76) *if and only if it is a weak solution to* (103).

Proof Suppose that the positive càdlàg process $\{y(t)\}$ is a weak solution to (103). By Itô's formula one can see $\{y(t)\}$ solves the martingale problem (96). By Theorem 7.2 we infer that $\{y(t)\}$ is a CBI-process with branching and immigration mechanisms given, respectively, by (19) and (76). Conversely, suppose that $\{y(t)\}$ is a càdlàg realization of the CBI-process with branching and immigration mechanisms given, respectively, by (19) and (76). By Theorem 7.2 the process has no negative jumps and the random measure

$$
N_0(\mathrm{d}s, \mathrm{d}z) := \sum_{s>0} 1_{\{\Delta y(s)>0\}}\delta_{(s,\Delta y(s))}(\mathrm{d}s, \mathrm{d}z)
$$

has predictable compensator

$$
\hat{N}_0(\mathrm{d}s, \mathrm{d}z) = y(s-)\mathrm{d}sm(\mathrm{d}z) + \mathrm{d}s\nu(\mathrm{d}z).
$$

Moreover, we have

$$
\begin{aligned}
y(t)=y(0) &+ t\Big[\beta + \int_0^\infty u\nu(\mathrm{d}u)\Big] - \int_0^t by(s-)\mathrm{d}s \\
&+ M^c(t) + \int_0^t \int_0^\infty z\tilde{N}_0(\mathrm{d}s, \mathrm{d}z),
\end{aligned}
$$

where $\tilde{N}_0(\mathrm{d}s, \mathrm{d}z) = N_0(\mathrm{d}s, \mathrm{d}z) - \hat{N}_0(\mathrm{d}s, \mathrm{d}z)$ and $t \mapsto M^c(t)$ is a continuous local martingale with quadratic variation $2cy(t-)\mathrm{d}t$. By Theorem III.7.1′ in Ikeda and Watanabe [26, p. 90], on an extension of the original probability space there is a standard Brownian motion $\{B(t)\}$ so that

$$
M^c(t) = \int_0^t \sqrt{2cy(s-)}\mathrm{d}B(s).
$$

By Theorem III.7.4 in Ikeda and Watanabe [26, p. 93], on a further extension of the probability space we can define independent Poisson random measures $M(ds, dz, du)$ and $N(ds, dz)$ with intensities $ds m(dz) du$ and $ds \nu(dz)$, respectively, so that

$$\int_0^t \int_0^\infty z\tilde{N}_0(ds, dz) = \int_0^t \int_0^\infty \int_0^{y(s-)} z\tilde{M}(ds, dz, du) + \int_0^t \int_0^\infty z\tilde{N}(ds, dz).$$

Then $\{y(t)\}$ is a weak solution to (103).

Theorem 8.2 *For any initial value $y(0) = x \geq 0$, there is a pathwise unique positive strong solution to* (103).

Proof By Theorem 8.1 there is a weak solution to (103). Then we only need to prove the pathwise uniqueness of the solution; see, e.g., Situ [49, p. 76 and p. 104]. Suppose that $\{x(t) : t \geq 0\}$ and $\{y(t) : t \geq 0\}$ are two positive solutions of (103) with deterministic initial states. By Theorem 8.1, both of them are CBI-processes. We may assume $x(0)$ and $y(0)$ are deterministic upon taking a conditional probability. In view of (79), the processes have locally bounded first moments. Let $\zeta(t) = x(t) - y(t)$ for $t \geq 0$. For each integer $n \geq 0$ define $a_n = \exp\{-n(n+1)/2\}$. Then $a_n \to 0$ decreasingly as $n \to \infty$ and

$$\int_{a_n}^{a_{n-1}} z^{-1} dz = n, \qquad n \geq 1.$$

Let $x \mapsto g_n(x)$ be a positive continuous function supported by (a_n, a_{n-1}) so that

$$\int_{a_n}^{a_{n-1}} g_n(x) dx = 1$$

and $g_n(x) \leq 2(nx)^{-1}$ for every $x > 0$. For $n \geq 0$ and $z \in \mathbb{R}$ let

$$f_n(z) = \int_0^{|z|} dy \int_0^y g_n(x) dx.$$

Then $f_n(z) \to |z|$ increasingly as $n \to \infty$. Moreover, we have $|f_n'(z)| \leq 1$ and

$$0 \leq |z| f_n''(z) = |z| g_n(|z|) \leq 2/n. \tag{104}$$

For $z, \zeta \in \mathbb{R}$ it is easy to see that

$$|f_n(\zeta + z) - f_n(\zeta) - z f_n'(\zeta)| \leq |f_n(\zeta + z) - f_n(\zeta)| + |z f_n'(\zeta)| \leq 2|z|.$$

By Taylor's expansion, when $z\zeta \geq 0$, there is η between ζ and $\zeta + z$ so that

$$|\zeta||f_n(\zeta + z) - f_n(\zeta) - z f_n'(\zeta)| \leq |\zeta||f_n''(\eta)|z^2/2 \leq |\eta||f_n''(\eta)|z^2/2 \leq z^2/n.$$

where we used (104) for the last inequality. It follows that, when $z\zeta \geq 0$,

$$|\zeta||f_n(\zeta + z) - f_n(\zeta) - zf_n'(\zeta)| \leq (2|z\zeta|) \wedge (z^2/n) \leq (1 + 2|\zeta|)[|z| \wedge (z^2/n)]. \quad (105)$$

From (103) we have

$$\begin{aligned}\zeta(t) = &\zeta(0) - b\int_0^t \zeta(s-)\mathrm{d}s + \sqrt{2c}\int_0^t \big(\sqrt{x(s)} - \sqrt{y(s)}\big)\mathrm{d}B(s)\\ &+ \int_0^t\int_0^\infty\int_{y(s-)}^{x(s-)} z1_{\{\zeta(s-)>0\}}\tilde{M}(\mathrm{d}s, \mathrm{d}z, \mathrm{d}u)\\ &- \int_0^t\int_0^\infty\int_{x(s-)}^{y(s-)} z1_{\{\zeta(s-)<0\}}\tilde{M}(\mathrm{d}s, \mathrm{d}z, \mathrm{d}u).\end{aligned}$$

By this and Itô's formula,

$$\begin{aligned}f_n(\zeta(t)) = &f_n(\zeta(0)) - b\int_0^t f_n'(\zeta(s))\zeta(s)\mathrm{d}s + c\int_0^t f_n''(\zeta(s))\big[\sqrt{x(s)} - \sqrt{y(s)}\,\big]^2\mathrm{d}s\\ &+ \int_0^t \zeta(s)1_{\{\zeta(s)>0\}}\mathrm{d}s\int_0^\infty [f_n(\zeta(s) + z) - f_n(\zeta(s)) - zf_n'(\zeta(s))]m(\mathrm{d}z)\\ &- \int_0^t \zeta(s)1_{\{\zeta(s)<0\}}\mathrm{d}s\int_0^\infty [f_n(\zeta(s) - z) - f_n(\zeta(s)) + zf_n'(\zeta(s))]m(\mathrm{d}z)\\ &+ \text{mart.}\end{aligned}$$

Taking the expectation in both sides of the above and using (104) and (105) we see

$$\mathbf{P}[f_n(\zeta(t))] \leq f_n(\zeta(0)) + |b|\int_0^t \mathbf{P}[|\zeta(s)|]\mathrm{d}s + \varepsilon_n(t), \quad (106)$$

where

$$\varepsilon_n(t) = 2cn^{-1}t + \int_0^t (1 + 2\mathbf{P}[|\zeta(s)|])\mathrm{d}s\int_0^\infty [z \wedge (n^{-1}z^2)]m(\mathrm{d}z).$$

Clearly, we have $\lim_{n\to\infty}\varepsilon_n(t) = 0$. Then letting $n \to \infty$ in (106) we get

$$\mathbf{P}[|x(t) - y(t)|] \leq |x(0) - y(0)| + |b|\int_0^t \mathbf{P}[|x(s) - y(s)|]\mathrm{d}s.$$

If $x(0) = y(0)$, we have $\mathbf{P}[|x(t) - y(t)|] = 0$ by Gronwall's inequality, and so $\mathbf{P}\{x(t) = y(t)\} = 1$ for $t \geq 0$. Then $\mathbf{P}\{x(t) = y(t) \text{ for } t \geq 0\} = 1$ by the right continuity of the processes. That gives the pathwise uniqueness for (103).

We can give a formulation of the CBI-process in terms of another stochastic integral equation weakly equivalent to (103). Let $\{M(\mathrm{d}s, \mathrm{d}z, \mathrm{d}u)\}$ and $\{\eta(s)\}$ be as in (103). Let $\{W(\mathrm{d}s, \mathrm{d}u)\}$ be a Gaussian time-space white noise on $(0, \infty)^2$ with intensity $\mathrm{d}s\mathrm{d}u$. We assume $\{W(\mathrm{d}s, \mathrm{d}u)\}$, $\{M(\mathrm{d}s, \mathrm{d}z, \mathrm{d}u)\}$, and $\{\eta(s)\}$ are defined on a complete probability space and are independent of each other. Consider the stochastic integral equation

$$\begin{aligned} y(t)=&y(0) + \sqrt{2c}\int_0^t\int_0^{y(s-)} W(\mathrm{d}s, \mathrm{d}u) - b\int_0^t y(s-)\mathrm{d}s \\ &+\int_0^t\int_0^\infty\int_0^{y(s-)} z\tilde{M}(\mathrm{d}s, \mathrm{d}z, \mathrm{d}u) + \eta(t). \end{aligned} \tag{107}$$

The reader may refer to Li [40, Sect. 7.3] and Walsh [50, Chap. 2] for discussions of stochastic integration with respect to Gaussian time-space white noises.

Theorem 8.3 *A positive càdlàg process $\{y(t) : t \geq 0\}$ is a CBI-process with branching and immigration mechanisms (ϕ, ψ) given, respectively, by* (19) *and* (76) *if and only if it is a weak solution to* (107).

Proof Suppose that $\{y(t)\}$ is a CBI-process with branching and immigration mechanisms given, respectively, by (19) and (76). By Theorem 8.1, the process is a weak solution to (103). By El Karoui and Méléard [15, Theorem III.6], on an extension of the probability space we can define a Gaussian time-space white noise $W(\mathrm{d}s, \mathrm{d}u)$ with intensity $\mathrm{d}s\mathrm{d}u$ so that

$$\int_0^t \sqrt{y(s-)}\mathrm{d}B(s) = \int_0^t\int_0^{y(s-)} W(\mathrm{d}s, \mathrm{d}u).$$

Then $\{y(t)\}$ is a weak solution to (107). Conversely, suppose that $\{y(t)\}$ is a weak solution to (107). By Itô's formula one can see $\{y(t)\}$ solves the martingale problem (96). By Theorem 7.2 we infer that $\{y(t)\}$ is a CBI-process with branching and immigration mechanisms given, respectively, by (19) and (76).

Theorem 8.4 *Suppose that $\{y_1(t) : t \geq 0\}$ and $\{y_2(t) : t \geq 0\}$ are two positive solutions to* (107) *with $\mathbf{P}\{y_1(0) \leq y_2(0)\} = 1$. Then we have $\mathbf{P}\{y_1(t) \leq y_2(t)$ for all $t \geq 0\} = 1$.*

Proof Let $\zeta(t) = y_1(t) - y_2(t)$ for $t \geq 0$. For $n \geq 0$ let f_n be the function defined as in the proof of Theorem 8.2. Let $h_n(z) = f_n(z \vee 0)$ for $z \in \mathbb{R}$. Then $h_n(z) \to z_+ := z \vee 0$ increasingly as $n \to \infty$. From (107) it follows that

$$\begin{aligned} \zeta(t)=&\zeta(0) - b\int_0^t \zeta(s-)\mathrm{d}s + \sqrt{2c}\int_0^t\int_{y_2(s-)}^{y_1(s-)} 1_{\{\zeta(s-)>0\}} W(\mathrm{d}s, \mathrm{d}u) \\ &- \sqrt{2c}\int_0^t\int_{y_1(s-)}^{y_2(s-)} 1_{\{\zeta(s-)<0\}} W(\mathrm{d}s, \mathrm{d}u) \\ &+ \int_0^t\int_0^\infty\int_{y_2(s-)}^{y_1(s-)} 1_{\{\zeta(s-)>0\}} z\tilde{M}(\mathrm{d}s, \mathrm{d}z, \mathrm{d}u) \\ &- \int_0^t\int_0^\infty\int_{y_1(s-)}^{y_2(s-)} 1_{\{\zeta(s-)<0\}} z\tilde{M}(\mathrm{d}s, \mathrm{d}z, \mathrm{d}u). \end{aligned}$$

Since $h_n(z) = 0$ for $z \leq 0$, by Itô's formula we have

$$\begin{aligned}
h_n(\zeta(t)) =& -b\int_0^t h_n'(\zeta(s-))\zeta(s-)\mathrm{d}s + c\int_0^t h_n''(\zeta(s-))|\zeta(s-)|\mathrm{d}s \\
&+ \int_0^t \zeta(s-)1_{\{\zeta(s-)>0\}}\mathrm{d}s \int_0^\infty \Big[h_n(\zeta(s-)+z) - h_n(\zeta(s-)) \\
&- zh_n'(\zeta(s-))\Big]m(\mathrm{d}z) - \int_0^t \zeta(s-)1_{\{\zeta(s-)<0\}}\mathrm{d}s \int_0^\infty \Big[h_n(\zeta(s-)-z) \\
&- h_n(\zeta(s-)) + zh_n'(\zeta(s-))\Big]m(\mathrm{d}z) + \text{local mart.} \\
=& -b\int_0^t h_n'(\zeta(s-))\zeta(s-)_+\mathrm{d}s + c\int_0^t h_n''(\zeta(s-))\zeta(s-)_+\mathrm{d}s \\
&+ \int_0^t \zeta(s-)_+\mathrm{d}s \int_0^\infty \Big[h_n(\zeta(s-)+z) - h_n(\zeta(s-)) \\
&- zh_n'(\zeta(s-))\Big]m(\mathrm{d}z) + \text{local mart.}
\end{aligned}$$

For any $k \geq 1$ define $\tau_k = \inf\{t \geq 0 : \zeta(t)_+ \geq k\}$. Taking the expectation in the above equality at time $t \wedge \tau_k$ and using (104) and (105) we have

$$\mathbf{P}[h_n(\zeta(t \wedge \tau_k))] \leq |b|\mathbf{P}\bigg[\int_0^{t\wedge\tau_k} \zeta(s-)_+\mathrm{d}s\bigg] + \varepsilon_n(t),$$

where

$$\varepsilon_n(t) = 2cn^{-1}t + \mathbf{P}\bigg[\int_0^{t\wedge\tau_k}(1+2\zeta(s-)_+)\mathrm{d}s\bigg]\int_0^\infty (z \wedge n^{-1}z^2)m(\mathrm{d}z).$$

Then we let $n \to \infty$ to obtain

$$\mathbf{P}[\zeta(t\wedge\tau_k)_+] \leq |b|\mathbf{P}\bigg[\int_0^{t\wedge\tau_k}\zeta(s-)_+\mathrm{d}s\bigg] \leq |b|\int_0^t \mathbf{P}[\zeta(s\wedge\tau_k)_+]\mathrm{d}s.$$

By Gronwall's inequality, for each $t \geq 0$ we have

$$\mathbf{P}[(y_1(t\wedge\tau_k) - y_2(t\wedge\tau_k))_+] = \mathbf{P}[\zeta(t\wedge\tau_k)_+] = 0.$$

By letting $k \to \infty$ and using Fatou's lemma we see $\mathbf{P}[(y_1(t) - y_2(t))_+] = 0$ for $t \geq 0$, and so $\mathbf{P}\{y_1(t) \leq y_2(t) \text{ for all } t \geq 0\} = 1$ by the right continuity of the processes.

Theorem 8.5 *For any initial value $y(0) = x \geq 0$, there is a pathwise unique positive strong solution to* (107).

Proof By Theorem 8.3 there is a weak solution $\{y(t)\}$ to (107). The pathwise uniqueness of the solution follows from Theorem 8.4. Then $\{y(t)\}$ is a strong solution to (107). See, e.g., Situ [49, p. 76 and p. 104].

From (103) or (107) we see that the immigration of the CBI-process $\{y(t)\}$ is represented by the increasing Lévy process $\{\eta(t)\}$. By the Lévy–Itô decomposition, there is a Poisson time-space random measure $\{N(\mathrm{d}s, \mathrm{d}z)\}$ with intensity $\mathrm{d}s\nu(\mathrm{d}z)$ such that

$$\eta(t) = \beta t + \int_0^t \int_0^\infty z N(\mathrm{d}s, \mathrm{d}z), \qquad t \geq 0.$$

Then the immigration of $\{y(t)\}$ involves two parts: the *continuous part* determined by the drift coefficient β and the *discontinuous part* given by the Poisson random measure $\{N(\mathrm{d}s, \mathrm{d}z)\}$.

Now let us consider a special CBI-process. Let $c, q \geq 0, b \in \mathbb{R}$ and $1 < \alpha < 2$ be given constants. Let $\{B(t)\}$ be a standard Brownian motion. Let $\{z(t)\}$ be a spectrally positive α-stable Lévy process with Lévy measure

$$\gamma(\mathrm{d}z) := (\alpha - 1)\Gamma(2-\alpha)^{-1} z^{-1-\alpha}\mathrm{d}z, \qquad z > 0$$

and $\{\eta(t)\}$ an increasing Lévy process with $\eta(0) = 0$ and with Laplace exponent ψ. We assume that $\{B(t)\}$, $\{z(t)\}$, and $\{\eta(t)\}$ are defined on a complete probability space and are independent of each other. Consider the stochastic differential equation

$$\mathrm{d}y(t) = \sqrt{2cy(t-)}\mathrm{d}B(t) + \sqrt[\alpha]{\alpha q y(t-)}\mathrm{d}z(t) - by(t-)\mathrm{d}t + \mathrm{d}\eta(t). \qquad (108)$$

Theorem 8.6 *A positive càdlàg process $\{y(t) : t \geq 0\}$ is a CBI-process with branching mechanism $\phi(z) = bz + cz^2 + qz^\alpha$ and immigration mechanism ψ given by* (76) *if and only if it is a weak solution to* (108).

Proof Suppose that $\{y(t)\}$ is a weak solution to (108). By Itô's formula one can see that $\{y(t)\}$ solves the martingale problem (96) associated with the generator L defined by (94) with $m(\mathrm{d}z) = \alpha q\gamma(\mathrm{d}z)$. Then $\{y(t)\}$ is a CBI-process with branching mechanism $\phi(z) = bz + cz^2 + qz^\alpha$ and immigration mechanism ψ given by (76). Conversely, suppose that $\{y(t)\}$ is a CBI-process with branching mechanism $\phi(z) = bz + cz^2 + qz^\alpha$ and immigration mechanism ψ given by (76). Then $\{y(t)\}$ is a weak solution to (103) with $\{M(\mathrm{d}s, \mathrm{d}z, \mathrm{d}u)\}$ being a Poisson random measure on $(0, \infty)^3$ with intensity $\alpha q \mathrm{d}s\gamma(\mathrm{d}z)\mathrm{d}u$. Let us assume $q > 0$, for otherwise the proof is easier. Define the random measure $\{N_0(\mathrm{d}s, \mathrm{d}z)\}$ on $(0, \infty)^2$ by

$$N_0((0,t] \times B) = \int_0^t \int_0^\infty \int_0^{y(s-)} 1_{\{y(s-)>0\}} 1_B\Big(\frac{z}{\sqrt[\alpha]{\alpha q y(s-)}}\Big) M(\mathrm{d}s, \mathrm{d}z, \mathrm{d}u)$$
$$+ \int_0^t \int_0^\infty \int_0^{1/\alpha q} 1_{\{y(s-)=0\}} 1_B(z) M(\mathrm{d}s, \mathrm{d}z, \mathrm{d}u).$$

It is easy to compute that $\{N_0(\mathrm{d}s, \mathrm{d}z)\}$ has predictable compensator

$$\begin{aligned}\hat{N}_0((0,t]\times B)=&\int_0^t\int_0^\infty 1_{\{y(s-)>0\}}1_B\Big(\frac{z}{\sqrt[\alpha]{\alpha q y(s-)}}\Big)\frac{\alpha q y(s-)(\alpha-1)\mathrm{d}s\mathrm{d}z}{\Gamma(2-\alpha)z^{1+\alpha}}\\&+\int_0^t\int_0^\infty 1_{\{y(s-)=0\}}1_B(z)\frac{(\alpha-1)\mathrm{d}s\mathrm{d}z}{\Gamma(2-\alpha)z^{1+\alpha}}\\=&\int_0^t\int_0^\infty 1_B(z)\frac{(\alpha-1)\mathrm{d}s\mathrm{d}z}{\Gamma(2-\alpha)z^{1+\alpha}}.\end{aligned}$$

Thus $\{N_0(\mathrm{d}s,\mathrm{d}z)\}$ is a Poisson random measure with intensity $\mathrm{d}s\gamma(\mathrm{d}z)$; see, e.g., Theorem III.7.4 in Ikeda and Watanabe [26, p. 93]. Now define the Lévy processes

$$z(t)=\int_0^t\int_0^\infty z\tilde{N}_0(\mathrm{d}s,\mathrm{d}z)\ \text{ and }\ \eta(t)=\beta t+\int_0^t\int_0^\infty zN(\mathrm{d}s,\mathrm{d}z),$$

where $\tilde{N}_0(\mathrm{d}s,\mathrm{d}z)=N_0(\mathrm{d}s,\mathrm{d}z)-\hat{N}_0(\mathrm{d}s,\mathrm{d}z)$. It is easy to see that

$$\begin{aligned}\int_0^t\sqrt[\alpha]{\alpha q y(s-)}\mathrm{d}z(s)=&\int_0^t\int_0^\infty\sqrt[\alpha]{\alpha q y(s-)}\,z\tilde{N}_0(\mathrm{d}s,\mathrm{d}z)\\=&\int_0^t\int_0^\infty\int_0^{y(s-)}z\tilde{M}(\mathrm{d}s,\mathrm{d}z,\mathrm{d}u).\end{aligned}$$

Then $\{y(t)\}$ is a weak solution to (108).

Theorem 8.7 *For any initial value* $y(0)=x\geq 0$, *there is a pathwise unique positive strong solution to* (108).

Proof By Theorem 8.6 there is a weak solution $\{y(t)\}$ to (108), so it suffices to prove the pathwise uniqueness of the solution. We first recall that the one-sided α-stable process $\{z(t)\}$ can be represented as

$$z(t)=\int_0^t\int_0^\infty z\tilde{M}(\mathrm{d}s,\mathrm{d}z),$$

where $M(\mathrm{d}s,\mathrm{d}z)$ is a Poisson random measure on $(0,\infty)^2$ with intensity $\mathrm{d}s\gamma(\mathrm{d}z)$. Let

$$z_1(t)=\int_0^t\int_0^1 z\tilde{M}(\mathrm{d}s,\mathrm{d}z)\quad\text{and}\quad z_2(t)=\int_0^t\int_1^\infty zM(\mathrm{d}s,\mathrm{d}z).$$

Since $t\mapsto z_2(t)$ has at most finitely many jumps in each bounded interval, we only need to prove the pathwise uniqueness of

$$\begin{aligned}\mathrm{d}y(t)=&\sqrt{2cy(t-)}\mathrm{d}B(t)+\sqrt[\alpha]{\alpha q y(t-)}\mathrm{d}z_1(t)-by(t-)\mathrm{d}t\\&-\alpha^{-1}(\alpha-1)\Gamma(2-\alpha)^{-1}\sqrt[\alpha]{\alpha q y(t-)}\mathrm{d}t+\mathrm{d}\eta(t),\end{aligned}\tag{109}$$

Suppose that $\{x(t)\}$ and $\{y(t)\}$ are two positive solutions to (109) with deterministic initial values. Let $\zeta_\theta(t)=\sqrt[\theta]{x(t)}-\sqrt[\theta]{y(t)}$ for $0<\theta\leq 2$ and $t\geq 0$. Then we have

$$\begin{aligned}\mathrm{d}\zeta_1(t)=&\sqrt{2c}\zeta_2(t-)\mathrm{d}B(t)+\sqrt[\alpha]{\alpha q}\zeta_\alpha(t-)\mathrm{d}z_1(t)-b\zeta_1(t-)\mathrm{d}t\\&-\alpha^{-1}(\alpha-1)\Gamma(2-\alpha)^{-1}\sqrt[\alpha]{\alpha q}\zeta_\alpha(t-)\mathrm{d}t.\end{aligned}$$

For $n \geq 0$ let f_n be the function defined as in the proof of Theorem 8.2. By Itô's formula,

$$\begin{aligned}f_n(\zeta_1(t))=&f_n(\zeta_1(0))+c\int_0^t f_n''(\zeta_1(s-))\zeta_1(s-)^2\mathrm{d}s-b\int_0^t f_n'(\zeta_1(s-))\zeta_1(s-)\mathrm{d}s\\&-\alpha^{-1}(\alpha-1)\Gamma(2-\alpha)^{-1}\sqrt[\alpha]{\alpha q}\int_0^t f_n'(\zeta_1(s-))\zeta_\alpha(s-)\mathrm{d}s\\&+\int_0^t ds\int_0^1\Big[f_n(\zeta_1(s-)+\sqrt[\alpha]{\alpha q}\zeta_\alpha(s-)z)-f_n(\zeta_1(s-))\\&-\sqrt[\alpha]{\alpha q}\zeta_\alpha(s-)zf_n'(\zeta_1(s-))\Big]\gamma(\mathrm{d}z)+\text{local mart.}\end{aligned}\qquad(110)$$

For any $k \geq 1+x(0)\vee y(0)$ let $\tau_k=\inf\{s\geq 0: x(s)\geq k \text{ or } y(s)\geq k\}$. For $0\leq t\leq \tau_k$ we have $|\zeta_1(t-)|\leq k$, $|\zeta_\alpha(t-)|\leq \sqrt[\alpha]{k}$ and

$$|\zeta_1(t)|\leq|\zeta_1(t-)|+|\zeta_1(t)-\zeta_1(t-)|\leq k+\sqrt[\alpha]{\alpha qk}.$$

By Taylor's expansion, there exists $0<\xi<z$ so that

$$\begin{aligned}&[f_n(\zeta_1(s-)+\sqrt[\alpha]{\alpha q}\zeta_\alpha(s-)z)-f_n(\zeta_1(s-))-\sqrt[\alpha]{\alpha q}\zeta_\alpha(s-)zf_n'(\zeta_1(s-))]\\&\quad=2^{-1}(\alpha q)^{2/\alpha}f_n''(\zeta_1(s-)+\sqrt[\alpha]{\alpha q}\zeta_\alpha(s-)\xi)\zeta_\alpha(s-)^2z^2\\&\quad\leq 2^{-1}(\alpha q)^{2/\alpha}\sqrt[\alpha]{k}f_n''(\zeta_1(s-)+\sqrt[\alpha]{\alpha q}\zeta_\alpha(s-)\xi)|\zeta_1(s-)+\sqrt[\alpha]{\alpha q}\zeta_\alpha(s-)\xi|z^2\\&\quad\leq n^{-1}(\alpha q)^{2/\alpha}\sqrt[\alpha]{k}z^2,\end{aligned}$$

where we have used (104) and the fact $\zeta_1(s-)\zeta_\alpha(s-)\geq 0$. Taking the expectation in both sides of (110) gives

$$\begin{aligned}\mathbf{P}[f_n(\zeta_1(t\wedge\tau_k))]\leq&\mathbf{P}[f_n(\zeta_1(0))]+|b|\int_0^t\mathbf{P}[|\zeta_1(s\wedge\tau_k)|]\mathrm{d}s+2cn^{-1}kt\\&+n^{-1}(\alpha q)^{2/\alpha}\sqrt[\alpha]{k}\int_0^t\mathrm{d}s\int_0^1 z^2\gamma(\mathrm{d}z).\end{aligned}$$

Now, if $x(0)=y(0)$, we can let $n\to\infty$ in the inequality above to get

$$\mathbf{P}[|x(t\wedge\tau_k)-y(t\wedge\tau_k)|]\leq|b|\int_0^t\mathbf{P}[|x(s\wedge\tau_k)-y(s\wedge\tau_k)|]\mathrm{d}s.$$

Then $\mathbf{P}[|x(t\wedge\tau_k)-y(t\wedge\tau_k)|]=0$ for $t\geq 0$ by Gronwall's inequality. By letting $k\to\infty$ and using Fatou's lemma, we obtain the pathwise uniqueness for (108).

Example 8.8 The stochastic integral equation (107) can be thought as a continuous time-space counterpart of the definition (68) of the GWI-process. In fact, assuming $\mu = \mathbf{E}(\xi_{1,1}) < \infty$, from (68) we have

$$y(n) - y(n-1) = \sum_{i=1}^{y(n-1)} (\xi_{n,i} - \mu) - (1-\mu)y(n-1) + \eta_n. \tag{111}$$

It follows that

$$y(n) - y(0) = \sum_{k=1}^{n} \sum_{i=1}^{y(k-1)} (\xi_{k,i} - \mu) - (1-\mu)\sum_{k=1}^{n} y(k-1) + \sum_{k=1}^{n} \eta_k. \tag{112}$$

The exact continuous time-state counterpart of (112) would be the stochastic integral equation

$$y(t) = y(0) + \int_0^t \int_0^\infty \int_0^{y(s-)} \xi \tilde{M}(\mathrm{d}s, \mathrm{d}\xi, \mathrm{d}u) - \int_0^t by(s)\mathrm{d}s + \eta(t), \tag{113}$$

which is a typical special form of (107); see Bertoin and Le Gall [5] and Dawson and Li [10]. Here the ξ's selected by the Poisson random measure $M(\mathrm{d}s, \mathrm{d}\xi, \mathrm{d}u)$ are distributed in a i.i.d. fashion and the compensation of the measure corresponds to the centralization in (112). The increasing Lévy process $t \mapsto \eta(t)$ in (113) corresponds to the increasing random walk $n \mapsto \sum_{k=1}^{n} \eta_k$ in (112). The additional term in (107) involving the stochastic integral with respect to the Gaussian white noise is just a continuous time-space parallel of that with respect to the compensated Poisson random measure.

Example 8.9 The stochastic differential equation (108) captures the structure of the CBI-process in a typical special case. Let $1 < \alpha \le 2$. Under the condition $\mu := \mathbf{E}(\xi_{1,1}) < \infty$, from (111) we have

$$y(n) - y(n-1) = \sqrt[\alpha]{y(n-1)} \sum_{i=1}^{y(n-1)} \frac{\xi_{n,i} - \mu}{\sqrt[\alpha]{y(n-1)}} - (1-\mu)y(n-1) + \eta_n.$$

Observe that the partial sum on the right-hand side corresponds to a one-sided α-stable type central limit theorem. Then a continuous time-state counterpart of the above equation would be

$$\mathrm{d}y(t) = \sqrt[\alpha]{\alpha q y(t-)}\mathrm{d}z(t) - by(t)\mathrm{d}t + \beta\mathrm{d}t, \qquad t \ge 0, \tag{114}$$

where $\{z(t) : t \ge 0\}$ is a standard Brownian motion if $\alpha = 2$ and a spectrally positive α-stable Lévy process with Lévy measure $(\alpha - 1)\Gamma(2-\alpha)^{-1}z^{-1-\alpha}\mathrm{d}z$ if $1 < \alpha < 2$. This is a typical special form of (108).

Example 8.10 When $\alpha = 2$ and $\beta = 0$, the CB-process defined by (114) is a diffusion process, which is known as *Feller's branching diffusion*. This process was first studied by Feller [18].

Example 8.11 In the special case of $\alpha = 2$, the CBI-process defined by (114) is known in mathematical finance as the *Cox–Ingersoll–Ross model* (CIR-model), which was used by Cox et al. [8] to describe the evolution of interest rates. The asymptotic behavior of the estimators of the parameters in the CIR-model was studied by Overbeck and Rydén [45]. In the general case, the solution to (114) is called a *α-stable Cox–Ingersoll–Ross model* (α-stable CIR-model); see, e.g., Jiao et al. [28] and Li and Ma [43].

As a simple application of the stochastic equation (103) or (107), we can give a simple derivation of the joint Laplace transform of the CBI-process and its positive integral functional. The next theorem extends the results in Sect. 4.

Theorem 8.12 *Let $Y = (\Omega, \mathscr{F}, \mathscr{F}_t, y(t), \mathbf{P}_x)$ be a Hunt realization of the CBI-process. Then for $t, \lambda, \theta \geq 0$ we have*

$$\mathbf{P}_x \exp\Big\{ - \lambda y(t) - \theta \int_0^t y(s)\mathrm{d}s \Big\} = \exp\Big\{ - x v(t) - \int_0^t \psi(v(s))\mathrm{d}s \Big\},$$

where $t \mapsto v(t) = v(t, \lambda, \theta)$ is the unique positive solution to (64).

Proof We can construct the process $\{y(t) : t \geq 0\}$ as the solution to (103) or (107) with $y(0) = x \geq 0$. Let

$$z(t) = \int_0^t y(s)\mathrm{d}s, \qquad t \geq 0.$$

Consider a function $G = G(t, y, z)$ on $[0, \infty)^3$ with bounded continuous derivatives up to the first-order relative to $t \geq 0$ and $z \geq 0$ and up to the second-order relative to $x \geq 0$. By Itô's formula,

$$\begin{aligned} G(t, y(t), z(t)) =& G(0, y(0), 0) + \text{local mart.} + \int_0^t \Big\{ G'_t(s, y(s), z(s)) \\ & + y(s)G'_z(s, y(s), z(s)) + [\beta - by(s)]G'_y(s, y(s), z(s)) \\ & + cy(s)G''_{yy}(s, y(s), z(s)) + y(s)\int_0^\infty \Big[G(s, y(s) + z, z(s)) \\ & - G(s, y(s), z(s)) - zG'_y(s, y(s), z(s)) \Big] m(\mathrm{d}z) \Big\} \mathrm{d}s \\ & + \int_0^t \mathrm{d}s \int_0^\infty \Big[G(s, y(s) + z, z(s)) - G(s, y(s), z(s)) \Big] \nu(\mathrm{d}z). \end{aligned}$$

We can apply the above formula to the function

$$G_T(t, y, z) = \exp\bigg\{ - v(T-t)x - \theta z - \int_0^{T-t} \psi(v(s))\mathrm{d}s \bigg\}.$$

Using (64) we see $t \mapsto G_T(t \wedge T, y(t \wedge T), z(t \wedge T))$ is a local martingale, and hence a martingale by the boundedness. From the relation $\mathbf{P}_x[G_T(t, y(t), z(t))] = G_T(0, x, 0)$ with $T = t$ we get the desired result.

The existence and uniqueness of strong solution to (103) were first established in Dawson and Li [39]. The moment condition (78) was removed in Fu and Li [19]. A stochastic flow of discontinuous CB-processes with critical branching mechanism was constructed in Bertoin and Le Gall [5] by using weak solutions of a special case of (103). The existence and uniqueness of strong solution to (108) were proved in Fu and Li [19] and those for (107) were given in Dawson and Li [11] and Li and Ma [42]. The results of Bertoin and Le Gall [5] were extended to flows of CBI-processes in Dawson and Li [11] and Li [41] using strong solutions. Although the study of branching processes has a long history, the stochastic equations (103), (107), and (108) were not established until the works mentioned above.

A natural generalization of the CBI-process is the so-called *affine Markov process*; see Duffie et al. [13] and the references therein. Those authors defined the regularity property of affine processes and gave a number of characterizations of those processes under the regularity assumption. By a result of Kawazu and Watanabe [29], a stochastically continuous CBI-process is automatically regular. Under the first moment assumption, the regularity of affine processes was proved in Dawson and Li [10]. The regularity problem was settled in Keller-Ressel et al. [30], where it was proved that any stochastically continuous affine process is regular. This problem is related to Hilbert's fifth problem; see Keller-Ressel et al. [30] for details.

9 Local and Global Maximal Jumps

In this section, we use stochastic equations of the CB- and CBI-processes to derive several characterizations of the distributions of their local and global maximal jumps. Let us consider a branching mechanism ϕ given by (19). Let $\{B(t)\}$ be a standard Brownian motion and $\{M(\mathrm{d}s, \mathrm{d}z, \mathrm{d}u)\}$ a Poisson time-space random measure on $(0, \infty)^3$ with intensity $\mathrm{d}s m(\mathrm{d}z)\mathrm{d}u$. By Theorem 8.2, for any $x \geq 0$, there is a pathwise unique positive strong solution to

$$\begin{aligned} x(t) =& x + \int_0^t \sqrt{2cx(s-)}\mathrm{d}B(s) - b\int_0^t x(s-)\mathrm{d}s \\ &+ \int_0^t \int_0^\infty \int_0^{x(s-)} z\tilde{M}(\mathrm{d}s, \mathrm{d}z, \mathrm{d}u). \end{aligned} \tag{115}$$

By Theorem 8.1, the solution $\{x(t) : t \geq 0\}$ is a CB-process with branching mechanism ϕ. For $t \geq 0$ and $r > 0$ let

$$N_r(t) = \int_0^t \int_r^\infty \int_0^{x(s-)} M(\mathrm{d}s, \mathrm{d}z, \mathrm{d}u),$$

which denotes the number of jumps with sizes in (r, ∞) of the trajectory $t \mapsto x(t)$ on the interval $(0, t]$. By (33) we have

$$\mathbf{P}[N_r(t)] = m(r, \infty)\mathbf{P}\bigg[\int_0^t x(s)\mathrm{d}s\bigg] = xb^{-1}(1 - \mathrm{e}^{-bt})m(r, \infty),$$

where $b^{-1}(1 - \mathrm{e}^{-bt}) = t$ for $b = 0$ by convention. In particular, we have $\mathbf{P}\{N_r(t) < \infty\} = 1$. For $r > 0$ we can define another branching mechanism by

$$\phi_r(z) = b_r z + cz^2 + \int_0^r (\mathrm{e}^{-zu} - 1 + zu)m(\mathrm{d}u), \tag{116}$$

where

$$b_r = b + \int_r^\infty um(\mathrm{d}u).$$

For $\theta \geq 0$ let $t \mapsto u(t, \theta)$ be the unique positive solution to (67). Let $t \mapsto u_r(t, \theta)$ be the unique positive solution to

$$\frac{\partial}{\partial t}u(t, \theta) = \theta - \phi_r(u(t, \theta)), \quad u(0, \theta) = 0. \tag{117}$$

The following theorem gives a characterization of the distribution of the local maximal jump of the CB-process.

Theorem 9.1 *Let $\Delta x(t) = x(t) - x(t-)$ for $t \geq 0$. Then for any $r > 0$ we have*

$$\mathbf{P}_x\bigg\{\max_{0<s\leq t} \Delta x(s) \leq r\bigg\} = \exp\{-xu_r(t)\},$$

where $u_r(t) = u_r(t, m(r, \infty))$.

Proof Let $M_r(\mathrm{d}s, \mathrm{d}z, \mathrm{d}u)$ and $M^r(\mathrm{d}s, \mathrm{d}z, \mathrm{d}u)$ denote the restrictions of $M(\mathrm{d}s, \mathrm{d}z, \mathrm{d}u)$ to $(0, \infty) \times (0, r] \times (0, \infty)$ and $(0, \infty) \times (r, \infty) \times (0, \infty)$, respectively. We can rewrite (115) into

$$\begin{aligned} x(t) = x &+ \int_0^t \sqrt{2cx(s-)}\mathrm{d}B(s) + \int_0^t \int_0^r \int_0^{x(s-)} z\tilde{M}_r(\mathrm{d}s, \mathrm{d}z, \mathrm{d}u) \\ &- \int_0^t b_r x(s-)\mathrm{d}s + \int_0^t \int_r^\infty \int_0^{x(s-)} zM^r(\mathrm{d}s, \mathrm{d}z, \mathrm{d}u), \end{aligned}$$

where the last term collects the jumps with sizes in (r, ∞) of $\{x(t)\}$. Let $\{x_r(t)\}$ be the unique positive strong solution to

$$x_r(t)=x - \int_0^t b_r x_r(s-)\mathrm{d}s + \int_0^t \sqrt{2cx_r(s-)}\mathrm{d}B(s) + \int_0^t \int_0^r \int_0^{x_r(s-)} z\tilde{M}_r(\mathrm{d}s, \mathrm{d}z, \mathrm{d}u).$$

Then $\{x_r(t)\}$ is a CB-process with branching mechanism ϕ_r. Let $\tau_r = \inf\{s \geq 0 : \Delta x(s) > r\}$. We have $x_r(s) = x(s)$ for $0 \leq s < \tau_r$ and

$$\Big\{\max_{0<s\leq t} \Delta x(s) \leq r\Big\} = \Big\{\int_0^t \int_r^\infty \int_0^{x(s-)} M^r(\mathrm{d}s, \mathrm{d}z, \mathrm{d}u) = 0\Big\} = \Big\{\int_0^t \int_r^\infty \int_0^{x_r(s-)} M^r(\mathrm{d}s, \mathrm{d}z, \mathrm{d}u) = 0\Big\}.$$

Since the strong solution $\{x_r(t)\}$ is progressively measurable with respect to the filtration generated by $\{B(t)\}$ and $\{M_r(\mathrm{d}s, \mathrm{d}z, \mathrm{d}u)\}$, it is independent of $\{M^r(\mathrm{d}s, \mathrm{d}z, \mathrm{d}u)\}$. Then $\{M^r(\mathrm{d}s, \mathrm{d}z, \mathrm{d}u)\}$ is still a Poisson random measure conditionally upon $\{x_r(t)\}$. It follows that

$$\mathbf{P}_x\Big\{\max_{0<s\leq t} \Delta x(s) \leq r\Big\} = \mathbf{P}_x\Big[\exp\Big\{-m(r, \infty)\int_0^t x_r(s)\mathrm{d}s\Big\}\Big].$$

Then the desired result follows by Corollary 4.4.

Corollary 9.2 *Suppose that the measure $m(\mathrm{d}u)$ has unbounded support. Then we have, as $r \to \infty$,*

$$\mathbf{P}_x\Big\{\max_{0<s\leq t} \Delta x(s) > r\Big\} \sim xb^{-1}(1 - \mathrm{e}^{-bt})m(r, \infty).$$

Proof Recall that $t \mapsto u(t, \theta)$ is defined by (67) and $t \mapsto u_r(t, \theta)$ is defined by (117). It is easy to see that $u(t, 0) = u_r(t, 0) = 0$. Moreover, by (67) we have

$$\frac{\partial}{\partial t}\frac{\partial}{\partial \theta}u(t, 0) = 1 - b\frac{\partial}{\partial \theta}u(t, 0), \quad \frac{\partial}{\partial \theta}u(0, 0) = 0.$$

We can solve the above equation to get

$$\frac{\partial}{\partial \theta}u(t, 0) = b^{-1}(1 - \mathrm{e}^{-bt}), \tag{118}$$

where $b^{-1}(1 - \mathrm{e}^{-bt}) = t$ for $b = 0$ by convention. Similarly we have

$$\frac{\partial}{\partial \theta}u_r(t, 0) = b_r^{-1}(1 - \mathrm{e}^{-b_r t}). \tag{119}$$

By Theorem 9.1 it follows that

$$\mathbf{P}_x\left\{\max_{0<s\le t}\Delta x(s)>r\right\}=1-\exp\{-xu_r(t,m(r,\infty))\},$$

For $r>q>0$, we have obviously $\phi\le\phi_r\le\phi_q$. By Corollary 4.6 we see

$$u_q(t,m(r,\infty))\le u_r(t,m(r,\infty))\le u(t,m(r,\infty)).$$

It follows that

$$\begin{aligned}1-\exp\{-xu_q(t,m(r,\infty))\}\le&\mathbf{P}_x\left\{\max_{0<s\le t}\Delta x(s)>r\right\}\\ \le&1-\exp\{-xu(t,m(r,\infty))\}.\end{aligned}$$

By (118) and (119), as $r\to\infty$ we have

$$\begin{aligned}1-\exp\{-xu(t,m(r,\infty))\}\sim&xu(t,m(r,\infty))\\ \sim&xb^{-1}(1-\mathrm{e}^{-bt})m(r,\infty),\end{aligned}$$

and

$$\begin{aligned}1-\exp\{-xu_q(t,m(r,\infty))\}\sim&xu_q(t,m(r,\infty))\\ \sim&xb_q^{-1}(1-\mathrm{e}^{-b_qt})m(r,\infty).\end{aligned}$$

The proof is completed as we notice $\lim_{q\to\infty}b_q=b$.

We can also give some characterizations of the global maximal jump of the CB-process. Let $\phi_r^{-1}(\theta):=\inf\{z\ge 0:\phi_r(z)>\theta\}$ for $\theta\ge 0$. It is easy to see that $\phi_r^{-1}(m(r,\infty))\to 0$ as $r\to\infty$ if and only if $b\ge 0$. Let $\phi'(\infty)$ be given by (30). By Theorems 4.8 and 9.1 we have:

Corollary 9.3 *Suppose that $\phi'(\infty)>0$. Then for any $r>0$ with $m(r,\infty)>0$ we have*

$$\mathbf{P}_x\left\{\sup_{s>0}\Delta x(s)\le r\right\}=\exp\{-x\phi_r^{-1}(m(r,\infty))\}.$$

Corollary 9.4 *Suppose that $b>0$ and the measure $m(\mathrm{d}u)$ has unbounded support. Then as $r\to\infty$ we have*

$$\mathbf{P}_x\left\{\sup_{s>0}\Delta x(s)>r\right\}\sim xb^{-1}m(r,\infty).$$

The results on local maximal jumps obtained above can be generalized to the case of a CBI-process. Let (ϕ,ψ) be the branching and immigration mechanisms given, respectively, by (19) and (76) with $\nu(du)$ satisfying (78). Let $\{y(t) : t \geq 0\}$ be the CBI-process defined by (103) with $y(0) = x \geq 0$. For $r > 0$ let

$$\psi_r(z) = \beta z + \int_0^r (1 - e^{-zu})\nu(du).$$

Based on Theorem 8.12, the following theorem can be proved by modifying the arguments in the proof of Theorem 9.1.

Theorem 9.5 *Let $\Delta y(t) = y(t) - y(t-)$ for $t \geq 0$. Then for any $r > 0$ we have*

$$\mathbf{P}_x\Big\{\max_{0<s\leq t} \Delta y(s) \leq r\Big\} = \exp\Big\{ - xu_r(t) - \nu(r, \infty)t - \int_0^t \psi_r(u_r(s))ds\Big\},$$

where $u_r(t) = u_r(t, m(r, \infty))$.

The results given in this section were adopted from He and Li [24]. We refer the reader to Bernis and Scotti [4] and Jiao et al. [28] for more careful analysis of the jumps of CBI-processes. In particular, the distributions of the numbers of large jumps in intervals were characterized in Jiao et al. [28]. The analysis is important for the study in mathematical finance as it allows one to describe in a unified way several recent observations on the bond markets such as the persistence of low interest rates together with the presence of large jumps.

10 A Coupling of CBI-Processes

In this section, we give some characterizations of a coupling of CBI-processes constructed by the stochastic equation (107). Using this coupling, we prove the strong Feller property and the exponential ergodicity of the CBI-process under suitable conditions. We shall follow the arguments of Li and Ma [43]. Suppose that (ϕ, ψ) are the branching and immigration mechanisms given, respectively, by (19) and (76) with $\nu(du)$ satisfying (78). Let $(P_t)_{t\geq 0}$ be the transition semigroup of the corresponding CBI-process defined by (25) and (77).

Theorem 10.1 *If $\{x(t) : t \geq 0\}$ and $\{y(t) : t \geq 0\}$ are positive solutions to (107) with $\mathbf{P}\{x(0) \leq y(0)\} = 1$, then $\{y(t) - x(t) : t \geq 0\}$ is a CB-process with branching mechanism ϕ.*

Proof By Theorem 8.4 we have $\mathbf{P}\{x(t) \leq y(t)$ for all $t \geq 0\} = 1$. Let $z(t) = y(t) - x(t)$. From (107) we have

$$z(t)=z(0) + \sqrt{2c}\int_0^t \int_{x(s-)}^{y(s-)} W(ds, du) - b\int_0^t z(s-)ds$$

$$
\begin{aligned}
&+\int_0^t \int_0^\infty \int_{x(s-)}^{y(s-)} z\tilde{M}(\mathrm{d}s, \mathrm{d}z, \mathrm{d}u) \\
=&z(0) + \sqrt{2c}\int_0^t \int_0^{z(s-)} W(\mathrm{d}s, x(s-) + \mathrm{d}u) - b\int_0^t z(s-)\mathrm{d}s \\
&+\int_0^t \int_0^\infty \int_0^{z(s-)} z\tilde{M}(\mathrm{d}s, \mathrm{d}z, x(s-) + \mathrm{d}u),
\end{aligned}
$$

where $W(\mathrm{d}s, x(s-) + \mathrm{d}u)$ is a Gaussian time-space white noise with intensity $\mathrm{d}s\mathrm{d}u$ and $M(\mathrm{d}s, \mathrm{d}z, x(s-) + \mathrm{d}u)$ is a Poisson time-space random measure with intensity $\mathrm{d}sm(\mathrm{d}z)\mathrm{d}u$. That shows $\{z(t)\}$ is a weak solution to (107) with $\eta(t) \equiv 0$. Then it is a CB-process with branching mechanism ϕ.

For $x \geq 0$ and $y \geq 0$, let $\{x(t) : t \geq 0\}$ and $\{y(t) : t \geq 0\}$ be the positive strong solutions to (107) with $x(0) = x$ and $y(0) = y$. This construction gives a natural *coupling* of the CBI-processes. Let $\tau(x, y) = \inf\{t \geq 0 : x(t) = y(t)\}$ be the *coalescence time* of the coupling. The distribution of this stopping time is given in the following theorem.

Theorem 10.2 *Suppose that Condition 3.5 holds. Then for any $t \geq 0$ we have*

$$
\mathbf{P}\{\tau(x, y) \leq t\} = \mathbf{P}\{y(t) = x(t)\} = \exp\{-|x - y|\bar{v}_t\}, \tag{120}
$$

where $t \mapsto \bar{v}_t$ is the unique solution to (47) *with singular initial condition $\bar{v}_{0+} = \infty$.*

Proof It suffices to consider the case of $y \geq x \geq 0$. By Theorem 10.1 the difference $\{y(t) - x(t) : t \geq 0\}$ is a CB-process with branching mechanism ϕ. By Theorem 3.4 the probability $\mathbf{P}\{\tau(x, y) \leq t\} = \mathbf{P}\{y(t) = x(t)\}$ is given by (120).

Theorem 10.3 *Suppose that Condition 3.5 holds. Then for $t > 0$ and $x, y \geq 0$ we have*

$$
\| P_t(x, \cdot) - P_t(y, \cdot)\|_{\mathrm{var}} \leq 2(1 - \mathrm{e}^{-\bar{v}_t|x-y|}) \leq 2\bar{v}_t|x - y|, \tag{121}
$$

where $\| \cdot \|_{\mathrm{var}}$ denotes the total variation norm.

Proof Let $\{x(t) : t \geq 0\}$ and $\{y(t) : t \geq 0\}$ be given as above. Since $\{y(t) - x(t) : t \geq 0\}$ is a CB-process with branching mechanism ϕ, for any bounded Borel function f on $[0, \infty)$, we have

$$
\begin{aligned}
|P_t f(x) - P_t f(y)| =& \big|\mathbf{P}[f(x(t))] - \mathbf{P}[f(y(t))]\big| \\
\leq& \mathbf{P}\big[|f(x(t)) - f(y(t))|1_{\{y(t) \neq x(t)\}}\big] \\
\leq& \mathbf{P}\big[(|f(x(t))| + |f(y(t))|)1_{\{y(t) \neq x(t)\}}\big] \\
\leq& 2\|f\|\mathbf{P}\{y(t) - x(t) \neq 0\} \\
=& 2\|f\|(1 - \mathrm{e}^{-\bar{v}_t|x-y|}).
\end{aligned}
$$

where the last equality follows by Theorem 10.2. Then we get (121) by taking the supremum over f with $\|f\| \leq 1$.

By Theorem 10.3, for each $t > 0$ the operator P_t maps any bounded Borel function on $[0, \infty)$ into a bounded continuous function, that is, the transition semigroup $(P_t)_{t\geq 0}$ satisfies the *strong Feller property*.

Theorem 10.4 *Suppose that $b > 0$. Then the transition semigroup $(P_t)_{t\geq 0}$ has a unique stationary distribution η given by*

$$L_\eta(\lambda) = \exp\bigg\{-\int_0^\infty \psi(v_s(\lambda))\mathrm{d}s\bigg\} = \exp\bigg\{-\int_0^\lambda \frac{\psi(z)}{\phi(z)}\mathrm{d}z\bigg\}, \tag{122}$$

and $P_t(x, \cdot) \to \eta$ weakly on $[0, \infty)$ as $t \to \infty$ for every $x \geq 0$. Moreover, we have

$$\int_{[0,\infty)} y\eta(\mathrm{d}y) = b^{-1}\psi'(0), \tag{123}$$

where $\psi'(0)$ is given by (80).

Proof Since $b > 0$, we have $\phi(z) \geq 0$ for all $z \geq 0$, so $t \mapsto v_t(\lambda)$ is decreasing. By Corollary 3.2 we have $\lim_{t\to\infty} v_t(\lambda) = 0$. From (34) it follows that

$$\int_0^t \psi(v_s(\lambda))\mathrm{d}s = \int_{v_t(\lambda)}^\lambda \frac{\psi(z)}{\phi(z)}\mathrm{d}z.$$

In view of (77), we have

$$\begin{aligned}\lim_{t\to\infty}\int_{[0,\infty)} \mathrm{e}^{-\lambda y}P_t(x, \mathrm{d}y) &= \exp\bigg\{-\int_0^\infty \psi(v_s(\lambda))\mathrm{d}s\bigg\}\\ &= \exp\bigg\{-\int_0^\lambda \frac{\psi(z)}{\phi(z)}\mathrm{d}z\bigg\}.\end{aligned}$$

By Theorem 1.2 there is a probability measure η on $[0, \infty)$ defined by (122). It is easy to show that η is the unique stationary distribution for $(P_t)_{t\geq 0}$. The expression (123) for its first moment follows by differentiating both sides of (122) at $\lambda = 0$.

Theorem 10.5 *Suppose that $b > 0$ and Condition 3.5 holds. Then for any $x \geq 0$ and $t \geq r > 0$ we have*

$$\|P_t(x, \cdot) - \eta(\cdot)\|_{\mathrm{var}} \leq 2[x + b^{-1}\psi'(0)]\bar{v}_r\mathrm{e}^{b(r-t)}, \tag{124}$$

where η is given by (122).

Proof Since η is a stationary distribution for $(P_t)_{t\geq 0}$, by Theorem 10.3 one can see

$$\begin{aligned}\|P_t(x,\cdot)-\eta(\cdot)\|_{\mathrm{var}}=&\left\|\int_{[0,\infty)}[P_t(x,\cdot)-P_t(y,\cdot)]\eta(\mathrm{d}y)\right\|_{\mathrm{var}}\\ \leq&\int_{[0,\infty)}\|P_t(x,\cdot)-P_t(y,\cdot)\|_{\mathrm{var}}\eta(\mathrm{d}y)\\ \leq&2\bar{v}_t\int_{[0,\infty)}|x-y|\eta(\mathrm{d}y)\\ \leq&2\bar{v}_t\int_{[0,\infty)}(x+y)\eta(\mathrm{d}y)\\ =&2[x+b^{-1}\psi'(0)]\bar{v}_t,\end{aligned}$$

where the last equality follows by (123). The semigroup property of $(v_t)_{t\geq 0}$ implies $\bar{v}_t = v_{t-r}(\bar{v}_r)$ for any $t \geq r > 0$. By (32) we see $\bar{v}_t = v_{t-r}(\bar{v}_r) \leq \mathrm{e}^{b(r-t)}\bar{v}_r$. Then (124) holds.

The result of Theorem 10.1 was used to construct flows of CBI-processes in Dawson and Li [11]. Clearly, the right-hand side of (124) decays exponentially fast as $t \to \infty$. This property is called the *exponential ergodicity* of the transition semigroup $(P_t)_{t\geq 0}$. It has played an important role in the study of asymptotics of the estimators for the α-stable CIR-model in Li and Ma [43].

References

1. Aliev, S.A.: A limit theorem for the Galton–Watson branching processes with immigration. Ukr. Math. J. **37**, 535–538 (1985)
2. Aliev, S.A., Shchurenkov, V.M.: Transitional phenomena and the convergence of Galton–Watson processes to Jiřina processes. Theory Probab. Appl. **27**, 472–485 (1982)
3. Athreya, K.B., Ney, P.E.: Branching Processes. Springer, Berlin (1972)
4. Bernis, G., Scotti, S.: Clustering effects through Hawkes processes. From Probability to Finance – Lecture Note of BICMR Summer School on Financial Mathematics. Series of Mathematical Lectures from Peking University. Springer, Berlin (2018+)
5. Bertoin, J., Le Gall, J.-F.: Stochastic flows associated to coalescent processes III: limit theorems. Ill. J. Math. **50**, 147–181 (2006)
6. Bienaymé, I.J.: De la loi de multiplication et de la durée des families. Soc. Philomat. Paris Extr. **5**, 37–39 (1845)
7. Chung, K.L.: Lectures from Markov Processes to Brownian Motion. Springer, Heidelberg (1982)
8. Cox, J., Ingersoll, J., Ross, S.: A theory of the term structure of interest rate. Econometrica **53**, 385–408 (1985)
9. Dawson, D.A., Li, Z.: Construction of immigration superprocesses with dependent spatial motion from one-dimensional excursions. Probab. Theory Relat. Fields **127**, 37–61 (2003)
10. Dawson, D.A., Li, Z.: Skew convolution semigroups and affine Markov processes. Ann. Probab. **34**, 1103–1142 (2006)
11. Dawson, D.A., Li, Z.: Stochastic equations, flows and measure-valued processes. Ann. Probab. **40**, 813–857 (2012)
12. Dellacherie, C., Meyer, P.A.: Probabilities and Potential. North-Holland, Amsterdam (1982)
13. Duffie, D., Filipović, D., Schachermayer, W.: Affine processes and applications in finance. Ann. Appl. Probab. **13**, 984–1053 (2003)

14. Duquesne, T., Labbé, C.: On the Eve property for CSBP. Electron. J. Probab. **19**, Paper No. 6, 1–31 (2014)
15. El Karoui, N., Méléard, S.: Martingale measures and stochastic calculus. Probab. Theory Relat. Fields **84**, 83–101 (1990)
16. El Karoui, N., Roelly, S.: Propriétés de martingales, explosion et representation de Lévy–Khintchine d'une classe de processus de branchement à valeurs mesures. Stoch. Process. Appl. **38**, 239–266 (1991)
17. Ethier, S.N., Kurtz, T.G.: Markov Processes: Characterization and Convergence. Wiley, New York (1986)
18. Feller, W.: Diffusion processes in genetics. In: Proceedings of the 2nd Berkeley Symposium on Mathematical Statistics and Probability (1951), University of California Press, Berkeley and Los Angeles, pp. 227–246 (1950)
19. Fu, Z., Li, Z.: Stochastic equations of nonnegative processes with jumps. Stoch. Process. Appl. **120**, 306–330 (2010)
20. Galton, F., Watson, H.W.: On the probability of the extinction of families. J. Anthropol. Inst. G. B. Irel. **4**, 138–144 (1874)
21. Grey, D.R.: Asymptotic behaviour of continuous time, continuous state-space branching processes. J. Appl. Probab. **11**, 669–677 (1974)
22. Grimvall, A.: On the convergence of sequences of branching processes. Ann. Probab. **2**, 1027–1045 (1974)
23. Harris, T.E.: The Theory of Branching Processes. Springer, Berlin (1963)
24. He, X., Li, Z.: Distributions of jumps in a continuous-state branching process with immigration. J. Appl. Probab. **53**, 1166–1177 (2016)
25. Hewitt, E., Stromberg, K.: Real and Abstract Analysis. Springer, Heidelberg (1965)
26. Ikeda, N., Watanabe, S.: Stochastic Differential Equations and Diffusion Processes, 2nd edn. North-Holland, Amsterdam (1989)
27. Jacod, J., Shiryaev, A.N.: Limit Theorems for Stochastic Processes, 2nd edn. Springer, Heidelberg (2003)
28. Jiao, Y., Ma, C., Scotti, S.: Alpha-CIR model with branching processes in sovereign interest rate modeling. Financ. Stoch. **21**, 789–813 (2017)
29. Kawazu, K., Watanabe, S.: Branching processes with immigration and related limit theorems. Theory Probab. Appl. **16**, 36–54 (1971)
30. Keller-Ressel, M., Schachermayer, W., Teichmann, J.: Affine processes are regular. Probab. Theory Relat. Fields **151**, 591–611 (2011)
31. Kyprianou, A.E.: Fluctuations of Lévy Processes with Applications, 2nd edn. Springer, Heidelberg (2014)
32. Lambert, A.: The branching process with logistic growth. Ann. Appl. Probab. **15**, 1506–1535 (2005)
33. Lamberton, D., Lapeyre, B.: Introduction to Stochastic Calculus Applied to Finance. Chapman and Hall, London (1996)
34. Lamperti, J.: The limit of a sequence of branching processes. Z. Wahrsch. verw. Geb. **7**, 271–288 (1967)
35. Lamperti, J.: Continuous state branching processes. Bull. Am. Math. Soc. **73**, 382–386 (1967)
36. Li, Z.: Immigration structures associated with Dawson–Watanabe superprocesses. Stoch. Process. Appl. **62**, 73–86 (1996)
37. Li, Z.: Asymptotic behavior of continuous time and state branching processes. J. Aust. Math. Soc. Ser. A **68**, 68–84 (2000)
38. Li, Z.: Skew convolution semigroups and related immigration processes. Theory Probab. Appl. **46**, 274–296 (2003)
39. Li, Z.: A limit theorem for discrete Galton–Watson branching processes with immigration. J. Appl. Probab. **43**, 289–295 (2006)
40. Li, Z.: Measure-Valued Branching Markov Processes. Springer, Heidelberg (2011)
41. Li, Z.: Path-valued branching processes and nonlocal branching superprocesses. Ann. Probab. **42**, 41–79 (2014)

42. Li, Z., Ma, C.: Catalytic discrete state branching models and related limit theorems. J. Theor. Probab. **21**, 936–965 (2008)
43. Li, Z., Ma, C.: Asymptotic properties of estimators in a stable Cox–Ingersoll–Ross model. Stoch. Process. Appl. **125**, 3196–3233 (2015)
44. Li, Z., Shiga, T.: Measure-valued branching diffusions: immigrations, excursions and limit theorems. J. Math. Kyoto Univ. **35**, 233–274 (1995)
45. Overbeck, L., Rydén, T.: Estimation in the Cox–Ingersoll–Ross model. Econom. Theory **13**, 430–461 (1997)
46. Pardoux, E.: Probabilistic Models of Population Evolution: Scaling Limits, Genealogies and Interactions. Springer, Switzerland (2016)
47. Parthasarathy, K.R.: Probability Measures on Metric Spaces. Academic, New York (1967)
48. Pitman, J., Yor, M.: A decomposition of Bessel bridges. Z. Wahrsch. verw. Geb. **59**, 425–457 (1982)
49. Situ, R.: Theory of Stochastic Differential Equations with Jumps and Applications. Springer, Heidelberg (2005)
50. Walsh, J.B.: An introduction to stochastic partial differential equations. Ecole d'Eté de Probabilités de Saint-Flour XIV-1984. Lecture Notes in Mathematics, vol. 1180, pp. 265–439. Springer, Berlin (1986)

Enlargement of Filtration in Discrete Time

Christophette Blanchet-Scalliet and Monique Jeanblanc

Abstract In this lecture, we study enlargement of filtration in a discrete time setting. In a discrete time setting, considering two filtrations $\mathbb{F}$ and $\mathbb{G}$ with $\mathbb{F} \subset \mathbb{G}$, any $\mathbb{F}$-martingale is a $\mathbb{G}$-semimartingale. We give the decomposition of $\mathbb{F}$-martingales in $\mathbb{G}$-semimartingales in the case of initial (and progressive) enlargement. We study progressive enlargement with pseudo-stopping times and honest times.

Keywords Enlargement of filtration · Initial enlargement · Progressive enlargement · Azéma's supermartingale · Discrete time · Doob decomposition · Pseudo-stopping times · Honest times

Mathematics Subject Classification (2010) 60G05 · 60G42 · 60G99

Introduction

In continuous time, considering two filtrations $\mathbb{F}$ and $\mathbb{G}$ such that $\mathbb{F} \subset \mathbb{G}$, it may happen that some $\mathbb{F}$-martingales are not $\mathbb{G}$-semimartingales and one of the problems in enlargement of filtration framework is to give conditions so that all $\mathbb{F}$-martingales are $\mathbb{G}$-semimartingales. We refer to [6] for references and results. The conditions are not simple and one has to work carefully. In these notes, we present classical results on enlargement of filtration, in a discrete time framework. In such a setting, any $\mathbb{F}$-martingale is a $\mathbb{G}$-semimartingale for any filtration $\mathbb{G}$ larger than $\mathbb{F}$, and one can think that there are not so many things to do. From our point of view, one interest of this lecture is that the proofs of the semimartingale decomposition formula are

C. Blanchet-Scalliet (✉)
Université de Lyon - CNRS, UMR 5208, Institut Camille Jordan - Ecole Centrale de Lyon, 36 avenue Guy de Collongue, 69134 Ecully Cedex, France
e-mail: christophette.blanchet@ec-lyon.fr

M. Jeanblanc
LaMME, UMR CNRS 8071, Univ Evry, Université Paris Saclay, 23 boulevard de France, 91037 Evry, France
e-mail: monique.jeanblanc@univ-evry.fr

Y. Jiao (ed.), *From Probability to Finance*, Mathematical Lectures from Peking University, https://doi.org/10.1007/978-981-15-1576-7_2

simple and give a pedagogical support to understand the general formulae obtained in the literature in continuous time. It can be noted that many results are established in continuous time under the hypothesis that all $\mathbb{F}$-martingales are continuous or, in the progressive enlargement case, that the random time avoids the $\mathbb{F}$-stopping times and the extension to the general case is difficult. In discrete time, one cannot make any of such assumptions, since all non-constant martingales are discontinuous and none of the random times valued in the set of integers avoids $\mathbb{F}$-stopping times.

Many books are devoted to discrete time in finance, among them Föllmer and Schied [15], Shreve [27] and Chaps. V and VI of Shiryaev [28]. The lecture notes of Privault [22] and Spreij [29] are available online. See also Neveu [23] for martingales in discrete time.

In the first section, we recall some well-known facts on processes in discrete time and notion of arbitrages in a financial market. We give a definition of viable markets, linked with no-arbitrage condition in the context of enlargement of filtration. Section 2 is devoted to the case of initial enlargement, and we give some examples. Section 3 presents the case of progressive enlargement with a random time τ and we study in particular the immersion property. The study of progressive enlargement before τ is done in Sect. 4. Section 5 deals with enlargement after τ; in particular, we present the case of honest times (which are a standard example in continuous time). In Sect. 6, we give various characterizations of pseudo-stopping times. Then, Sect. 7 is devoted to an optional representation theorem for martingales. In Sect. 8, we consider enlargement with a process, and in Sect. 9, we study applications to credit risk.

We thank the anonymous referee for pointing out some misprints and for interesting comments and Marek Rutkowski for useful comments on a preliminary version.

1 Some Well-Known Results and Definitions

1.1 Basic Definitions

In these notes, we are working in a discrete time setting: $X = (X_n, n \geq 0)$ is a process on a probability space $(\Omega, \mathcal{G}, \mathbb{P})$, and $\mathbb{H} = (\mathcal{H}_n, n \geq 0)$ is a filtration, i.e. a family of σ-algebra such that $\mathcal{H}_n \subset \mathcal{H}_{n+1} \subset \mathcal{G}$. In that setting, $\mathcal{H}_\infty$ is the smallest σ algebra which contains all the $\mathcal{H}_n$. We note $\Delta X_n := X_n - X_{n-1}, n \geq 1$ the increment of X at time n and we set $\Delta X_0 = X_0$. An inequality (or equality) between two random variables is always $\mathbb{P}$ a.s. We note $X \overset{\text{law}}{=} Y$ to indicate that the random variables X and Y have the same law.

We recall, for the ease of the reader some basic definitions.

A process X is $\mathbb{H}$-**adapted** if, for any $n \geq 0$, the random variable X_n is $\mathcal{H}_n$-measurable.

A process X is $\mathbb{H}$-**predictable** if, for any $n \geq 1$, the random variable X_n is $\mathcal{H}_{n-1}$-measurable and X_0 is a constant.

A process X is **integrable** (resp. square integrable) if $\mathbb{E}[|X_n|] < \infty$ (resp. $\mathbb{E}[X_n^2] < \infty$) for all $n \geq 0$. A process X is **uniformly integrable** (in short u.i.) if $\sup_n \mathbb{E}[|X_n| \mathbb{1}_{\{|X_n| \geq c\}}]$ converges to 0 when c goes to infinity.

The process X_- is defined as the process equal to X_{n-1} at time n and to 0 for $n = 0$, if X is $\mathbb{H}$-adapted, X_- is $\mathbb{H}$-predictable. A random variable ζ is said to be **positive** if $\zeta > 0$ a.s.; a process X is positive if the r.v. X_n is positive for any $n \geq 0$ and a process A is **increasing** (resp. decreasing) if $A_n \geq A_{n-1}$ (resp. $A_n \leq A_{n-1}$) a.s. , for all $n \geq 1$. For two r.v. ζ and ξ, we write $\xi > \zeta$ for $\xi - \zeta > 0$, a.s. (resp. $\xi \geq \zeta$ for $\xi - \zeta \geq 0$, a.s.); for two processes X and Y we write $X < Y$ for $X_n < Y_n$, for any $n \geq 0$. For a probability measure $\mathbb{P}$ defined on $\mathscr{G}$, we note $\mathbb{P}_n$ (or $\mathbb{P}_n^{\mathbb{H}}$ in case of ambiguity) the restriction of $\mathbb{P}$ to $\mathscr{H}_n$.

We recall that, if $\zeta > 0$, and $\mathscr{G}$ a sigma-algebra, then $\mathbb{E}[\zeta|\mathscr{G}] > 0$. Indeed, for $A := \{\mathbb{E}[\zeta|\mathscr{G}] = 0\}$, one has $\mathbb{E}[\zeta \mathbb{1}_A] = 0$.

The **natural filtration** of a process X is the smallest filtration which makes it adapted, i.e. it is the filtration $\mathbb{F}^X$ defined as $\mathscr{F}_n^X = \sigma(X_0, X_1, \cdots, X_n)$, $\forall n \geq 0$.

The **tower property** states that, if $\mathscr{F}$ and $\mathscr{G}$ are sigma-algebra, with $\mathscr{F} \subset \mathscr{G}$, and ζ is an integrable r.v. then

$$\mathbb{E}[\mathbb{E}[\zeta|\mathscr{G}]|\mathscr{F}] = \mathbb{E}[\mathbb{E}[\zeta|\mathscr{F}]|\mathscr{G}] = \mathbb{E}[\zeta|\mathscr{F}].$$

For a process X, we denote $X_\infty = \lim_{n\to\infty} X_n$, if the a.s. limit exists.

Let $\mathbb{P}$ and $\mathbb{Q}$ be two probability measures on the same σ-algebra $\mathscr{G}$. One says that $\mathbb{Q}$ is absolutely continuous w.r.t. $\mathbb{P}$, and we note $\mathbb{Q} << \mathbb{P}$ if $\mathbb{P}(A) = 0$ implies $\mathbb{Q}(A) = 0$. In that case, there exists a $\mathscr{G}$-measurable non-negative r.v. L such that $\mathbb{E}_{\mathbb{Q}}(X) = \mathbb{E}_{\mathbb{P}}(XL)$ for any bounded $\mathscr{G}$-measurable r.v. X. This r.v. L is called Radon–Nikodym density and we note $d\mathbb{Q} = Ld\mathbb{P}$. If $\mathbb{P}(A) = 0$ is equivalent to $\mathbb{Q}(A) = 0$, one says that $\mathbb{P}$ and $\mathbb{Q}$ are equivalent ($\mathbb{P} \sim \mathbb{Q}$), and L is positive.

1.2 $\mathbb{H}$-Martingales

An integrable $\mathbb{H}$-adapted process X is an $\mathbb{H}$-**martingale** (resp. an $\mathbb{H}$-supermartingale) if, for any $n \geq 1$, $\mathbb{E}[X_n|\mathscr{H}_{n-1}] = X_{n-1}$, or equivalently $\mathbb{E}[\Delta X_n|\mathscr{H}_{n-1}] = 0$ (resp. $\mathbb{E}[X_n|\mathscr{H}_{n-1}] \leq X_{n-1}$). A process $X = (X^1, \cdots, X^d)$ is a d-dimensional martingale if X^i is a martingale for any $i = 1, \cdots, d$.

We give some obvious results on the form of $\mathbb{H}$-martingales. In what follows, $\sum_{k=1}^0 \bullet = 0$ and $\prod_{k=1}^0 \bullet = 1$.

Proposition 1.1 *(a) The set of processes of the form*

$$\psi_0 + \sum_{k=1}^{n} \left(\psi_k - \mathbb{E}[\psi_k|\mathscr{H}_{k-1}]\right),\ n \geq 0$$

where ψ is an $\mathbb{H}$-adapted integrable process is equal to the set of all $\mathbb{H}$-martingales.
(b) The set of processes of the form

$$\psi_0 \prod_{k=1}^{n} \frac{\psi_k}{\mathbb{E}[\psi_k|\mathscr{H}_{k-1}]}, \, n \geq 0$$

where ψ is a positive integrable $\mathbb{H}$-adapted process is equal to the set of all positive $\mathbb{H}$-martingales.

Proof (a) Let X be a process such that $X_n = \psi_0 + \sum_{k=1}^{n} \left(\psi_k - \mathbb{E}[\psi_k|\mathscr{H}_{k-1}]\right)$ for all $n \geq 0$ where ψ is an $\mathbb{H}$-adapted integrable process. Then, X is integrable, as a linear combination of integrable r.v., and

$$\mathbb{E}[X_n - X_{n-1}|\mathscr{H}_{n-1}] = \mathbb{E}[\psi_n - \mathbb{E}[\psi_n|\mathscr{H}_{n-1}]|\mathscr{H}_{n-1}] = 0 \, , \forall n \geq 1.$$

Therefore X is an $\mathbb{H}$-martingale.

Let X be an $\mathbb{H}$-martingale. Then, since $\mathbb{E}[\Delta X_k|\mathscr{H}_{k-1}] = 0$, one has $X_n = \Delta X_0 + \sum_{k=1}^{n} \Delta X_k - \mathbb{E}[\Delta X_k|\mathscr{H}_{k-1}]$. Choosing $\psi = \Delta X$, we obtain the result.

(b) Let $X_n = \psi_0 \prod_{k=1}^{n} \frac{\psi_k}{\mathbb{E}[\psi_k|\mathscr{H}_{k-1}]}$, $n \geq 0$. Then, for each $n \geq 1$ fixed, the positive random variable X_n is integrable since, by recurrence,

$$\mathbb{E}[X_n] = \mathbb{E}\big[X_{n-1}\mathbb{E}\big[\frac{\psi_n}{\mathbb{E}[\psi_n|\mathscr{H}_{n-1}]}|\mathscr{H}_{n-1}\big]\big] = \mathbb{E}\big[X_{n-1}\big] = X_0,$$

and one has

$$\mathbb{E}[X_n|\mathscr{H}_{n-1}] = X_{n-1}\, \mathbb{E}[\frac{\psi_n}{\mathbb{E}[\psi_n|\mathscr{H}_{n-1}]}|\mathscr{H}_{n-1}] = X_{n-1} \, .$$

If X is a positive martingale, setting $\psi_k = X_k$ leads to the result.

Lemma 1.2 *A predictable martingale X is constant.*

Proof Indeed, the fact that X is predictable implies $\mathbb{E}[X_n|\mathscr{F}_{n-1}] = X_n$, then the martingale property $\mathbb{E}[X_n|\mathscr{F}_{n-1}] = X_{n-1}$ leads to $X_n = X_0$, $\forall n \geq 0$.

Lemma 1.3 *If X is a square integrable martingale, then*

$$\mathbb{E}[(\Delta X_n)^2|\mathscr{H}_{n-1}] = \mathbb{E}[\Delta (X_n)^2|\mathscr{H}_{n-1}] \, .$$

Proof This is easy since, using the tower property

$$\begin{aligned}\mathbb{E}[(\Delta X_n)^2|\mathscr{H}_{n-1}] &= \mathbb{E}[X_n^2 - 2X_n X_{n-1} + X_{n-1}^2|\mathscr{H}_{n-1}] = \mathbb{E}[X_n^2 - 2X_{n-1}^2 + X_{n-1}^2|\mathscr{H}_{n-1}] \\ &= \mathbb{E}[X_n^2 - X_{n-1}^2|\mathscr{H}_{n-1}] \, .\end{aligned}$$

An $\mathbb{H}$-martingale X is **closed** if there exists an integrable random variable ζ such that $X_n = \mathbb{E}[\zeta|\mathscr{H}_n]$. If X is a u.i. martingale, then X_n converges to X_∞ (which exists) when n goes to infinity and it is closed: $X_n = \mathbb{E}[X_\infty|\mathscr{H}_n]$. Furthermore, X_n converges to X_∞ in L^1.

1.3 Doob's Decomposition and Applications

1.3.1 Doob's Decomposition

Lemma 1.4 *Any integrable $\mathbb{H}$-adapted process X admits a (unique) decomposition $X = M^{X,\mathbb{H}} + X^{p,\mathbb{H}}$ where $M^{X,\mathbb{H}}$ is an $\mathbb{H}$-martingale and $X^{p,\mathbb{H}}$ is an $\mathbb{H}$-predictable process with $X_0^{p,\mathbb{H}} = 0$. Furthermore,*

$$\boxed{(\Delta X^{p,\mathbb{H}})_n = \mathbb{E}[\Delta X_n|\mathscr{H}_{n-1}],\ \forall n \geq 1\,.}$$

The process $M^{X,\mathbb{H}}$ is called the $\mathbb{H}$-martingale part of X and $X^{p,\mathbb{H}}$ the $\mathbb{H}$-predictable part of X (also denoted X^p if no confusion occurs).

Proof In the proof, $X^p := X^{p,\mathbb{H}}$ and $M := M^{X,\mathbb{H}}$. Setting $X_0^p = 0$ and, for $n \geq 1$

$$X_n^p - X_{n-1}^p = \mathbb{E}[X_n - X_{n-1}|\mathscr{H}_{n-1}],$$

we construct an $\mathbb{H}$-predictable process. This leads to

$$\Delta M_n = \Delta X_n - \Delta X_n^p = X_n - \mathbb{E}[X_n|\mathscr{H}_{n-1}],\ \forall n \geq 1\,.$$

Setting $M_0 = X_0$, the process M is an $\mathbb{H}$-martingale from Proposition 1.1.

As an immediate corollary, we obtain the Doob decomposition of supermartingales and submartingales:

Corollary 1.5 *If X is an $\mathbb{H}$-adapted supermartingale, it admits a unique decomposition*

$$X = M^X - A^X$$

where M^X is an $\mathbb{H}$-martingale and A^X is an increasing $\mathbb{H}$-predictable process with $A_0^X = 0$.

If X is an increasing $\mathbb{H}$-adapted process, the increasing $\mathbb{H}$-predictable process A^p such that $X - A^p$ is a martingale is called the predictable compensator of X.

Proof The supermartingale property of X implies that $\Delta X^p \leq 0$, and hence X^p is decreasing. It remains to set $A^X = -X^p$.

If X is increasing, $-X$ is a supermartingale, and we can use its Doob decomposition and set $A = A^{-X}$.

Of course, a process of the form $M - A$ where M is a martingale and A an increasing adapted process is a supermartingale.

1.3.2 Predictable Brackets

Proposition 1.6 *If X and Y are square integrable $\mathbb{H}$-martingales, there exists a unique $\mathbb{H}$-predictable process $(XY)^p$ such that $(XY)^p_0 = 0$ and $XY - (XY)^p$ is an $\mathbb{H}$-martingale. Furthermore*

$$\Delta(XY)^p_n = \mathbb{E}[Y_n \Delta X_n | \mathcal{H}_{n-1}] = \mathbb{E}[\Delta Y_n \Delta X_n | \mathcal{H}_{n-1}], \ n \geq 1.$$

The process $\langle X, Y\rangle := (XY)^p$ is called the predictable bracket of the two martingales X and Y.

Proof From Lemma 1.4, the r.v. XY being integrable, we have, for $n \geq 1$,

$$\begin{aligned}\Delta(XY)^p_n &= (XY)^p_n - (XY)^p_{n-1} = \mathbb{E}[X_n Y_n - X_{n-1}Y_{n-1} | \mathcal{H}_{n-1}]\\ &= \mathbb{E}[Y_n \Delta X_n | \mathcal{H}_{n-1}] + \mathbb{E}[X_{n-1}\Delta Y_n | \mathcal{H}_{n-1}]\\ &= \mathbb{E}[Y_n \Delta X_n | \mathcal{H}_{n-1}] = \mathbb{E}[\Delta Y_n \Delta X_n | \mathcal{H}_{n-1}].\end{aligned}$$

In the fourth equality, we have used that, from the martingale property of Y,

$$\mathbb{E}[X_{n-1}\Delta Y_n | \mathcal{H}_{n-1}] = X_{n-1}\mathbb{E}[\Delta Y_n | \mathcal{H}_{n-1}] = 0.$$

1.3.3 Stochastic Integral of Adapted Processes and Martingale Property

Definition 1.7 The **stochastic integral** of a process Y w.r.t. a process X is the process $Y \centerdot X$ defined as $(Y \centerdot X)_n := \sum_{k=1}^n Y_k \Delta X_k, \ n \geq 0$. If $X = (X^1, \cdots, X^d)$ is a d-dimensional process, for a d-dimensional process Y, one defines the stochastic integral by $(Y \centerdot X)_n := \sum_{k=1}^n \sum_{j=1}^d Y^j_k \Delta X^j_k, \ n \geq 0$. The process $Y \centerdot X$ is also called d-martingale transform.

The **covariation** process of two processes X and Y is

$$[X, Y]_0 = 0, \ [X, Y]_n := \sum_{k=1}^n \Delta X_k \, \Delta Y_k, n \geq 1.$$

The covariation process of X is $[X] := [X, X]$.

Proposition 1.8 ***Integration by parts formula.*** *For two processes X and Y*

$$XY = X_0Y_0 + X_- \centerdot Y + Y_- \centerdot X + [X, Y] = X_0Y_0 + X_- \centerdot Y + Y \centerdot X\,.$$

Proof This equality is based on the fact that, $\forall n \geq 1$,

$$\Delta(XY)_n = X_{n-1}\Delta Y_n + Y_{n-1}\Delta X_n + \Delta X_n\, \Delta Y_n = X_{n-1}\Delta Y_n + Y_n\Delta X_n.$$

Lemma 1.9 *If X is a square integrable $\mathbb{H}$-martingale and H an $\mathbb{H}$-predictable square integrable process, then the process $H \centerdot X$ is an $\mathbb{H}$-martingale.*

Proof For H predictable,

$$\mathbb{E}[H_n\Delta X_n|\mathscr{H}_{n-1}] = H_n\mathbb{E}[\Delta X_n|\mathscr{H}_{n-1}] = 0$$

and the result follows.

Lemma 1.10 *If X and Y are two square integrable $\mathbb{H}$-martingales then $XY - [X, Y]$ is an $\mathbb{H}$-martingale. In particular, $X^2 - [X]$ is an $\mathbb{H}$-martingale for a square integrable $\mathbb{H}$-martingale X.*

Proof This is a direct consequence of integration by parts formula and the fact that X_- and Y_- are predictable.

Definition 1.11 Two square integrable martingales X and Y are said to be **orthogonal** if their product is a martingale, i.e. if $\mathbb{E}[\Delta(XY)_n|\mathscr{H}_{n-1}] = 0\,, \forall n \geq 1$. This condition is equivalent to any of the following assertions:

(a) $\mathbb{E}[\Delta Y_n\Delta X_n|\mathscr{H}_{n-1}] = 0,\ \forall n \geq 1$
(b) $\mathbb{E}[Y_n\Delta X_n|\mathscr{H}_{n-1}] = 0,\ \forall n \geq 1$
(c) $[X, Y]$ is a martingale
(d) $\langle X, Y\rangle = 0$.

Proof *(of the various equivalence conditions)* From integration by parts formula, the orthogonality equivalent to $[X, Y]$ is a martingale, which is equivalent to the two other conditions, due to $\Delta(XY)_n = X_{n-1}\Delta Y_n + Y_n\Delta X_n$, and the fact that $X_- \centerdot Y$ and $Y_- \centerdot X$ are martingales.

Lemma 1.12 *If X is a square integrable $\mathbb{H}$-martingale and H an $\mathbb{H}$-adapted square integrable process with $\mathbb{H}$-martingale part orthogonal to X, then the process $H \centerdot X$ is an $\mathbb{H}$-martingale.*

Proof Let $H = M^H + H^p$. Since, $\forall n \geq 1$,

$$\mathbb{E}[H_n\Delta X_n|\mathscr{H}_{n-1}] = \mathbb{E}[M^H_n\,\Delta X_n|\mathscr{H}_{n-1}] + H^p_n\mathbb{E}[\Delta X_n|\mathscr{H}_{n-1}] = \mathbb{E}[M^H_n\,\Delta X_n|\mathscr{H}_{n-1}] = 0$$

the result is obvious. If H is predictable, $M^H = 0$, therefore as we have already proven $H \centerdot X$ is an $\mathbb{H}$-martingale.

Definition 1.13 Let X, Y be two square integrable $\mathbb{H}$-adapted processes. The $\mathbb{H}$-predictable bracket of the processes X and Y is the predictable part of the process $[X, Y]$.

Lemma 1.14 *Let X, Y be two square integrable $\mathbb{H}$-adapted processes. Then, the $\mathbb{H}$-predictable bracket of the processes X and Y, is the process $\langle X, Y\rangle$ defined as*

$$\langle X, Y\rangle_0 = 0\,,\ \Delta\langle X, Y\rangle_n = \mathbb{E}[\Delta X_n\, \Delta Y_n|\mathscr{H}_{n-1}]\,,\ n \geq 1\,.$$

Proof From Doob's decomposition (Lemma 1.4), for $n \geq 1$, denoting by $[X, Y]^p$ the predictable part of the process $[X, Y]$, one has

$$(\Delta[X, Y]^p)_n = \mathbb{E}[[X, Y]_n - [X, Y]_{n-1}|\mathscr{H}_{n-1}] = \mathbb{E}[\Delta X_n\, \Delta Y_n|\mathscr{H}_{n-1}]\,.$$

In the case where X and Y are martingales, we recover the result of Proposition 1.6.

Note that the predictable bracket depends on the filtration, which is not the case for continuous semimartingales in a continuous time setting, since in that case, the predictable bracket is equal to the limit in probability of quadratic variation (see [26, Th 1.8]).

We now give the isometry formula, similar to the one in continuous time.

Proposition 1.15 *If X is a martingale, Y an adapted process such that XY is square integrable and $Y \centerdot X$ is a martingale, then $\forall n \geq 1$, $\mathbb{E}[((Y \centerdot X)_n)^2] = \sum_{k=1}^n \mathbb{E}[Y_k^2 (\Delta X_k)^2] = \mathbb{E}[(Y^2 \centerdot \langle X, X\rangle)_n]$.*

Proof Since $Y \centerdot X$ is a martingale, Lemma 1.3 implies that $\mathbb{E}\big[\Delta((Y \centerdot X)^2)_k \big| \mathscr{H}_{k-1}\big] = \mathbb{E}\big[(\Delta(Y \centerdot X)_k)^2 \big| \mathscr{H}_{k-1}\big]$. Then, using $\Delta(Y \centerdot X)_k = Y_k \Delta X_k$ and taking expectations, we get

$$\mathbb{E}\big[\Delta((Y \centerdot X)^2)_k\big] = \mathbb{E}\big[Y_k^2 (\Delta X_k)^2\big]\,.$$

Finally, taking the sum for k from 0 to n, we obtain

$$\mathbb{E}\big[(Y \centerdot X)_n^2\big] = \mathbb{E}\Big[\sum_{k=0}^n Y_k^2 (\Delta X_k)^2\Big]\,,\quad \forall n \geq 0\,.$$

We end this section by giving a characterization of martingales.

Proposition 1.16 *Let X an adapted integrable process. Then X is a martingale iff for all predictable processes Y, $\mathbb{E}[(Y \centerdot X)_\infty] = 0$.*

Proof The proof is left to the reader.

1.4 Projections

In this section, $\mathbb{H}$ is a filtration, and X is a process, *not assumed to be $\mathbb{H}$-adapted.* For future use, we mimic two definitions which are important in continuous time. In particular, we introduce optional projections, even if in discrete time, optional means adapted. We keep the continuous time denomination to make easier the comparison between both results for readers aware about continuous time.

Definition 1.17 The $\mathbb{H}$-**optional projection** of an integrable process X is the $\mathbb{H}$-adapted process ${}^oX$ defined as ${}^oX_n := \mathbb{E}[X_n|\mathscr{H}_n]$, $\forall n \geq 0$.

The $\mathbb{H}$-**predictable projection** of an integrable process X is the $\mathbb{H}$-predictable process ${}^pX$ defined as ${}^pX_n := \mathbb{E}[X_n|\mathscr{H}_{n-1}]$, $\forall n \geq 1$ and ${}^pX_0 := \mathbb{E}[X_0|\mathscr{H}_0]$.

Definition 1.18 The **dual** $\mathbb{H}$-**optional projection** of an increasing integrable process X is the increasing $\mathbb{H}$-adapted process X^o defined by

$$\Delta X^o_n := \mathbb{E}[\Delta X_n|\mathscr{H}_n],\ n \geq 1 \text{ and } X^o_0 := \mathbb{E}[X_0|\mathscr{H}_0]\,.$$

The **dual** $\mathbb{H}$-**predictable projection** of an increasing integrable process X is the increasing $\mathbb{H}$-predictable process X^p defined by $\Delta X^p_n = \mathbb{E}[\Delta X_n|\mathscr{H}_{n-1}]$ for $n \geq 1$ and $X^p_0 = 0$. If X is an integrable $\mathbb{H}$-adapted process, $X^o = X$ and X^p is the $\mathbb{H}$-predictable part in the Doob decomposition of the process X, and, as we already mentioned, is also called predictable compensator.

If useful, we shall denote by $X^{o,\mathbb{H}}$ the $\mathbb{H}$-optional projection, with similar notation for the other projections.

The dual $\mathbb{H}$-optional projection of X satisfies

$$\mathbb{E}\Big[(Y \centerdot X)_\infty\Big] = \mathbb{E}\Big[(Y \centerdot X^o)_\infty\Big] \tag{1}$$

for any non-negative bounded $\mathbb{H}$-adapted process Y, where Z_∞ is defined for adequate processes Z in Sect. 1.1. The dual $\mathbb{H}$-predictable projection of X satisfies

$$\mathbb{E}\Big[(Y \centerdot X)_\infty\Big] = \mathbb{E}\Big[(Y \centerdot X^p)_\infty\Big] \tag{2}$$

for any non-negative bounded $\mathbb{H}$-predictable process Y.

Using that any integrable process X can be written as $X = X^\uparrow - X^\downarrow$ where $X^\uparrow$ and $X^\downarrow$ are increasing processes, one can also define dual predictable projection for any process X as $(X^\uparrow)^p - (X^\downarrow)^p$ (same construction for dual optional projection). The processes $X^\uparrow$ and $X^\downarrow$ can be constructed as follows:

$$X^\uparrow_n = \sum_{k=1}^n (X_k - X_{k-1})^+,\ X^\uparrow_0 = X^+_0,\quad X^\downarrow_n = \sum_{k=1}^n (X_k - X_{k-1})^-,\ X^\downarrow_0 = X^-_0\,.$$

These processes are integrable and increasing and one has

$$(X_k - X_{k-1})^+ - (X_k - X_{k-1})^- = \Delta X_k\,.$$

Exercise 1.19 Prove that for a pair of processes (X, Y), the following duality formulae hold:

$$\mathbb{E}[(Y \centerdot X^o)_\infty] = \mathbb{E}[({}^oY \centerdot X)_\infty]$$
$$\mathbb{E}[(Y \centerdot X^p)_\infty] = \mathbb{E}[({}^pY \centerdot X)_\infty] .$$

Proposition 1.20 *Let X be an integrable process, and $\mathbb{H}$ a filtration. The processes ${}^{o,\mathbb{H}}X - X^{o,\mathbb{H}}$ and ${}^{o,\mathbb{H}}X - X^{p,\mathbb{H}}$ are $\mathbb{H}$-martingales.*

Proof In the proof, we forget the superscript $\mathbb{H}$. Using the tower property

$$\begin{aligned}\mathbb{E}[\Delta({}^oX - X^o)_n|\mathcal{H}_{n-1}] &= \mathbb{E}\Big[\,\mathbb{E}[X_n|\mathcal{H}_n] - \mathbb{E}[X_{n-1}|\mathcal{H}_{n-1}] - \mathbb{E}[X_n - X_{n-1}|\mathcal{H}_n]\,|\mathcal{H}_{n-1}\Big] \\ &= \mathbb{E}\Big[-\mathbb{E}[X_{n-1}|\mathcal{H}_{n-1}] + \mathbb{E}[X_{n-1}|\mathcal{H}_n]|\mathcal{H}_{n-1}\Big] = 0 .\end{aligned}$$

The proof that ${}^oX - X^p$ is a martingale is left to the reader.

1.5 Multiplicative Decomposition

Theorem 1.21 *Let X be an $\mathbb{H}$-adapted integrable positive process, then X can be represented in a unique form as*

$$X = K^X N^X,$$

where K^X is an $\mathbb{H}$-predictable process with $K_0^X = 1$ and N^X is an $\mathbb{H}$-martingale. More precisely,

$$N_0^X = X_0, \quad N_n^X = X_0 \prod_{k=1}^{n} \frac{X_k}{\mathbb{E}[X_k|\mathcal{H}_{k-1}]}, \quad \forall n \geq 1 ,$$
$$K_0^X = 1, \quad K_n^X = \prod_{k=1}^{n} \frac{\mathbb{E}[X_k|\mathcal{H}_{k-1}]}{X_{k-1}}, \quad \forall n \geq 1.$$

Proof From Proposition 1.1, N^X is a martingale. On the other hand, the process K^X, defined by

$$K_n^X = \frac{X_n}{X_0 \prod_{k=1}^n \frac{X_k}{\mathbb{E}[X_k|\mathcal{H}_{k-1}]}} = \prod_{k=1}^{n} \frac{\mathbb{E}[X_k|\mathcal{H}_{k-1}]}{X_{k-1}} ,$$

is an $\mathbb{H}$-predictable process.

Remark 1.22 In terms of Doob's decomposition $X = M^X + X^p$, one has

$$K_n^X = \frac{M_{n-1}^X + X_n^p}{X_{n-1}} K_{n-1}^X = X_0 \prod_{k=1}^{n} \frac{M_{k-1}^X + X_k^p}{X_{k-1}}, \quad N_n^X = X_n / K_n^X, \ n \geq 1.$$

Corollary 1.23 *Any positive $\mathbb{H}$-supermartingale Y admits a unique multiplicative predictable decomposition $Y = N^Y D^Y$ where N^Y is a positive $\mathbb{H}$-martingale and D^Y is an $\mathbb{H}$-predictable positive decreasing process with $D_0^Y = 1$. Conversely, any process of the form $Y = ND$ where N is a positive $\mathbb{H}$-martingale and D is an $\mathbb{H}$-adapted positive decreasing process is a positive supermartingale.*

Proof The process $D^Y := K^Y$ is indeed decreasing if Y is a supermartingale and N^Y is positive if Y is positive. The proof of the converse is left to the reader.

1.6 Stochastic Exponential Process

Given a process X, we define the stochastic exponential of X denoted by $\mathcal{E}(X)$ as the solution Y of the following equation in differences:

$$\begin{cases} \Delta Y_n = Y_{n-1} \Delta X_n, & \forall n \geq 1, \\ Y_0 = 1. \end{cases} \tag{3}$$

In other terms, $\mathcal{E}(X)$ satisfies

$$\begin{cases} \Delta \mathcal{E}(X)_n = \mathcal{E}(X)_{n-1} \Delta X_n, & \forall n \geq 1, \\ \mathcal{E}(X)_0 = 1. \end{cases}$$

Proposition 1.24 *The solution of (3) is given by*

$$\mathcal{E}(X)_n := \Pi_{k=1}^n (\Delta X_k + 1), \quad \forall n \geq 1. \tag{4}$$

If X is an $\mathbb{H}$-martingale, $\mathcal{E}(X)$ is an $\mathbb{H}$-martingale. If $\Delta X_n > -1$, for all $n \geq 1$, then $\mathcal{E}(X)$ is positive.

Proof The equality (4) is obtained by recurrence. The martingale property is easily checked from (3).

Proposition 1.25 *If Y is a positive process with $Y_0 = 1$, there exists a unique process X such that $Y = \mathcal{E}(X)$.*

Proof Set $\Delta X_n = \frac{\Delta Y_n}{Y_{n-1}}$, $X_0 = 0$.

We now give the obvious relation between stochastic exponential and exponential.

Proposition 1.26 *If X is a process such that $\Delta X > -1$ then $\mathcal{E}(X) = e^U$ where the process U is defined as $U_0 = 0$ and $\Delta U_n = \log(1 + \Delta X_n)$, $\forall n \geq 1$.*

Proof Note that $e^{U_n} = e^{U_{n-1}}(1 + \Delta X_n)$.

Lemma 1.27 *Let ψ, γ, M and N be four processes. Then*

$$\mathcal{E}(\psi \centerdot M)\mathcal{E}(\gamma \centerdot N) = \mathcal{E}(\psi \centerdot M + \gamma \centerdot N + \psi\gamma \centerdot [M, N]) .$$

If ψ, γ are predictable and M, N are orthogonal martingales, the product of martingales $\mathcal{E}(\psi \centerdot M)\mathcal{E}(\gamma \centerdot N)$ is a martingale.

Proof By definition, the two sides are equal to 1 at time 0. For $n \geq 1$, the left-hand side $K_n := \mathcal{E}(\psi \centerdot M)_n\mathcal{E}(\gamma \centerdot N)_n$ satisfies $K_n = K_{n-1}(1 + \psi_n\Delta M_n)(1 + \gamma_n\Delta N_n)$. The right-hand side $J_n := \mathcal{E}(\psi \centerdot M + \gamma \centerdot N + \psi\gamma \centerdot [M, N])_n$ satisfies

$$J_n = J_{n-1}(1 + \psi_n\Delta M_n + \gamma_n\Delta N_n + \psi_n\gamma_n\Delta M_n\Delta N_n) .$$

Assuming by recurrence that $K_{n-1} = J_{n-1}$, the result follows.

This result is known in continuous time as Yor's equality, see [28, Page 691].

1.7 Stopping Times and Local Martingales

1.7.1 Random and Stopping Times

A random time is a random variable valued in $\mathbb{N} \cup \{+\infty\}$.

An $\mathbb{H}$-stopping time is a random variable valued in $\mathbb{N} \cup \{+\infty\}$ such that $\{\tau \leq n\} \in \mathcal{H}_n$, for any $n \geq 0$, or equivalently $\{\tau = n\} \in \mathcal{H}_n$, for any $n \geq 0$. A stopping time τ is called an $\mathbb{H}$-predictable stopping time if $\{\tau = 0\} \in \mathcal{H}_0$ and $\{\tau \leq n\} \in \mathcal{H}_{n-1}$, for any $n \geq 0$. If τ is an $\mathbb{H}$-stopping time, we define the two σ-algebra of events before τ and strictly before τ

$$\mathcal{H}_\tau = \{F \in \mathcal{H}_\infty \,:\, F \cap \{\tau \leq n\} \in \mathcal{H}_n, \,\forall n\}$$
$$\mathcal{H}_{\tau-} = \sigma\{F \cap \{n < \tau\} \text{ for } F \in \mathcal{H}_n, n \geq 0\}.$$

Note that, for $\tau \equiv n$, one has $\mathcal{H}_\tau = \mathcal{H}_n$ and $\mathcal{H}_{\tau-} = \mathcal{H}_{n-1}$.

If τ is a random time, the **stopped** process X^τ is defined as $X_n^\tau = X_{\tau\wedge n}$. If τ is an $\mathbb{H}$-stopping time and X is $\mathbb{H}$-adapted, X^τ is $\mathbb{H}$-adapted.

If T and S are two random times, we define the stochastic intervals

$$[\![T, S]\!] : = \{(\omega, n), T(\omega) \leq n \leq S(\omega)\}$$
$$[\![T, S[\![: = \{(\omega, n), T(\omega) \leq n < S(\omega)\} .$$

The stochastic interval $[\![T]\!] := \{(\omega, n), T(\omega) = n\}$ is called the **graph** of T.

Lemma 1.28 *If $A \in \mathcal{H}_\infty$, then $A \cap \{\tau = +\infty\}$ is $\mathcal{H}_{\tau-}$-measurable.*

Proof For n fixed and $B \in \mathcal{H}_n$, one has $B \cap \{\tau = +\infty\} \in \mathcal{H}_{\tau-}$ as $\{\tau = +\infty\} = \bigcap_{m \geq n}\{\tau > m\}$. Moreover, $\{B \in \mathcal{H}_\infty, B \cap \{\tau = +\infty\} \in \mathcal{H}_{\tau-}\}$ is a σ-algebra. So the result follows.

The **restriction** of a random time to a given set $F \in \mathcal{H}_\infty$ is defined as

$$\tau_F(\omega) := \begin{cases} \tau(\omega) & \omega \in F \\ \infty & \omega \notin F\,. \end{cases} \tag{5}$$

In particular $0_A = 0$ on A and ∞ elsewhere. For a stopping time τ, the following proposition gives some relationships between the properties of a process X and the measurability of the random variable $X_\tau \mathbb{1}_{\{\tau<\infty\}}$.

Proposition 1.29 *(a) If X is an $\mathbb{H}$-adapted process, and τ an $\mathbb{H}$-stopping time, the random variable $X_\tau \mathbb{1}_{\{\tau<\infty\}}$ is $\mathcal{H}_\tau$-measurable.*
(b) If X is an $\mathbb{H}$-predictable process, and τ an $\mathbb{H}$-stopping time, the random variable $X_\tau \mathbb{1}_{\{\tau<\infty\}}$ is $\mathcal{H}_{\tau-}$-measurable.
(c) The process X is $\mathbb{H}$-predictable iff $X_\tau \mathbb{1}_{\{\tau<\infty\}}$ is $\mathcal{H}_{\tau-}$-measurable for any $\mathbb{H}$-predictable stopping time τ.
(d) Let τ be an $\mathbb{H}$-stopping time. A random variable ζ in $\mathcal{H}_\infty$ is $\mathcal{H}_{\tau-}$-measurable if and only if there exists a predictable process X such that $X_\tau = \zeta$ on $\{\tau < +\infty\}$.

Proof (a) The assertion follows from $X_\tau \mathbb{1}_{\{\tau<\infty\}} = \sum_n X_n \mathbb{1}_{\{\tau=n\}}$ and from $X_n \mathbb{1}_{\{\tau=n\}} \in \mathcal{H}_\tau$.
(b) In a first step, note that τ is $\mathcal{H}_{\tau\_}$ measurable. Indeed $\{\tau > n\} \in \mathcal{H}_{\tau\_}$ (taking $F = \Omega$ in the definition of $\mathcal{H}_{\tau-}$), hence $\{\tau \leq n\} \in \mathcal{H}_{\tau\_}$, and $\{\tau = n\} = \{\tau \leq n\} \cap \{n-1 < \tau\} \in \mathcal{H}_{\tau\_}$. The assertion follows from $X_n \mathbb{1}_{\{\tau=n\}} \in \mathcal{H}_{\tau\_}$ when X is predictable. Indeed, $X_n \mathbb{1}_{\{\tau=n\}} = X_n \mathbb{1}_{\{n-1<\tau\}} \mathbb{1}_{\{\tau=n\}}$ where $X_n \mathbb{1}_{\{n-1<\tau\}}$ is $\mathcal{H}_{\tau\_}$ measurable (due to $X_n \in \mathcal{H}_{n-1}$) and $\{\tau = n\}$ belongs to $\mathcal{H}_{\tau\_}$.
(c) For the sufficient condition, we can choose $\tau \equiv n$ for a fixed n, τ is predictable and $X_\tau \mathbb{1}_{\{\tau<+\infty\}} = X_n \in \mathcal{H}_{n-1}$. Then X is predictable.
(d) Let ζ in $\mathcal{H}_\infty$. We suppose that there exists a predictable process X such that $X_\tau = \zeta$ on $\{\tau < +\infty\}$. Then $\zeta = \zeta \mathbb{1}_{\{\tau=+\infty\}} + X_\tau \mathbb{1}_{\{\tau<+\infty\}}$. The random variable $\zeta \mathbb{1}_{\{\tau=+\infty\}}$ is $\mathcal{H}_{\tau-}$-measurable. The result follows from Lemma 1.28 and the first assertion. Conversely, it is sufficient to prove the assertion for $\zeta = \mathbb{1}_B$ with $B \in \mathcal{H}_0$ and for $\zeta = \mathbb{1}_{B\cap\{n<\tau\}}$ with $B \in \mathcal{H}_n$. In this case, we can choose $X = \mathbb{1}_{[\![0_B]\!]}$ in the first case and $X = \mathbb{1}_{[\![n_B]\!]}$ in the second.

Exercise 1.30 Prove that, for any $\mathbb{H}$-stopping time,

$$\begin{aligned} \mathcal{H}_\tau &= \{X_\tau \ : \ X \text{ is an adapted process}\} \\ \mathcal{H}_{\tau-} &= \{X_\tau \ : \ X \text{ is a predictable process}\}. \end{aligned}$$

Lemma 1.31 *One has $\mathscr{H}_{\tau-} \subset \mathscr{H}_\tau$. In general the inclusion is strict.*

Proof The inclusion is obvious. For the strict inclusion, take $X_n = \sum_{i=0}^n Y_i$ where Y_i are i.i.d. random variables, non-constant, and $\mathbb{F}$ the natural filtration of X. For $\tau = \inf\{n : X_n \geq a\}$, the r.v. $\zeta = X_\tau \mathbb{1}_{\{\tau<\infty\}}$ is $\mathscr{F}_\tau$-measurable but is not $\mathscr{F}_{\tau-}$ measurable.

Exercise 1.32 Prove that, for any $\mathbb{H}$-stopping time T, and any process X (not necessarily $\mathbb{H}$-adapted),

$$\mathbb{E}[X_T \mathbb{1}_{\{T<\infty\}}] = \mathbb{E}[{}^o X_T \mathbb{1}_{\{T<\infty\}}]$$

and that for any predictable stopping time S,

$$\mathbb{E}[X_S \mathbb{1}_{\{S<\infty\}}] = \mathbb{E}[{}^p X_S \mathbb{1}_{\{S<\infty\}}] .$$

1.7.2 Local Martingales

We present the definition of local martingales and some classical results (see [28, 29]).

Definition 1.33 The process X is said to be an $\mathbb{H}$-local martingale if it is $\mathbb{H}$-adapted and if there exists an increasing sequence $(\tau_k, k \geq 1)$ of $\mathbb{H}$-stopping times such that $\mathbb{P}(\lim_{k\to\infty} \tau_k = \infty) = 1$ and for any k, the stopped process X^{τ_k} is an $\mathbb{H}$-martingale.

We denote by $\mathscr{M}_{loc}$ the set of local martingales.

We borrow from [25] an example of a process which is a local martingale but not a martingale.

Example 1.34 Let Y be a random variable which is integrable but not square integrable and U a random variable taking the value 1 and the value -1 with probability $1/2$, independent of Y. We define $\mathscr{F}_0 = \sigma(Y)$ and $\mathscr{F}_n = \sigma(Y, U)$ for all $n \geq 1$. Let us consider $M_0 = Y$ and $M_n = Y + UY^2$ if $n \geq 1$. Then M is a local martingale which is not a martingale since M_1 is not integrable.

Let τ_k be the $\mathbb{F}$-stopping time defined as $\tau_k = \min\{n, |M_n| \geq k\}$. Then, the sequence (τ_k) goes to infinity a.s when k goes to infinity (as usual the minimum of an empty set is $+\infty$). We show that M^{τ_k} is a martingale.

Let k fixed, by definition, $M_0 = Y$ is integrable, and hence $M_0^{\tau_k} = M_0$ is integrable and

$$\begin{aligned}\mathbb{E}[|M_1^{\tau_k}|] &= \mathbb{E}[|M_1^{\tau_k}|\mathbb{1}_{\{\tau_k=0\}}] + \mathbb{E}[|M_1^{\tau_k}|\mathbb{1}_{\{\tau_k>0\}}]\\ &= \mathbb{E}[|M_0|\mathbb{1}_{\{\tau_k=0\}}] + \mathbb{E}[|M_1|\mathbb{1}_{\{\tau_k>0\}}] \leq \mathbb{E}[|M_0|] + k + k^2\end{aligned}$$

since $M_1 = M_0 + U(M_0)^2$ and $M_0 < k$ on $\tau_k > 0$. Then $M_n^{\tau_k}$ is integrable for all n. Moreover, to prove the martingale property, we just have to check it for $n = 1$

$$\begin{aligned}\mathbb{E}[M_1^{\tau_k}|\mathscr{F}_0] &= \mathbb{E}[M_0^{\tau_k}\mathbb{1}_{\{\tau_k=0\}}|\mathscr{F}_0] + \mathbb{E}[M_1^{\tau_k}\mathbb{1}_{\{\tau_k>0\}}|\mathscr{F}_0] \\ &= M_0\mathbb{1}_{\{\tau_k=0\}} + \mathbb{1}_{\{\tau_k>0\}}(M_0 + M_0^2\mathbb{E}[U]) = M_0 = M_0^{\tau_k}\end{aligned}$$

where we have used the fact that $\{\tau_k = 0\} \in \mathscr{F}_0$ and $\{\tau_k > 0\} \in \mathscr{F}_0$ and that $\mathbb{E}[U|\mathscr{F}_0] = \mathbb{E}[U] = 0$.

1.7.3 Generalized Martingales

In this subsection, we follow closely [28, Chap. II, Sect. 1c]. Let $\mathscr{F}$ and $\mathscr{G}$ be two σ-algebra with $\mathscr{F} \subset \mathscr{G}$. If ζ is a positive $\mathscr{G}$-measurable random variable, one can define $\mathbb{E}[\zeta|\mathscr{F}]$ even if ζ is not integrable. Indeed, there exists an $\mathscr{F}$-measurable random variable $\widehat{\zeta}$ (which can take $+\infty$ value) such that $\mathbb{E}[\zeta\mathbb{1}_F] = \mathbb{E}[\widehat{\zeta}\mathbb{1}_F]$ for all $F \in \mathscr{F}$: it suffices to take $\widehat{\zeta} = \lim_{n\to\infty}\mathbb{E}[\zeta \wedge n|\mathscr{F}]$ which is defined as the limit of an increasing sequence. We write $\widehat{\zeta} = \mathbb{E}[\zeta|\mathscr{F}]$. Let now X be a random variable, decomposed as $X = X^+ - X^-$. We can define $\mathbb{E}[X^\pm|\mathscr{F}]$, and assuming that $\inf(\mathbb{E}[X^+|\mathscr{F}], \mathbb{E}[X^-|\mathscr{F}]) < \infty$, we set $\mathbb{E}[X|\mathscr{F}] = \mathbb{E}[X^+|\mathscr{F}] - \mathbb{E}[X^-|\mathscr{F}]$. One can then in particular define the $\mathscr{F}$-conditional expectation of X under the hypothesis $\mathbb{E}[|X||\mathscr{F}] < \infty$.

Definition 1.35 The process X is a generalized martingale if it is adapted and if $\mathbb{E}[|X_{n+1}||\mathscr{H}_n] < \infty$ for any n and satisfies $\mathbb{E}[X_{n+1}|\mathscr{H}_n] = X_n$.

Here $\mathbb{E}[X_{n+1}|\mathscr{H}_n] = \mathbb{E}[X_{n+1}^+|\mathscr{H}_n] - \mathbb{E}[X_{n+1}^-|\mathscr{H}_n] < \infty$.

Proposition 1.36 *Let X be an $\mathbb{H}$-adapted process. The following conditions are equivalent: (a) X is a local martingale (b) X is a generalized martingale (c) There exists an $\mathbb{H}$-predictable process φ and an $\mathbb{H}$-martingale M such that $X = X_0 + \varphi \centerdot M$.*

Proof We refer the reader to [28, Chap. II, page 98].

Proposition 1.37 *Let X be a local martingale such that $\mathbb{E}[X_n^-] < \infty$ for all n, or that $\mathbb{E}[X_n^+] < \infty$ for all n, then X is a martingale. In particular, if X is an integrable local martingale, it is a martingale.*

Proof We reproduce the proof given in [28, p.100]. If $\mathbb{E}[X_n^-] < \infty$ (or $\mathbb{E}[X_n^+] < \infty$), then $\mathbb{E}[|X_n|] < \infty$. Indeed, if $\mathbb{E}[X_n^-] < \infty$, then, for a localizing sequence (T_k),

$$\begin{aligned}\mathbb{E}[X_n^+] &= \mathbb{E}[\liminf_k X_{n\wedge T_k}^+] \le \liminf_k \mathbb{E}[X_{n\wedge T_k}^+] = \liminf_k \mathbb{E}[X_{n\wedge T_k}] + \mathbb{E}[X_{n\wedge T_k}^-] \\ &= \mathbb{E}[X_0] + \liminf_k \mathbb{E}[X_{n\wedge T_k}^-] \le \mathbb{E}[X_0] + \sum_{i=0}^n \mathbb{E}[X_i^-] < \infty\,.\end{aligned}$$

Then, $\mathbb{E}[|X_n|] < \infty$. From

$$\mathbb{E}[X_{T_k\wedge(n+1)}|\mathscr{F}_n] = X_{T_k\wedge n}\,,$$

using the fact that $|X_{T_k\wedge(n+1)}| \leq \sum_{i=0}^{n+1} |X_i|$ and that $\mathbb{E}[|X_i|] < \infty$, applying the dominated convergence Lebesgue theorem, we get $\mathbb{E}[X_{n+1}|\mathscr{F}_n] = X_n$.

Corollary 1.38 *If X is a local martingale bounded from below with $\mathbb{E}[X_0] < \infty$, then it is a true martingale.*

In particular, if X is a local martingale and $\mathscr{E}(X)$ is non-negative, $\mathscr{E}(X)$ is a martingale.

Proposition 1.39 *If H is a predictable process and M a local martingale, then $H \centerdot M$ is a local martingale.*

Proof We leave the (easy) proof to the reader.

Definition 1.40 Two local martingales X and Y are orthogonal if XY is a local martingale.

Definition 1.41 A semimartingale is a process X such that $X = M + V$ where M is a local martingale and V a finite variation process. A special semimartingale is a process X such that $X = M + V$ where M is a local martingale and V a predictable finite variation process.

Proposition 1.42 *Let X be an adapted process and assume that for all n, $\mathbb{E}(|X_n|\,|\mathscr{F}_n) < \infty$ and $\mathbb{E}(|X_n|\,|\mathscr{F}_{n-1}) < \infty$. Then, the process V defined as $V_0 = 0$, $\Delta V_n = \mathbb{E}(X_n - X_{n-1}|\mathscr{F}_{n-1})$ is predictable and $M := X - V$ is a generalized martingale.*

Proof We leave the proof to the reader.

1.8 Change of Probability

Let $\mathbb{P}$ and $\mathbb{Q}$ be two probability measures defined on $\mathscr{H}_\infty$. The probability measure $\mathbb{Q}$ is **locally absolutely continuous** w.r.t. the probability measure $\mathbb{P}$ if $\mathbb{Q}_n << \mathbb{P}_n$ for all $n \geq 0$ (the notation $\mathbb{P}_n$ is defined in Sect. 1.1). In this case, we can define the Radon–Nikodym density $L_n = d\mathbb{Q}_n/d\mathbb{P}_n$. The process L is a $\mathbb{P}$-martingale called the density of $\mathbb{Q}$ with respect to $\mathbb{P}$. This process can vanish, however, $L_n = 0$ on the set $\{L_{n-1} = 0\}$; indeed $\mathbb{E}_{\mathbb{P}}[L_n \mathbb{1}_{\{L_{n-1}=0\}}] = \mathbb{E}_{\mathbb{P}}[L_{n-1}\mathbb{1}_{\{L_{n-1}=0\}}] = 0$. We take the convention that, on the set $\{L_{n-1} = 0\}$, one has $L_n/L_{n-1} = 0$.

We now recall the standard Bayes' formula.

Lemma 1.43 ***Bayes' formula.*** *Let $\mathbb{Q} \ll \mathbb{P}$ on a σ-algebra $\mathscr{G}$ and $L := d\mathbb{Q}/d\mathbb{P}$ the Radon–Nikodym derivative. Let $\mathscr{F}$ be a sub-σ-algebra of $\mathscr{G}$ and X a $\mathscr{G}$ measurable, $\mathbb{Q}$-integrable random variable. Then, the random variable L is positive $\mathbb{Q}$-a.s and one has*

$$\mathbb{E}_{\mathbb{Q}}[X|\mathscr{F}] = \frac{\mathbb{E}_{\mathbb{P}}[XL|\mathscr{F}]}{\mathbb{E}_{\mathbb{P}}[L|\mathscr{F}]}, \mathbb{Q} - a.s$$

Proof In a first step, we prove that the random variable L is positive $\mathbb{Q}$-a.s. Indeed,

$$\mathbb{Q}(L=0)=\mathbb{E}_{\mathbb{P}}[L\mathbb{1}_{\{L=0\}}]=0.$$

As $\mathscr{F}$ is a sub-σ-algebra, this implies that $\mathbb{Q}(\mathbb{E}[L|\mathscr{F}]>0)=1$ (see Sect. 1.1).

In our setting, the Bayes formula states that, for any integrable $\mathscr{H}_N$-measurable random variable Y and $n\le N$,

$$\mathbb{E}_{\mathbb{Q}}[Y|\mathscr{H}_n]=\frac{1}{L_n}\mathbb{E}_{\mathbb{P}}[YL_N|\mathscr{H}_n]\,,\mathbb{Q}-a.s.$$

If $\mathbb{P}_n\sim\mathbb{Q}_n$ for all n, $\mathbb{P}$ and $\mathbb{Q}$ are called **locally equivalent**. In that case, L^{-1} is a $\mathbb{Q}$-martingale.

Proposition 1.44 *Let X be an adapted process and assume that $\mathbb{Q}$ is a probability measure locally absolutely continuous w.r.t. $\mathbb{P}$ with density process L.*
(a) The process X is a martingale under $\mathbb{Q}$ if the process XL is a martingale under $\mathbb{P}$.
(b) The process X is a local martingale under $\mathbb{Q}$ if the process XL is a local martingale under $\mathbb{P}$.
(c) If moreover $\mathbb{P}$ is also locally absolutely continuous w.r.t. $\mathbb{Q}$ (i.e. $\mathbb{P}$ and $\mathbb{Q}$ are locally equivalent), then the process XL is a local martingale under $\mathbb{P}$ iff the process X is a local martingale under $\mathbb{Q}$.

Proof (a) From $\mathbb{E}_{\mathbb{Q}}[|X_n|]=\mathbb{E}_{\mathbb{P}}[|X_nL_n|]$ we obtain that XL is $\mathbb{P}$-integrable. Bayes's formula $\mathbb{E}_{\mathbb{Q}}[X_n|\mathscr{H}_{n-1}]=\dfrac{\mathbb{E}_{\mathbb{P}}[X_nL_n|\mathscr{H}_{n-1}]}{L_{n-1}}$ leads to the result.
(b) This assertion is obtained by localization.
(c) If $\mathbb{P}$ and $\mathbb{Q}$ are locally equivalent, L^{-1} is a $\mathbb{Q}$-martingale. The process XL being a $\mathbb{P}$ local martingale and $\mathbb{P}\ll\mathbb{Q}$ with Radon–Nikodym density L^{-1}, part (b) applied to the $\mathbb{P}$ local martingale XL, leads to $(XL)L^{-1}$ is a $\mathbb{Q}$ local martingale.

Proposition 1.45 ***Girsanov's Theorem*** *Let X be an $\mathbb{H}$-local martingale under $\mathbb{P}$ and let $\mathbb{Q}$ be locally absolutely continuous w.r.t. $\mathbb{P}$ with density process L. The process $X^{\mathbb{Q}}$ defined as*

$$X_n^{\mathbb{Q}}=X_n-\sum_{k=0}^{n}\frac{1}{L_k}\Delta[X,L]_k=X_n-\sum_{k=0}^{n}\frac{1}{L_k}\Delta X_k\Delta L_k,\ \forall n\ge 0\,,$$

or in a more concise way

$$X^{\mathbb{Q}}:=X-\frac{1}{L}\centerdot[X,L]$$

and the process $\widetilde{X}^{\mathbb{Q}}$ defined as

$$\widetilde{X}_n^{\mathbb{Q}} := X_n - \sum_{k=0}^{n} \frac{1}{L_{k-1}} \Delta\langle X, L\rangle_k \,,$$

are well defined under $\mathbb{Q}$ *and*
(a) if $\mathbb{E}_{\mathbb{P}}[L_{n-1}|\Delta X_n||\mathcal{H}_{n-1}] < \infty$*, the process* $X^{\mathbb{Q}}$ *is a local martingale under* $\mathbb{Q}$*,*
(b) if $\mathbb{E}[|\Delta X_n|L_n\,|\mathcal{H}_{n-1}] < \infty$ *a.s. for all n, then* $\widetilde{X}^{\mathbb{Q}}$ *is a local martingale under* $\mathbb{Q}$*.*

Proof (a) Assume that X is an $(\mathbb{P}, \mathbb{H})$-martingale. We have, $\forall n \geq 0$,

$$\Delta X_n^{\mathbb{Q}} = \Delta X_n - \frac{\Delta X_n \Delta L_n}{L_n} = \frac{\Delta X_n}{L_n}(L_n - \Delta L_n) = \frac{\Delta X_n}{L_n} L_{n-1} \,.$$

From Bayes' formula and the martingale property of L

$$\mathbb{E}_{\mathbb{Q}}[\Delta X_n^{\mathbb{Q}}|\mathcal{H}_{n-1}] = \frac{\mathbb{E}_{\mathbb{P}}[L_n \Delta X_n^{\mathbb{Q}}|\mathcal{H}_{n-1}]}{\mathbb{E}_{\mathbb{P}}[L_n|\mathcal{H}_{n-1}]} = \frac{\mathbb{E}_{\mathbb{P}}[L_{n-1}\Delta X_n|\mathcal{H}_{n-1}]}{L_{n-1}} = 0 \,.$$

The case of local martingales is obtained by standard localization.
(b) The result follows from the fact that the process $\widetilde{X}^{\mathbb{Q}}$ satisfies, $\forall n \geq 0$,

$$\widetilde{X}_n^{\mathbb{Q}} = X_n - \sum_{k=1}^{n} \frac{1}{L_{k-1}} \mathbb{E}_{\mathbb{P}}[L_k \Delta X_k|\mathcal{H}_{k-1}] = X_n - \sum_{k=1}^{n} \mathbb{E}_{\mathbb{Q}}[\Delta X_k|\mathcal{H}_{k-1}]$$

hence is a $\mathbb{Q}$-martingale if X is $\mathbb{Q}$-integrable. If $\mathbb{E}_{\mathbb{P}}[|\Delta X_n|L_n|\mathcal{H}_{n-1}] < \infty$, then $E_{\mathbb{Q}}[|\Delta X_n||\mathcal{H}_{n-1}] < +\infty$ and $\widetilde{X}^{\mathbb{Q}}$ is a $\mathbb{Q}$-local martingale.

Remark 1.46 If X is a martingale under $\mathbb{P}$ and is $\mathbb{Q}$-integrable, then $\widetilde{X}^{\mathbb{Q}}$ is a martingale under $\mathbb{Q}$. This result can be proved directly using Doob's decomposition. Moreover, if $\mathbb{E}_{\mathbb{P}}[|\Delta X_n|L_{n-1}] < \infty$, then $X^{\mathbb{Q}}$ is a martingale under $\mathbb{Q}$.

Proposition 1.47 *In discrete time, for any* $\mathbb{P}$*-local martingale* X *there exists a probability measure* $\mathbb{Q}$ *equivalent to* $\mathbb{P}$ *so that* X *is a* $\mathbb{Q}$*-martingale.*

Proof The proof of this result, based on functional analysis, can be found in [21].

Example 1.48 Let us come back to example (1.34) about a local martingale which is not a martingale and show how to obtain the equivalent probability measure stated in the previous proposition. One can define $d\mathbb{Q} = Ld\mathbb{P}$ with $L = \frac{\zeta}{\mathbb{E}_{\mathbb{P}}[\zeta]}$ and

$$\zeta = |Y|\mathbb{1}_{\{|Y|<1\}} + \frac{1}{|Y|}\mathbb{1}_{\{|Y|\geq 1\}} \,.$$

Then, Y is independent from U under $\mathbb{Q}$ and is $\mathbb{Q}$ square integrable and M is integrable under $\mathbb{Q}$ and is a $\mathbb{Q}$-martingale.

1.8.1 Predictable Representation Property

An important property for finance purpose is the existence of a martingale which enjoys the predictable representation property (PRP).

Definition 1.49 An $\mathbb{H}$-martingale X (which can be d-dimensional) enjoys the predictable representation property in the filtration $\mathbb{H}$ if any one-dimensional $\mathbb{H}$-martingale has the form $Y = Y_0 + \varphi \centerdot X$ for a predictable d-dimensional process φ and $Y_0 \in \mathscr{H}_0$.

We now give the Kunita–Watanabe decomposition.

Proposition 1.50 *If X is a given square integrable $\mathbb{H}$-martingale, then for any square integrable $\mathbb{H}$-martingale Y there exists an $\mathbb{H}$-predictable process φ and an $\mathbb{H}$-martingale $X^\perp$ orthogonal to X such that $Y = Y_0 + \varphi \centerdot X + X^\perp$.*

Proof Notice that on the set $\Delta\langle X, X\rangle_n = 0$, one has $\Delta\langle X, Y\rangle_n = 0$. Let φ be the $\mathbb{H}$-predictable process $\varphi_n = \frac{\Delta\langle X,Y\rangle_n}{\Delta\langle X,X\rangle_n}\mathbb{1}_{\{\Delta\langle X,X\rangle_n>0\}}$([1]). Then $\varphi \centerdot X$ is a local martingale, and $X^\perp = Y - Y_0 - \varphi \centerdot X$ is a local martingale as the difference of two local martingales. Since

$$\mathbb{E}[\Delta X_n^\perp \Delta X_n | \mathscr{H}_{n-1}] = \Delta\langle X, Y\rangle_n - \frac{\Delta\langle X, Y\rangle_n}{\Delta\langle X, X\rangle_n}\mathbb{1}_{\{\Delta\langle X,X\rangle_n>0\}}\Delta\langle X, X\rangle_n = 0$$

the orthogonality is proved.

We now check that the PRP is stable by change of probability measure. This property is well known in continuous time. The goal here is to give a direct proof.

Theorem 1.51 *Let $X = (X^1, \cdots, X^d)$ be a d-dimensional $(\mathbb{P}, \mathbb{H})$-martingale enjoying the $\mathbb{H}$-predictable representation property under $\mathbb{P}$. Let $\mathbb{Q}$ be a probability measure locally equivalent to $\mathbb{P}$, and $L_n := \frac{d\mathbb{Q}_n}{d\mathbb{P}_n}$. The $\mathbb{Q}$-martingale $X^\mathbb{Q}$ defined as*

$$X^\mathbb{Q} = X - \frac{1}{L} \centerdot [X, L]$$

where $[X, L]$ is the d-dimensional process $([X^i, L], i = 1, \cdots, d)$ and enjoys the $(\mathbb{Q}, \mathbb{H})$-predictable representation property.

Proof We give the proof for $d = 1$. The general case does not present any difficulty. For a $(\mathbb{Q}, \mathbb{H})$-martingale Y, we have to show that there exists an $\mathbb{H}$-predictable process ϑ such that

$$\Delta Y_n = \vartheta_n \Delta X_n^\mathbb{Q}.$$

In a first step, we suppose that Y is a $\mathbb{Q}$ u.i. martingale, then there exists ζ an $\mathscr{H}_\infty$-measurable $\mathbb{Q}$-integrable random variable s.t $Y_n = \mathbb{E}_\mathbb{Q}[\zeta | \mathscr{H}_n]$. Then, for all $N \in \mathbb{N}$ and $0 \le n \le N$,

---

[1] Note that $\frac{1}{b}\mathbb{1}_{\{b>0\}} = 0$ if $b = 0$.

$$Y_n = \mathbb{E}_{\mathbb{Q}}[\zeta|\mathcal{H}_n] = \frac{1}{L_n}\mathbb{E}_{\mathbb{P}}[\mathbb{E}_{\mathbb{P}}[\zeta|\mathcal{H}_N]L_N|\mathcal{H}_n] =: \frac{\zeta_n}{L_n}\,.$$

From integration by parts formula

$$\Delta Y_n = \Delta\Big(\frac{1}{L_n}\zeta_n\Big) = \zeta_{n-1}\Delta\Big(\frac{1}{L_n}\Big) + \frac{1}{L_n}\Delta\zeta_n\,. \tag{6}$$

By definition, the process $(\zeta_n, n \geq 0)$ is a $(\mathbb{P}, \mathbb{H})$-martingale and using the PRP for X, we have the existence of an $\mathbb{H}$-predictable process ϕ such that

$$\Delta\zeta_n = \phi_n \Delta X_n. \tag{7}$$

From Girsanov's theorem, the process $X^{\mathbb{Q}}$ defined as

$$\Delta X_n^{\mathbb{Q}} = \Delta X_n - \frac{1}{L_n}\Delta X_n \Delta L_n = \frac{L_{n-1}}{L_n}\Delta X_n, \quad X_0^{\mathbb{Q}} = X_0$$

is an $(\mathbb{Q}, \mathbb{H})$-local martingale. Inserting ΔX_n in (7), it follows that

$$\Delta\zeta_n = \phi_n \frac{L_n}{L_{n-1}}\Delta X_n^{\mathbb{Q}}. \tag{8}$$

Since X enjoys $(\mathbb{P}, \mathbb{H})$ PRP, and L is a $(\mathbb{P}, \mathbb{H})$-martingale, there exists an $\mathbb{H}$-predictable process ψ such that $\Delta L_n = \psi_n \Delta X_n$. Therefore,

$$\Delta\Big(\frac{1}{L}\Big)_n = -\frac{1}{L_n L_{n-1}}\Delta L_n = -\frac{1}{L_n L_{n-1}}\psi_n \Delta X_n = -\frac{\psi_n}{L_{n-1}^2}\Delta X_n^{\mathbb{Q}}. \tag{9}$$

Plugging (8) and (9) in (6) yields to

$$\Delta Y_n = -\zeta_{n-1}\frac{\psi_n}{L_{n-1}^2}\Delta X_n^{\mathbb{Q}} + \phi_n \frac{1}{L_{n-1}}\Delta X_n^{\mathbb{Q}} = \theta_n \Delta X_n^{\mathbb{Q}}$$

where

$$\theta_n = \frac{\phi_n}{L_{n-1}} - \zeta_{n-1}\frac{\psi_n}{L_{n-1}^2}$$

belongs to $\mathcal{H}_{n-1}$. If Y is not u.i., or is a local martingale, we localize it with a sequence of stopping times (T_j) so that the stopped processes are u.i. We apply PRP to the stopped processes and we pass to the limit as follows.

Let j fixed, then there is a predictable process φ s.t.

$$Y_{n\wedge T_j} = \sum_{k=1}^{n} \varphi_{k\wedge T_j}\Delta X_k^{\mathbb{Q}}\,. \tag{10}$$

Indeed, since this equality is pathwise, and $Y_{n\wedge T_j} = Y_{n\wedge T_{j+1}\wedge T_j}$, the process φ associated to T_{j+1} stopped at T_j is equal to the process associated to φ for T_j. Then passing to the limit in Eq. (10), we obtain that $Y_n = \sum_{k=1}^n \varphi_k \Delta X_k^{\mathbb{Q}}$. Finally, $X^{\mathbb{Q}}$ enjoys PRP.

1.9 Some Facts on Finance

One of the goals of these notes is to present applications to finance and to answer the following question: Does some new information can lead the financial agent to make benefit without risk (i.e. to realize an arbitrage).

In this subsection, we present basic definitions and results in mathematical finance. We recall in particular the definition of self-financing strategies and the results on arbitrage opportunities.

We now consider a model $(\Omega, \mathbb{H}, \mathbb{P}, S)$ where $(\Omega, \mathbb{H}, \mathbb{P})$ is a filtered probability space, and S is a d-dimensional positive process which represents the prices of d risky assets. We assume that there exists a risk-free asset (the savings account) S^0 defined as $S_n^0 = (1+r)^n$, $\forall n \geq 0$, where r is the constant interest rate. A portfolio is a vector of processes (α, π) where $\pi = (\pi^i, i = 1, \cdots, d)$, with α adapted and π predictable. The wealth associated to this portfolio is the process X such that

$$X_n = \alpha_n S_n^0 + \sum_{i=1}^d \pi_n^i S_n^i = \alpha_n S_n^0 + \pi_n S_n, \ \forall n \geq 0.$$

The portfolio is said to be self-financing if $S_{n-1}^0 \Delta\alpha_n + S_{n-1}\Delta\pi_n = 0$ or, since $\Delta(xy)_n = x_{n-1}\Delta y_n + y_n \Delta x_n$, if $\Delta X_n = \alpha_n^0 \Delta S_n^0 + \pi_n \Delta S_n$.

It is standard to work with the discounted prices $\bar{S} = S/S^0$ and discounted wealth $\bar{X} = X/S^0$. Then, the self-financing condition can be written as $\Delta\bar{X} = \pi\Delta\bar{S}$. This allows us to consider only the part π of the portfolio and the initial wealth of the portfolio. The vector π being known, the process α such that (α, π) satisfies the self-financing condition is then $\alpha_n S_n^0 = X_n^{\pi,x} - \sum_{i=1}^d \pi_n^i S_n^i$, where $X^{\pi,x}$ is the wealth with initial value x, i.e. $X_0^{x,\pi} = x$, $\Delta\bar{X}^{x,\pi} = \pi\Delta\bar{S}$.

We consider a model $(\Omega, \mathbb{F}, \mathbb{P}, S)$ with a given horizon N: processes are defined only up to time N, and there are no more transactions after time N.

We now follow Jacod and Shiryaev [18].

Definition 1.52 (a) The financial market is **complete** if for every bounded $\mathscr{F}_N$-measurable random variable ζ, there exists a strategy π and an initial value x such that $\bar{X}_N^{x,\pi} = \bar{\zeta} = \zeta/S^0$ a.s. One says that the strategy π hedges ζ, meaning that the associated self-financing strategy (α, π) satisfies $\alpha_N S_N^0 + \pi_N S_N = \zeta$.

The financial market is strongly complete if every finite-valued $\mathscr{F}_N$-measurable variable ζ can be hedged.

(b) The model $(\Omega, \mathbb{F}, \mathbb{P}, S)$ is **arbitrage-free** if for any π with $X_0^{\pi,0} = 0$ and $X_N^{\pi,0} \geq 0$, one has $X_N^{\pi,0} = 0$.

The model is weakly arbitrage-free if for any π with $X_0^{\pi,0} = 0$ and $X_n^{\pi,0} \geq 0$, $\forall n \geq 0$, one has $X_N^{\pi,0} = 0$.

The model is strongly arbitrage-free if for any π with $X_0^{\pi,0} = 0$ and $X_N^{\pi,0} \geq 0$, one has $X_n^{\pi,0} = 0, \forall n \geq 0$.

We denote by $\mathcal{Q}$ the class of all probability measures which are equivalent to $\mathbb{P}$ and under which the discounted process $\bar{S}$ is a martingale, and by $\mathcal{Q}_{loc}$ the set of all probability measures which are equivalent to $\mathbb{P}$ and under which the discounted process $\bar{S}$ is a local martingale. Finally, $\mathcal{Q}_b$ is the set of all $\mathbb{Q}$ in $\mathcal{Q}$ such that the Radon–Nikodym density $L = d\mathbb{Q}/d\mathbb{P}$ is bounded.

Proposition 1.53 *There is equivalence between:*

(a) The model is arbitrage-free.
(b) The model is weakly arbitrage-free.
(c) The model is strongly arbitrage-free.
(d) The set $\mathcal{Q}$ is non-empty.
(e) The set $\mathcal{Q}_b$ is non-empty.
(f) The set $\mathcal{Q}_{loc}$ is non-empty.
In that case, we shall write that S satisfies NA($\mathbb{H}$).

Proof It is obvious that $c \Rightarrow a \Rightarrow b$ and $e \Rightarrow d$. Moreover, as a non-negative local martingale is a true martingale, then $d \Leftrightarrow f$. It remains to prove $d \Rightarrow c$ and $b \Rightarrow d$ and $f \Rightarrow e$.

$d \Rightarrow c$: Let $\mathbb{P} \in \mathcal{Q}$ such as $\bar{S}$ is a martingale. Let π with $X_0^{\pi,0} = 0$ and $X_N^{\pi,0} \geq 0$. Then, the discounted wealth process $\bar{X}^{\pi,0}$ is a local martingale with non-negative terminal value. Using the local martingale property, we show that the process $\bar{X}^{\pi,0}$ is non-negative. Hence, it is a true martingale. It implies that $\mathbb{E}[\bar{X}_0^{\pi,0}] = 0 = \mathbb{E}[\bar{X}_n^{\pi,0}]$ for all $n \in \mathbb{N}$. So the result follows.

For $b \Rightarrow d$ and $f \Rightarrow e$, see [18, Th. 3].

If $\mathcal{Q}$ is not empty, any Radon–Nikodym density is called a **deflator**.

Proposition 1.54 *Assume that the model is arbitrage-free, then the following assertions are equivalent:*

(a) The model is complete.
(b) The model is strongly complete.
(c) The set $\mathcal{Q}$ contains exactly one measure.
(d) The set $\mathcal{Q}_{loc}$ contains exactly one measure.
(e) There exists $\mathbb{P}^* \in \mathcal{Q}_{loc}$ such that S enjoys the $\mathbb{P}^*$-martingale representation property.

Proof We refer to [18, Theorem 6].

Assertion (c) of Proposition 1.54 is called the second fundamental asset pricing theorem.

Proposition 1.55 *In the case where the discrete time market with d assets is arbitrage-free and complete, the σ-field $\mathscr{F}_N$ is purely atomic relative to the measure $\mathbb{P}$, with at most $(d+1)^N$ atoms.*

Proof We refer to [18, Theorem 6].

1.10 Enlargement of Filtration

In continuous time, a difficult problem is to give conditions such that an $\mathbb{F}$-martingale is a $\mathbb{G}$-semimartingale for two filtrations satisfying $\mathbb{F} \subset \mathbb{G}$, and, if it is the case, to give the $\mathbb{G}$-semimartingale decomposition of an $\mathbb{F}$-martingale. In discrete time, the following proposition is an easy consequence of Doob's decomposition and states that if $\mathbb{F} \subset \mathbb{G}$, then any $\mathbb{F}$-martingale is a $\mathbb{G}$-semimartingale and gives explicitly the decomposition of this semimartingale.

Proposition 1.56 *In a discrete time setting, any integrable process is a special semimartingale in any filtration with respect to which it is adapted. In particular, if $\mathbb{F} \subset \mathbb{G}$, and if X is an $\mathbb{F}$-martingale, it is a $\mathbb{G}$ special semimartingale with decomposition*

$$X = M^{\mathbb{G}} + X^{p,\mathbb{G}}$$

where $M^{\mathbb{G}}$ is a $\mathbb{G}$-martingale and $X^{p,\mathbb{G}}$ is $\mathbb{G}$-predictable, $X_0^{p,\mathbb{G}} = 0$. This decomposition is unique and

$$\Delta X_n^{p,\mathbb{G}} = \mathbb{E}[\Delta X_n | \mathscr{G}_{n-1}],\ n \geq 1.$$

*The process $X^{p,\mathbb{G}}$ is often called the **information drift** of X relative to $\mathbb{G}$.*

Definition 1.57 Immersion is satisfied between the filtration $\mathbb{F}$ and a larger filtration $\mathbb{G}$ (or $\mathbb{F}$ is immersed in $\mathbb{G}$), if any $\mathbb{F}$-martingale is a $\mathbb{G}$-martingale. If this property holds, we will denote it by $\mathbb{F} \hookrightarrow \mathbb{G}$. In order to specify that the immersion is achieved under the probability measure $\mathbb{P}$, we will denote it by $\mathbb{F} \overset{\mathbb{P}}{\hookrightarrow} \mathbb{G}$.

Immersion is also called $(\mathscr{H})$-hypothesis in the literature. The following result is well known and useful (see [9, Th. 3]).

Proposition 1.58 *Immersion hypothesis is equivalent to any of the following properties, where for a set A, and a σ-algebra $\mathscr{K}$, we denote $\mathbb{P}(A|\mathscr{K}) = \mathbb{E}[\mathbb{1}_A|\mathscr{K}]$:*

(a) $\forall n \geq 0$, the σ-fields $\mathscr{F}_\infty$ and $\mathscr{G}_n$ are conditionally independent given $\mathscr{F}_n$, that is, if $\forall n \geq 0, \forall G_n \in \mathscr{G}_n, F \in \mathscr{F}_\infty$, one has $\mathbb{P}(F \cap G_n|\mathscr{F}_n) = \mathbb{P}(F|\mathscr{F}_n)\mathbb{P}(G_n|\mathscr{F}_n)$.
(b) $\forall n \geq 0,\ G_n \in \mathscr{G}_n,\ \mathbb{P}(G_n|\mathscr{F}_n) = \mathbb{P}(G_n|\mathscr{F}_\infty)$.
(c) $\forall n \geq 0,\ F \in \mathscr{F}_\infty,\ \mathbb{P}(F|\mathscr{F}_n) = \mathbb{P}(F|\mathscr{G}_n)$.

Proof We recall the proof (given in [9]) for the ease of the reader.

- Immersion $\Rightarrow$ (a). Let $F \in \mathcal{F}_\infty$. Under immersion, the $\mathbb{F}$-martingale defined by $\mathbb{E}[\mathbb{1}_F|\mathcal{F}_n]$ for all $n \geq 0$ is a $\mathbb{G}$-martingale. Hence, for any $n \geq 0$ and any $G_n \in \mathcal{G}_n$,

$$\begin{aligned}\mathbb{P}(F \cap G_n|\mathcal{F}_n) &= \mathbb{E}\big[\mathbb{P}(F|\mathcal{G}_n)\mathbb{1}_{G_n}|\mathcal{F}_n\big]=\mathbb{E}\big[\mathbb{P}(F|\mathcal{F}_n)\mathbb{1}_{G_n}|\mathcal{F}_n\big]\\ &= \mathbb{P}(F|\mathcal{F}_n)\mathbb{P}(G_n|\mathcal{F}_n)\ ,\end{aligned}$$

which is (a). The second equality comes from the fact that the $\mathbb{F}$-martingale $\mathbb{P}(F|\mathcal{F}_n)$, being a $\mathbb{G}$-martingale, is equal to $\mathbb{P}(F|\mathcal{G}_n)$.
- $(a) \Rightarrow (b)$. If (a) holds, then for any $F \in \mathcal{F}_\infty$ and any $G_n \in \mathcal{G}_n$

$$\mathbb{E}\big[\mathbb{1}_F\mathbb{P}(G_n|\mathcal{F}_n)\big] = \mathbb{E}\big[\mathbb{P}(F|\mathcal{F}_n)\mathbb{P}(G_n|\mathcal{F}_n)\big] \overset{(a)}{=} \mathbb{E}\big[\mathbb{P}(F \cap G_n|\mathcal{F}_n)\big] = \mathbb{P}(F \cap G_n)\ ,$$

which is exactly (b).
- $(b) \Rightarrow (c)$. Suppose (b) and let $F \in \mathcal{F}_\infty$ and $G_n \in \mathcal{G}_n$ for any $n \geq 0$, then

$$\mathbb{E}\big[\mathbb{P}(F|\mathcal{F}_n)\mathbb{1}_{G_n}\big] = \mathbb{E}\big[\mathbb{1}_F\mathbb{P}(G_n|\mathcal{F}_n)\big] \overset{(b)}{=} \mathbb{E}\big[\mathbb{1}_F\mathbb{P}(G_n|\mathcal{F}_\infty)\big] = \mathbb{P}(F \cap G_n)\ ,$$

which implies (c).
- $(c) \Rightarrow$ immersion. Consider $F \in \mathcal{F}_\infty$ and the $\mathbb{F}$-martingale $F_n := \mathbb{P}(F|\mathcal{F}_n), n \geq 0$. Then, $(G_n)_{n\geq 0}$ defined by $G_n := \mathbb{P}(F|\mathcal{G}_n)$ for all $n \geq 0$ is a $\mathbb{G}$-martingale. Then under (c), we have that $F_n = G_n$ for all $n \geq 0$; therefore, immersion is satisfied for u.i. martingales. The extension to all martingales is standard.

Our goal is to compute more explicitly the semimartingale decomposition in some specific cases, and to show, with elementary computations, that we recover the classical general formulae established in the literature in continuous time.

Comment 1.59 Note that results in continuous time can be directly applied to discrete time: if $\mathbb{F}$ is a discrete time filtration and X a discrete time process, one can study the continuous on right filtration $\widetilde{\mathbb{F}}$ defined in continuous time for $n \leq t < n+1$ as $\widetilde{\mathcal{F}}_t = \mathcal{F}_n$, and the càdlàg process $\widetilde{X}_t = \sum_n X_n \mathbb{1}_{\{n \leq t < n+1\}}$. One interest of our computations relies on the fact that we do not need hypotheses done in continuous time and that our proofs are simple.

Another goal of this paper is to study how enlarging the filtration may introduce arbitrages. We start with a general result, valid for any filtration $\mathbb{H}$.

Lemma 1.60 *Let Y be an integrable $\mathbb{H}$-semimartingale. If there exists a positive $\mathbb{H}$-adapted process ψ such that*

$$\mathbb{E}[Y_n\psi_n|\mathcal{H}_{n-1}] = Y_{n-1}\mathbb{E}[\psi_n|\mathcal{H}_{n-1}],\ \forall n \geq 1\ ,$$

there exists a positive $\mathbb{H}$-martingale L such that LY is an $\mathbb{H}$-martingale.

Proof Let Y be a $(\mathbb{P}, \mathbb{H})$-semimartingale with decomposition $Y = M + Y^p$, with $\Delta Y_n^p = \mathbb{E}[\Delta Y_n|\mathcal{H}_{n-1}]$ and where M is a $(\mathbb{P}, \mathbb{H})$-martingale. Define, for a given positive ψ, the positive $(\mathbb{P}, \mathbb{H})$-martingale L

$$L_0 = 1, \; L_n = \prod_{k=1}^{n} \frac{\psi_k}{\mathbb{E}[\psi_k | \mathcal{H}_{k-1}]} = L_{n-1} \frac{\psi_n}{\mathbb{E}[\psi_n | \mathcal{H}_{n-1}]}, n \geq 1,$$

then, setting $d\mathbb{Q}_n = L_n d\mathbb{P}_n$, the process M decomposes as $M = m + M^{p,\mathbb{Q}}$ where m is a $(\mathbb{Q}, \mathbb{H})$-martingale and $M^{p,\mathbb{Q}}$ a predictable process given by

$$\begin{aligned}\Delta M_n^{p,\mathbb{Q}} &= \mathbb{E}_{\mathbb{Q}}[\Delta M_n | \mathcal{H}_{n-1}] = \frac{1}{L_{n-1}} \mathbb{E}_{\mathbb{P}}[L_n \Delta M_n | \mathcal{H}_{n-1}] \\ &= \frac{1}{L_{n-1}} (\mathbb{E}_{\mathbb{P}}[L_n M_n | \mathcal{H}_{n-1}] - L_{n-1} M_{n-1}) = \frac{1}{\mathbb{E}_{\mathbb{P}}[\psi_n | \mathcal{H}_{n-1}]} \mathbb{E}_{\mathbb{P}}[\psi_n \Delta M_n | \mathcal{H}_{n-1}] .\end{aligned}$$

The semimartingale Y is a $(\mathbb{Q}, \mathbb{H})$-martingale if its predictable part is null, that is, $Y^p + M^{p,\mathbb{Q}} = 0$ or equivalently $\Delta Y^p + \Delta M^{p,\mathbb{Q}} = 0$, or

$$\mathbb{E}[\psi_n \Delta M_n | \mathcal{H}_{n-1}] + \mathbb{E}[\psi_n | \mathcal{H}_{n-1}] \mathbb{E}[\Delta Y_n | \mathcal{H}_{n-1}] = 0 .$$

We develop and use that $\Delta M_n = Y_n - \mathbb{E}[Y_n | \mathcal{H}_{n-1}]$ and obtain, after simplification

$$\mathbb{E}[\psi_n Y_n | \mathcal{H}_{n-1}] = \mathbb{E}[\psi_n | \mathcal{H}_{n-1}] Y_{n-1} .$$

In the setting of enlargement of filtration, we introduce the following definition of viable enlargement.

Definition 1.61 Let $\mathbb{F} \subset \mathbb{G}$, we say that the enlargement $(\mathbb{F}, \mathbb{G}, \mathbb{P})$ is viable if there exists a positive $(\mathbb{P}, \mathbb{G})$-martingale L with $L_0 = 1$ (called a **universal deflator**) such that, for any $(\mathbb{P}, \mathbb{F})$-martingale X, the process XL is a $(\mathbb{P}, \mathbb{G})$-martingale.

Our definition implies that, if there is a discounted price process S, which is a $(\mathbb{P}, \mathbb{F})$-martingale, then the market $(S, \mathbb{G})$ is arbitrage-free. The study of necessary and sufficient conditions in a progressive enlargement setting so that, for a given $(\mathbb{P}, \mathbb{F})$-martingale S, there exists a deflator can be found in Choulli and Deng [11].

2 Initial Enlargement

For a random variable ξ taking values in $\mathbb{R}$, the filtration $\mathbb{F}^{\sigma(\xi)} = (\mathcal{F}_n^{\sigma(\xi)}, n \geq 0)$ where $\mathcal{F}_n^{\sigma(\xi)} := \mathcal{F}_n \vee \sigma(\xi), n \geq 0$ is called the initial enlargement of $\mathbb{F}$ with ξ. We assume that $\mathcal{F}_0$ is trivial, so that $\mathcal{F}_0^{\sigma(\xi)} = \sigma(\xi)$. Any $\sigma(\xi)$-measurable random variable is on the form $h(\xi)$ where h is a Borel function. Any $\mathcal{F}_n^{\sigma(\xi)}$-measurable random variable is of the form $h_n(\xi)$ where the map $h_n(\omega, u)$ from $(\Omega, \mathbb{R})$ to $\mathbb{R}$ is $\mathcal{F}_n \times \mathcal{B}$ measurable, $\mathcal{B}$ being the Borelian σ algebra.

2.1 Bridge

We study the following particular example. Let $(Y_i, i \geq 1)$ be a sequence of i.i.d. random variables with zero mean and define the process X by $X_0 := 0,\ X_n := \sum_{i=1}^n Y_i,\ n \geq 1$. The process X is an $\mathbb{F}$-martingale, where $\mathbb{F}$ is the natural filtration of X. For N fixed, we choose $\xi := X_N$.

To obtain the Doob decomposition of X in the filtration $\mathbb{F}^{\sigma(X_N)}$, we compute $\mathbb{E}[\Delta X_n|\mathscr{F}_{n-1}^{\sigma(X_N)}] = \mathbb{E}[\Delta X_n|\mathscr{F}_{n-1} \vee \sigma(X_N)]$. Using the fact that $(Y_i, i \geq 1)$ are i.i.d, we have, for $n \leq j \leq N$,

$$(Y_j, X_1, \cdots, X_{n-1}, X_N) \overset{\text{law}}{=} (Y_n, X_1, \cdots, X_{n-1}, X_N),$$

hence

$$\begin{aligned}\mathbb{E}[Y_n|\mathscr{F}_{n-1} \vee \sigma(X_N)] &= \mathbb{E}[Y_j|\mathscr{F}_{n-1} \vee \sigma(X_N)] \\ &= \frac{1}{N-(n-1)}\mathbb{E}[Y_n + \cdots + Y_j + \cdots + Y_N|\mathscr{F}_{n-1} \vee \sigma(X_N)] \\ &= \frac{1}{N-(n-1)}\mathbb{E}[X_N - X_{n-1}|\mathscr{F}_{n-1} \vee \sigma(X_N)] = \frac{X_N - X_{n-1}}{N-(n-1)}.\end{aligned}$$

Therefore, the process $\widetilde{X}$ defined as

$$\widetilde{X}_n = X_n - \sum_{k=1}^n \frac{X_N - X_{k-1}}{N-(k-1)},\ n \geq 0,$$

is an $\mathbb{F}^{\sigma(X_N)}$-martingale.

Comment 2.1 This formula is similar to the one obtained for Lévy bridges: if X is an integrable Lévy process in continuous time (e.g. a Brownian motion) with natural filtration $\mathbb{F}^X$, setting $\mathbb{F}^{\sigma(X_T)} = \mathbb{F}^X \vee \sigma(X_T)$ leads to the fact that the process $\widetilde{X}$, defined as

$$\widetilde{X}_t = X_t - \int_0^t \frac{X_T - X_s}{T-s} ds,\ 0 \leq t \leq T,$$

is an $\mathbb{F}^{\sigma(X_T)}$-martingale. See [6, Chap. 4].

2.2 Viability

The following lemma establishes that if $\xi \in \mathscr{F}_N$ for some N and $\xi \notin \mathscr{F}_0$, the enlargement $(\mathbb{F}, \mathbb{F}^{\sigma(\xi)}, \mathbb{P})$ is not viable.

Lemma 2.2 *If ξ is an integrable $\mathscr{F}_N$-measurable r.v. for some N and ξ is not a constant, the enlargement $(\mathbb{F}, \mathbb{F}^{\sigma(\xi)}, \mathbb{P})$ is not viable.*

Proof Let $Y_n := \mathbb{E}[\xi|\mathscr{F}_n]$. If an $\mathbb{F}^{\sigma(\xi)}$-deflator L exists, the process YL would be an $\mathbb{F}^{\sigma(\xi)}$-martingale, and $Y_nL_n = \mathbb{E}[Y_NL_N|\mathscr{F}_n^{\sigma(\xi)}]$. Using the fact that $Y_N = \xi$ is $\mathscr{F}_n^{\sigma(\xi)}$-measurable for $0 \le n \le N$, we obtain $\mathbb{E}[Y_NL_N|\mathscr{F}_n^{\sigma(\xi)}] = Y_NL_n$, in particular, $Y_NL_0 = Y_0L_0$. It would follow that $Y_N = Y_0$, which is not possible since $Y_N = \xi$ is not $\mathscr{F}_0$-measurable.

2.3 *Initial Enlargement with a $\mathbb{Z}$-Valued Random Variable ξ*

Let X be an $\mathbb{F}$-martingale, ξ be a r.v. taking values in $\mathbb{Z}$ and, for any $j \in \mathbb{Z}$, let $p(j)$ be the $\mathbb{F}$-martingale defined as $p_n(j) = \mathbb{P}(\xi = j|\mathscr{F}_n)$, $\forall n \ge 0$. In what follows, the notation $\langle X, p(j)\rangle_k^{\mathbb{F}}|_{j=\xi}$ means that, in a first step, one computes the predictable bracket of the two $\mathbb{F}$-martingales X and $p(j)$ in the filtration $\mathbb{F}$, i.e.$\langle X, p(j)\rangle_k^{\mathbb{F}}$ which is an $\mathscr{F}_{k-1}$-measurable r.v., depending on the parameter j, and then, one replaces j by ξ.

Proposition 2.3 *(a) One has $p_n(\xi) > 0, \forall n \ge 0$.*
(b) The process $\widetilde{X}$ defined as

$$\widetilde{X}_n = X_n - \sum_{k=1}^{n} \frac{\Delta\langle X, p(j)\rangle_k^{\mathbb{F}}|_{j=\xi}}{p_{k-1}(\xi)}, \ \forall n \ge 0, \tag{11}$$

is an $\mathbb{F}^{\sigma(\xi)}$-martingale.

Proof (a) On the set $\{\xi = j\}$, one has $p_n(j) \neq 0, \forall n \ge 0$. Indeed,

$$\mathbb{E}[\mathbb{1}_{\{p_n(j)=0\}}\mathbb{1}_{\{\xi=j\}}] = \mathbb{E}[\mathbb{1}_{\{p_n(j)=0\}}\mathbb{E}[\xi = j|\mathscr{F}_n]] = \mathbb{E}[\mathbb{1}_{\{p_n(j)=0\}}p_n(j)] = 0\,.$$

Hence $p_n(\xi) = \sum_j \mathbb{1}_{\{\xi=j\}}p_n(j) > 0$.
(b) The Doob decomposition of X in $\mathbb{F}^{\sigma(\xi)}$ is $X = M + X^p$ where M is an $\mathbb{F}^{\sigma(\xi)}$-martingale and, for $n \ge 1$, $\Delta X_n^p = \mathbb{E}[\Delta X_n|\mathscr{F}_{n-1} \vee \sigma(\xi)]$ so that

$$\begin{aligned}(\Delta X_n^p)\mathbb{1}_{\{\xi=j\}} &= \mathbb{1}_{\{\xi=j\}}\frac{\mathbb{E}[\mathbb{1}_{\{\xi=j\}}\Delta X_n|\mathscr{F}_{n-1}]}{\mathbb{P}(\xi = j|\mathscr{F}_{n-1})} \\ &= \mathbb{1}_{\{\xi=j\}}\frac{\mathbb{E}[p_n(j)\Delta X_n|\mathscr{F}_{n-1}]}{p_{n-1}(j)} = \mathbb{1}_{\{\xi=j\}}\frac{\Delta\langle X, p(j)\rangle_n^{\mathbb{F}}}{p_{n-1}(j)},\end{aligned} \tag{12}$$

where we have used the tower property in the second equality.

Comment 2.4 From this formula, one can guess that, in continuous time the process $X^{\mathbb{F}^{\sigma(\xi)}}$ is an $\mathbb{F}^{\sigma(\xi)}$-martingale where

$$X_t^{\mathbb{F}^{\sigma(\xi)}} = X_t - \int_0^t \frac{d\langle X, p(u)\rangle_s^{\mathbb{F}}|_{u=\xi}}{p_{s-}(\xi)}, \quad \forall t \ge 0\,.$$

In order to give a meaning to that formula, one has to assume Jacod's hypothesis which stipulates that the conditional law of ξ w.r.t. $\mathscr{F}_t$ is absolutely continuous w.r.t. the law of ξ, i.e. there exists a family $p(u)$ such that

$$\mathbb{P}(\xi \in du|\mathscr{F}_t) = p_t(u)\mathbb{P}(\xi \in du)\,.$$

Then, the above decomposition formula is indeed true (see, e.g. [17] and [19, Chap. 3], [6, Chap. 4]). If ξ takes only discrete values, then Jacod's hypothesis is always true.

Comment 2.5 In what follows, we study the positive process $1/p(\xi)$. Note that this process is not integrable: in the case $p(k) > 0, \forall k$, indeed

$$\begin{aligned}\mathbb{E}[1/p_n(\xi)] &= \sum_{k=-\infty}^{\infty} \mathbb{E}\big[\mathbb{1}_{\{\xi=k\}}/p_n(\xi)\big] = \sum_{k=-\infty}^{\infty} \mathbb{E}\big[\mathbb{1}_{\{\xi=k\}}/p_n(k)\big] \\ &= \sum_{k=-\infty}^{\infty} \mathbb{E}\big[\mathbb{P}(\xi = k|\mathscr{F}_n)/p_n(k)\big] = \sum_{k=-\infty}^{\infty} \mathbb{E}\big[p_n(k)/p_n(k)\big] = \infty\end{aligned}$$

so, in the following proposition, we are dealing with generalized martingales. However, the process $p_0(\xi)/p(\xi)$ is integrable

$$\mathbb{E}[p_0(\xi)/p_n(\xi)] = \sum_{k=-\infty}^{\infty} \mathbb{E}\big[p_0(k)\big] = \sum_{k=-\infty}^{\infty} \mathbb{P}(\xi = k) = 1.$$

The following proposition establishes that the condition $p(k) > 0, \forall k \in \mathbb{Z}$ implies that, if a market $(S, \mathbb{F})$ has no arbitrages, the market $(S, \mathbb{F}^{\sigma(\xi)})$ has no arbitrages. Indeed, the enlargement $(\mathbb{F}, \mathbb{F}^{\sigma(\xi)})$ is viable, a universal deflator being $p_0(\xi)/p(\xi)$.

Proposition 2.6 *(a) The process* $\frac{1}{p(\xi)}$ *is an* $\mathbb{F}^{\sigma(\xi)}$*-generalized supermartingale. If* $p(k) > 0$ *for any* $k \in \mathbb{Z}$*, it is an* $\mathbb{F}^{\sigma(\xi)}$*-generalized martingale.*
(b) If X *is an* $\mathbb{F}$*-martingale and* $p(k) > 0, \forall k \in \mathbb{Z}$*, the process* $X/p(\xi)$ *is an* $\mathbb{F}^{\sigma(\xi)}$*-generalized martingale. If, moreover,* $Xp_0(\xi)/p(\xi)$ *is integrable, the process* $Xp_0(\xi)/p(\xi)$ *is an* $\mathbb{F}^{\sigma(\xi)}$*-martingale.*

Proof (a) We recall that, from Proposition 2.3, $p(\xi) > 0$. We now establish the supermartingale property.

$$\begin{aligned}\mathbb{E}[\frac{1}{p_n(\xi)}|\mathscr{F}_{n-1} \vee \sigma(\xi)] &= \sum_{k=-\infty}^{\infty} \mathbb{1}_{\{\xi=k\}} \frac{\mathbb{E}[\mathbb{1}_{\{\xi=k\}}\mathbb{1}_{\{p_n(k)>0\}}\frac{1}{p_n(k)}|\mathscr{F}_{n-1}]}{\mathbb{E}[\mathbb{1}_{\{\xi=k\}}|\mathscr{F}_{n-1}]} \\ &= \sum_{k=-\infty}^{\infty} \mathbb{1}_{\{\xi=k\}} \frac{\mathbb{E}[\mathbb{1}_{\{p_n(k)>0\}}|\mathscr{F}_{n-1}]}{p_{n-1}(k)} \le \sum_{k=-\infty}^{\infty} \mathbb{1}_{\{\xi=k\}} \frac{1}{p_{n-1}(k)} \\ &= \frac{1}{p_{n-1}(\xi)}.\end{aligned}$$

If $p(k) > 0$ for any k, the last inequality is in fact an equality, and $1/p(\xi)$ is a martingale.
(b) For a $\mathbb{P}$-martingale X such that $Xp_0(\xi)/p(\xi)$ is integrable, if $p(k) > 0, k \in \mathbb{Z}$, one has

$$\mathbb{E}[\frac{X_n p_0(\xi)}{p_n(\xi)}|\mathscr{F}_{n-1} \vee \sigma(\xi)] = \sum_{k=-\infty}^{\infty} \mathbb{1}_{\{\xi=k\}} \frac{\mathbb{E}[\mathbb{1}_{\{\xi=k\}}\frac{X_n}{p_n(k)}|\mathscr{F}_{n-1}]}{\mathbb{E}[\mathbb{1}_{\{\xi=k\}}|\mathscr{F}_{n-1}]} p_0(k)$$

$$= \sum_{k=-\infty}^{\infty} \mathbb{1}_{\{\xi=k\}} \frac{\mathbb{E}[X_n|\mathscr{F}_{n-1}]}{p_{n-1}(k)} p_0(k) = \frac{X_{n-1}p_0(\xi)}{p_{n-1}(\xi)} .$$

The general case is left to the reader.

We can now define a particular change of probability under which ξ and $\mathbb{F}$ will be independent.

Lemma 2.7 *If $p(k) > 0$ for any $k \in \mathbb{Z}$, the process $L_n := \frac{p_0(\xi)}{p_n(\xi)}$ is a positive martingale with expectation 1. Define, for any $n \geq 0$, $d\mathbb{Q} = L_n d\mathbb{P}$ on the σ-algebra $\mathscr{F}_n^{\sigma(\xi)}$. Then, for any n, ξ is independent from $\mathscr{F}_n$ under $\mathbb{Q}$, and $\mathbb{Q}|_{\mathscr{F}_n} = \mathbb{P}|_{\mathscr{F}_n}$, $\mathbb{Q}|_{\sigma(\xi)} = \mathbb{P}|_{\sigma(\xi)}$.*

Proof We have seen in part (b) of the previous proposition (take $X = 1$) that L is a martingale. For $F \in \mathscr{F}_n$,

$$\mathbb{E}_\mathbb{Q}[h(\xi)F] = \mathbb{E}[L_n h(\xi)F] = \mathbb{E}[\sum_k \frac{p_0(k)}{p_n(k)} h(k) F \mathbb{1}_{\{\xi=k\}}] = \mathbb{E}[\sum_k \frac{p_0(k)}{p_n(k)} h(k) F p_n(k)] .$$

In particular

$$\mathbb{E}_\mathbb{Q}[h(\xi)F] = \mathbb{E}[h(\xi)]\,\mathbb{E}[F] .$$

Taking $h \equiv 1$ (resp. $F = 1$), we see that

$$\mathbb{E}_\mathbb{Q}[F] = \mathbb{E}[F] \text{ (resp.} \, \mathbb{E}_\mathbb{Q}[h(\xi)] = \mathbb{E}[h(\xi)\,],$$

and

$$\mathbb{E}_\mathbb{Q}[h(\xi)F] = \mathbb{E}_\mathbb{Q}[h(\xi)]\mathbb{E}_\mathbb{Q}[F] .$$

The result follows.

Proposition 2.8 *Let $Y(k)$ be a family of $\mathbb{F}$-adapted processes. The process $Y(\xi)$ is an $\mathbb{F}^{\sigma(\xi)}$-martingale iff $Y(k)p(k)$ is an $\mathbb{F}$-martingale for any k.*

Proof Let $Y(\xi)$ be an $\mathbb{F}^{\sigma(\xi)}$-martingale. Then, for $n \geq m$,

$$\mathbb{E}[Y_n(\xi)|\mathscr{F}_m^{\sigma(\xi)}] = Y_m(\xi)$$

or, equivalently, for any $k \in \mathbb{Z}$

$$\mathbb{E}[Y_n(\xi)|\mathscr{F}_m^{\sigma(\xi)}]\mathbb{1}_{\{\xi=k\}} = Y_m(\xi)\mathbb{1}_{\{\xi=k\}} .$$

Taking the $\mathscr{F}_m$-conditional expectation of both sides, we obtain for $n \geq m$,

$$\mathbb{E}[Y_n(k)\mathbb{1}_{\{\xi=k\}}|\mathscr{F}_m] = Y_m(k)p_m(k)$$

hence, by tower property

$$\mathbb{E}[Y_n(k)p_n(k)|\mathscr{F}_m] = Y_m(k)p_m(k) .$$

Conversely,

$$\mathbb{E}[Y_n(\xi)|\mathscr{F}_m^{(\sigma(\xi)}] = \sum_k \mathbb{E}[Y_n(\xi)\mathbb{1}_{\{\xi=k\}}|\mathscr{F}_m^{\sigma(\xi)}] = \sum_k \mathbb{1}_{\{\xi=k\}}\mathbb{E}[Y_n(k)|\mathscr{F}_m^{\sigma(\xi)}] .$$

Then, if for any k, $Y(k)p(k)$ is an $\mathbb{F}$-martingale, for $m \leq n$, using the fact that

$$\mathbb{1}_{\{\xi=k\}}\mathbb{E}[Y_n(\xi)|\mathscr{F}_m^{\sigma(\xi)}] = \mathbb{1}_{\{\xi=k\}}\frac{\mathbb{E}[Y_n(k)p_n(k)|\mathscr{F}_m]}{p_m(k)} = \mathbb{1}_{\{\xi=k\}}Y_m(k)$$

we conclude that $Y(\xi)$ is an $\mathbb{F}^{\sigma(\xi)}$-martingale.

Proposition 2.9 *For any $\mathbb{Z}$-valued random variable ξ, the following assertions are equivalent.*
(a) The set $\{\omega : 0 = p_n(k)(\omega) < p_{n-1}(k)(\omega)\}$ is negligible, for all k and n.
(b) For any $\mathbb{F}$-adapted integrable non-negative process X satisfying NA($\mathbb{F}$), X satisfies NA($\mathbb{F}^{\sigma(\xi)}$).

Proof We refer the reader to [11].

2.4 *Example: Supremum of a Random Walk*

We consider a particular case of the previous setting where we can compute the probabilities $p_n(j)$. Let $(Y_i, i \geq 1)$ a sequence of i.i.d. random variables taking values in $\mathbb{Z}$ and the process X of the form $X_0 := 0,\ X_n := \sum_{i=1}^n Y_i,\ n \geq 1$. For N fixed, we put $\xi := \sup_{0\leq n\leq N} X_n$ and we denote by $\mathbb{F}$ the natural filtration of X.

We denote by $g(n,k) = \mathbb{P}(\sup_{1\le j\le n} X_j = k)$ and by $h(n,k) = \mathbb{P}(\sup_{1\le j\le n} X_j \le k)$. Then the probability $p_n(j)$ can be expressed as follows.
We note that

$$\{\sup_{k\le N} X_k = j\} = \{\sup_{k\le n} X_k = j, \sup_{1\le k\le N-n} X_{n+k} < j\} \cup \{\sup_{k\le n} X_k \le j, \sup_{1\le k\le N-n} X_{n+k} = j\}$$

and that, setting $\widetilde{X}_k = X_{n+k} - X_n = \sum_{i=1}^{N-n} \widetilde{Y}_i$ where $\widetilde{Y}_i$ are copies of Y_i, independent from $\mathscr{F}_n$

$$\begin{aligned}\mathbb{P}(\sup_{1\le k\le N-n} X_{n+k} < j|\mathscr{F}_n) &= \mathbb{P}(\sup_{1\le k\le N-n} X_{n+k} - X_n < j - X_n|\mathscr{F}_n)\\ &= \mathbb{P}(\sup_{1\le k\le N-n} \widetilde{X}_k < j - X_n|\mathscr{F}_n) = h(N-n, j-X_n)\end{aligned}$$

$$\mathbb{P}(\sup_{1\le k\le N-n} X_{n+k} = j|\mathscr{F}_n) = \mathbb{P}(\sup_{1\le k\le N-n} \widetilde{X}_k = j - X_n|\mathscr{F}_n) = g(N-n, j-X_n)).$$

Then,

$$p_n(j) = \mathbb{1}_{\{\sup_{0\le k\le n} X_k = j\}} h(N-n, j-X_n) + \mathbb{1}_{\{\sup_{0\le k\le n} X_k \le j\}} g(N-n, j-X_n).$$

3 Introduction to Progressive Enlargement

We assume that $\mathscr{F}_0$ is trivial. Let τ be a random time, i.e. a random variable valued in $\mathbb{N} \cup \{+\infty\}$, and introduce the filtration $\mathbb{G}$ where, for $n \ge 0$, we set $\mathscr{G}_n = \mathscr{F}_n \vee \sigma(\tau \wedge n)$. In other words, $\mathbb{G}$ is the smallest filtration which contains $\mathbb{F}$ and makes τ a stopping time. In particular $\{\tau = 0\} \in \mathscr{G}_0$, so that, in general $\mathscr{G}_0$ is not trivial.

In continuous time, many results are obtained under the hypothesis that τ avoids $\mathbb{F}$-stopping times, or that all $\mathbb{F}$-martingales are continuous, which is not the case in discrete time. We present here some basic results.

We define the process H as

$$\boxed{H_n = \mathbb{1}_{\{\tau\le n\}}\,.} \tag{13}$$

Then, $\mathbb{G} = \mathbb{F} \vee \mathbb{H}$, where $\mathbb{H}$ is the filtration generated by H.

3.1 General Results

3.1.1 Basic Properties

Lemma 3.1 *(a) If Y is a $\mathbb{G}$-adapted process, there exists an $\mathbb{F}$-adapted process y such that*

$$Y_n \mathbb{1}_{\{n<\tau\}} = y_n \mathbb{1}_{\{n<\tau\}},\ \forall n \geq 0. \tag{14}$$

(b) If $\widetilde{Y}$ is a $\mathbb{G}$-predictable process, there exists an $\mathbb{F}$-predictable process $\widetilde{y}$ such that

$$\widetilde{Y}_n \mathbb{1}_{\{n\leq\tau\}} = \widetilde{y}_n \mathbb{1}_{\{n\leq\tau\}},\ \forall n \geq 0. \tag{15}$$

(c) If $\widetilde{Y}$ is a $\mathbb{G}$-predictable process, there exists a family of $\mathbb{F}$-predictable process $y(k)$ such that

$$\widetilde{Y}_n \mathbb{1}_{\{n>\tau\}} = y_n(\tau) \mathbb{1}_{\{n>\tau\}}\,, \tag{16}$$

(d) If Y is a $\mathbb{G}$-adapted process, there exists a family $y(k)$ of $\mathbb{F}$-adapted process y such that

$$Y_n \mathbb{1}_{\{\tau\leq n\}} = y_n(\tau) \mathbb{1}_{\{\tau\leq n\}},\ \forall n \geq 0. \tag{17}$$

Proof (a) One proves (14) for $Y_n = X_n h(\tau \wedge n)$ where $X_n \in \mathscr{F}_n$ and h a bounded Borel function. In that case, the result is obvious, with $y_n = X_n h(n)$. The general case follows from the monotone class theorem.
(b) If $\widetilde{Y}$ is a $\mathbb{G}$-predictable process, then the process S defined by $S_0 = Y_0$ and $S_n = \widetilde{Y}_{n+1}$ is $\mathbb{G}$-adapted and so there exists an $\mathbb{F}$-adapted process s s.t.

$$\widetilde{Y}_n \mathbb{1}_{\{n\leq\tau\}} = S_{n-1} \mathbb{1}_{\{n-1<\tau\}} = s_{n-1} \mathbb{1}_{\{n-1<\tau\}} = \widetilde{y}_n \mathbb{1}_{\{n\leq\tau\}}$$

where $\widetilde{y}_n = s_{n-1}$.
(c) It suffices to give the proof for processes of the form $\widetilde{Y}_n = X_n h(\tau) \mathbb{1}_{\{n>\tau\}}$ where X is $\mathbb{F}$-predictable, which generate $\mathbb{G}$-predictable processes after τ. For this class, the result is straightforward.

Definition 3.2 Given τ, we define two supermartingales which will be important: the Azéma supermartingale Z defined as

$$\boxed{Z_n = \mathbb{P}(\tau > n|\mathscr{F}_n),\ \forall n \geq 0,}$$

and the second Azéma supermartingale $\widetilde{Z}$ defined as

$$\widetilde{Z}_n = \mathbb{P}(\tau \geq n|\mathscr{F}_n),\ \forall n \geq 0\,. \tag{18}$$

If several probability measures are involved, e.g. $\mathbb{Q}$, we denote by $Z^{\mathbb{Q}}$ the $\mathbb{Q}$-Azéma's supermartingale defined as

$$Z_n^{\mathbb{Q}} = \mathbb{Q}(\tau > n|\mathscr{F}_n), \forall n \geq 0.$$

In other terms, Z is the $\mathbb{F}$-optional projection of $1 - H$.

We shall often use the trivial equalities

$$\widetilde{Z}_n = \mathbb{P}(\tau > n-1|\mathscr{F}_n) = Z_n + \mathbb{P}(\tau = n|\mathscr{F}_n),$$
$$Z_n = \mathbb{P}(\tau \geq n+1|\mathscr{F}_n), \ \mathbb{E}[\widetilde{Z}_n|\mathscr{F}_{n-1}] = Z_{n-1}.$$

Proposition 3.3 *The Doob decomposition of the Azéma supermartingale Z is given by*

$$Z = M - H^{p,\mathbb{F}}, \tag{19}$$

where M is an $\mathbb{F}$-martingale. The Doob decomposition of the second Azéma supermartingale $\widetilde{Z}$ is

$$\widetilde{Z} = \widetilde{M} - \widetilde{A} \tag{20}$$

where $\widetilde{M}$ is an $\mathbb{F}$-martingale and $\widetilde{A}$ is defined as $\widetilde{A}_n = H_{n-1}^{o,\mathbb{F}}$.

Proof In the proof, the predictable (resp. optional) dual projections are w.r.t. $\mathbb{F}$ and so we omit the superscript $\mathbb{F}$. The Doob decomposition of Z is $M - Z^p$ where the predictable part Z^p satisfies

$$\Delta Z_n^p = -\mathbb{E}[\Delta Z_n|\mathscr{F}_{n-1}] = -\mathbb{E}[\Delta H_n|\mathscr{F}_{n-1}] = \mathbb{P}(\tau = n|\mathscr{F}_{n-1}) = H_n^{p,\mathbb{F}}, \ \forall n \geq 1, \tag{21}$$

where the second equality comes from tower property. Since $\mathscr{F}_0$ is assumed to be trivial, $Z_0 = M_0 = \mathbb{P}(\tau > 0)$. The predictable part of the Doob decomposition of $\widetilde{Z}$ is the $\mathbb{F}$-predictable increasing process $\widetilde{A}$ satisfying $\widetilde{A}_0 = 0$, $\Delta\widetilde{A}_n = \mathbb{P}(\tau = n-1|\mathscr{F}_{n-1}), \forall n \geq 1$. We show in the following that $\widetilde{A}_n = H_{n-1}^o$.

From Definition 1.18, H^o satisfies

$$H_n^o := \sum_{k=0}^{n} \mathbb{E}[\Delta H_k|\mathscr{F}_k] = \sum_{k=0}^{n} \mathbb{P}(\tau = k|\mathscr{F}_k), \quad \forall n \geq 0.$$

It follows that $\Delta\widetilde{A}_n = H_{n-1}^o - H_{n-2}^o$ and since $\widetilde{A}_1 = H_0^o$ we have $\widetilde{A}_n = H_{n-1}^o$, hence

$$Z_n + H_n^o = Z_n + \Delta H_n^o + H_{n-1}^o = \widetilde{Z}_n + \widetilde{A}_n = \widetilde{M}_n. \tag{22}$$

We recall that $H_\infty^o = \lim_{n\to\infty} H_n^o = \sum_{k=0}^{\infty} \mathbb{P}(\tau = k|\mathscr{F}_k)$. Since $\lim_{n\to\infty} Z_n = \mathbb{1}_{\{\tau=\infty\}}$, and $\mathbb{E}[H_\infty^o] = \lim \mathbb{E}[H_n^o] \leq 1$, one has

$$\widetilde{M}_n = Z_n + H_n^o = \mathbb{E}[\widetilde{M}_\infty | \mathscr{F}_n] = \mathbb{E}[\mathbb{1}_{\{\tau=\infty\}} + H_\infty^o | \mathscr{F}_n]\,. \tag{23}$$

We shall also need the following result:

$$\widetilde{Z}_n - Z_{n-1} = \Delta H_n^o + \Delta Z_n = \Delta \widetilde{M}_n, \tag{24}$$

where the first equality comes from the definition of the various processes and the second from (22).

Comment 3.4 In particular, if $\widetilde{Z}$ is predictable, $\widetilde{M} = 0$ and $\widetilde{Z}_n = Z_{n-1}$.

Note that, by definition of dual optional projection, for any $\mathbb{F}$-adapted process X

$$\mathbb{E}[X_\tau \mathbb{1}_{\{\tau<\infty\}}] = \mathbb{E}[\sum_n X_n \Delta H_n^o] \tag{25}$$

Proposition 3.5 *On the set $\{n \le \tau\}$, the random variables $\widetilde{Z}_n$ and Z_{n-1} are positive. On the set $\{n > \tau\}$, the random variables $\widetilde{Z}_n$ and Z_{n-1} are strictly smaller than 1. On the set $\{\tau < \infty\}$, one has $Z_\tau < 1$.*

Proof The first assertion is obtained from the two following equalities:

$$\mathbb{E}[\mathbb{1}_{\{n\le\tau\}}\mathbb{1}_{\{Z_{n-1}=0\}}] = \mathbb{E}[\mathbb{P}(n \le \tau | \mathscr{F}_{n-1})\mathbb{1}_{\{Z_{n-1}=0\}}] = \mathbb{E}[Z_{n-1}\mathbb{1}_{\{Z_{n-1}=0\}}] = 0\,,$$
$$\mathbb{E}[\mathbb{1}_{\{n\le\tau\}}\mathbb{1}_{\{\widetilde{Z}_n=0\}}] = \mathbb{E}[\mathbb{P}(n \le \tau | \mathscr{F}_n)\mathbb{1}_{\{\widetilde{Z}_n=0\}}] = \mathbb{E}[\widetilde{Z}_n\mathbb{1}_{\{\widetilde{Z}_n=0\}}] = 0\,.$$

The second assertion is left to the reader. By definition,

$$Z_\tau \mathbb{1}_{\{\tau<\infty\}} = \sum_n \mathbb{1}_{\{\tau=n\}} Z_n = \sum_n \mathbb{1}_{\{\tau=n\}} \mathbb{P}(\tau > n | \mathscr{F}_n) \le 1 - \sum_n \mathbb{1}_{\{\tau=n\}} \mathbb{P}(\tau \le n | \mathscr{F}_n)$$

and $\mathbb{P}(\tau \le n | \mathscr{F}_n) \ge \mathbb{P}(\tau = n | \mathscr{F}_n) = p_n(n)$. The quantity $p_n(n)$ being positive on $\{\tau = n\}$ the result follows.

We give a useful lemma known as key lemma. The proof of (a) is standard, and the proof of (b) can be found in Aksamit et al. [5] for continuous time. For the ease of the reader, we recall these proofs.

Lemma 3.6 *Let τ be a random time. Then,*
(a) for any integrable random variable ζ

$$\mathbb{E}[\zeta | \mathscr{G}_n]\mathbb{1}_{\{\tau>n\}} = \mathbb{1}_{\{\tau>n\}} \frac{\mathbb{E}[\zeta \mathbb{1}_{\{\tau>n\}} | \mathscr{F}_n]}{Z_n}\,, \forall n \ge 0\,.$$

(b) for any integrable $\mathscr{F}_n$-measurable r.v. Y_n,

$$\mathbb{E}[Y_n|\mathscr{G}_{n-1}]\mathbb{1}_{\{\tau\geq n\}} = \mathbb{1}_{\{\tau\geq n\}}\frac{1}{Z_{n-1}}\mathbb{E}[Y_n\widetilde{Z}_n|\mathscr{F}_{n-1}],\ \forall n\geq 1$$

$$\mathbb{E}[\frac{Y_n}{\widetilde{Z}_n}|\mathscr{G}_{n-1}]\mathbb{1}_{\{\tau\geq n\}} = \mathbb{1}_{\{\tau\geq n\}}\frac{1}{Z_{n-1}}\mathbb{E}[Y_n\mathbb{1}_{\{\widetilde{Z}_n>0\}}|\mathscr{F}_{n-1}],\ \forall n\geq 1. \tag{26}$$

Proof (a) For $Y_n = \mathbb{E}[\zeta|\mathscr{G}_n]$ in (14), taking expectation w.r.t. $\mathscr{F}_n$ we obtain

$$\mathbb{E}[\zeta\mathbb{1}_{\{\tau>n\}}|\mathscr{F}_n] =: \mathbb{E}[\mathbb{1}_{\{\tau>n\}}|\mathscr{F}_n]y_n = Z_n y_n\,,$$

hence, using the fact that Z_n is positive on the set $\{\tau > n\}$, one gets

$$y_n\mathbb{1}_{\{\tau>n\}} = \mathbb{1}_{\{\tau>n\}}\frac{\mathbb{E}[\zeta\mathbb{1}_{\{\tau>n\}}|\mathscr{F}_n]}{Z_n}\,.$$

(b) The first part follows from part (a) and tower property. Only (26) requires a proof. For $n\geq 1$, we have, using in the second equality, the fact that $\widetilde{Z}_n > 0$ on the set $\{\tau\geq n\}$,

$$\begin{aligned}\mathbb{E}[\frac{Y_n}{\widetilde{Z}_n}\mathbb{1}_{\{\tau\geq n\}}|\mathscr{G}_{n-1}] &= \mathbb{1}_{\{\tau\geq n\}}\frac{1}{Z_{n-1}}\mathbb{E}[Y_n\frac{1}{\widetilde{Z}_n}\mathbb{1}_{\{\tau\geq n\}}|\mathscr{F}_{n-1}]\\ &= \mathbb{1}_{\{\tau\geq n\}}\frac{1}{Z_{n-1}}\mathbb{E}[Y_n\frac{1}{\widetilde{Z}_n}\mathbb{1}_{\{\tau\geq n\}}\mathbb{1}_{\{\widetilde{Z}_n>0\}}|\mathscr{F}_{n-1}]\\ &= \mathbb{1}_{\{\tau\geq n\}}\frac{1}{Z_{n-1}}\mathbb{E}[Y_n\mathbb{1}_{\{\widetilde{Z}_n>0\}}|\mathscr{F}_{n-1}]\,.\end{aligned}$$

We give an immediate and important consequence in order to define in a canonical way[2] a process y which satisfies (14) and a process $\widetilde{y}$ which satisfies (15).

Lemma 3.7 *The process y, which satisfies (14) can be chosen as*

$$y_n = \frac{1}{Z_n}\mathbb{E}[Y_n\mathbb{1}_{\{n<\tau\}}|\mathscr{F}_n]\mathbb{1}_{\{Z_n>0\}},\ \forall n\geq 0\,.$$

This particular choice of y is called the $\mathbb{F}$-adapted reduction of Y.

The process $\widetilde{y}$, which satisfies (15), can be chosen as

$$\widetilde{y}_n = \frac{1}{Z_{n-1}}\mathbb{E}[\widetilde{Y}_n\mathbb{1}_{\{n\leq\tau\}}|\mathscr{F}_{n-1}]\mathbb{1}_{\{Z_{n-1}>0\}},\ \forall n\geq 1\,.$$

This particular choice of $\widetilde{y}$ is called the $\mathbb{F}$-predictable reduction of Y.

[2] We recall that $\frac{1}{b}\mathbb{1}_{\{b>0\}} = 0$ for $b = 0$.

Proof The proof is a consequence on Proposition 3.5 and Lemma 3.6.

Proposition 3.8 *The process $\Upsilon := (1-H)\frac{1}{Z}$ is a $\mathbb{G}$-supermartingale.*
Assume that Z is positive. If X is an $\mathbb{F}$-martingale, such that $X\Upsilon$ is integrable, $X\Upsilon$ is a $\mathbb{G}$-martingale. In particular, Υ is a martingale.

Proof

$$\begin{aligned}
\mathbb{E}[\mathbb{1}_{\{\tau>n\}}\frac{1}{Z_n}|\mathscr{G}_{n-1}] &= \mathbb{1}_{\{\tau>n-1\}}\frac{1}{Z_{n-1}}\mathbb{E}[\mathbb{1}_{\{\tau>n\}}\frac{1}{Z_n}|\mathscr{F}_{n-1}] \\
&= \mathbb{1}_{\{\tau>n-1\}}\frac{1}{Z_{n-1}}\mathbb{E}[\mathbb{1}_{\{\tau>n\}}\mathbb{1}_{\{Z_n>0\}}\frac{1}{Z_n}|\mathscr{F}_{n-1}] \\
&= \mathbb{1}_{\{\tau>n-1\}}\frac{1}{Z_{n-1}}\mathbb{E}[\mathbb{1}_{\{Z_n>0\}}\frac{1}{Z_n}\mathbb{E}[\mathbb{1}_{\{\tau>n\}}|\mathscr{F}_n]|\mathscr{F}_{n-1}] \\
&= \mathbb{1}_{\{\tau>n-1\}}\frac{1}{Z_{n-1}}\mathbb{E}[\mathbb{1}_{\{Z_n>0\}}|\mathscr{F}_{n-1}] \le \mathbb{1}_{\{\tau>n-1\}}\frac{1}{Z_{n-1}}\,.
\end{aligned}$$

In the first equality, we have used that, due to the fact that τ is a $\mathbb{G}$-stopping time, $\{\tau \le n-1\} \in \mathscr{G}_{n-1}$, hence $\mathbb{E}[\mathbb{1}_{\{\tau>n\}}\frac{1}{Z_n}|\mathscr{G}_{n-1}]\mathbb{1}_{\{\tau\le n-1\}} = 0$. In the second equality, we have used that $\{\tau > n\} \subset \{Z_n > 0\}$. In particular, one has equality in the last line if $Z > 0$. If X is an $\mathbb{F}$-martingale, the same kind of proof establishes that, for $Z > 0$,

$$\mathbb{E}[\mathbb{1}_{\{\tau>n\}}\frac{X_n}{Z_n}|\mathscr{G}_{n-1}] = \mathbb{1}_{\{\tau>n-1\}}\frac{X_{n-1}}{Z_{n-1}}\,.$$

The following lemma provides the Doob decomposition of H, and in particular its predictable compensator Λ^τ.

Lemma 3.9 *Let Λ be the $\mathbb{F}$-predictable process defined as*

$$\boxed{\Lambda_0 = 0,\ \Delta\Lambda_n := \frac{\Delta H_n^{p,\mathbb{F}}}{Z_{n-1}}\mathbb{1}_{\{Z_{n-1}>0\}},\ n \ge 1\,.}$$

The process N defined as

$$N_n := H_n - \Lambda_{n\wedge\tau} = H_n - \sum_{k=1}^{n\wedge\tau}\lambda_k,\ n \ge 0, \tag{27}$$

where $\lambda_n := \Delta\Lambda_n$, is a $\mathbb{G}$-martingale.
Moreover, if Z is predictable, then $\Delta\Lambda_n = -\frac{\Delta Z_n}{Z_{n-1}}\mathbb{1}_{\{Z_{n-1}>0\}}$.

Proof It suffices to find the Doob decomposition of the $\mathbb{G}$-semimartingale H. The predictable part of this decomposition is K with

$$\Delta K_n = \mathbb{E}[\Delta H_n|\mathscr{G}_{n-1}] = \mathbb{1}_{\{\tau\le n-1\}}0 + \mathbb{1}_{\{\tau>n-1\}}\frac{\mathbb{E}[\Delta H_n|\mathscr{F}_{n-1}]}{Z_{n-1}}$$
$$= \mathbb{1}_{\{\tau\ge n\}}\frac{H_n^{p,\mathbb{F}} - H_{n-1}^{p,\mathbb{F}}}{Z_{n-1}},\ n\ge 1.$$

Note that $\mathbb{E}[\Delta H_n|\mathscr{F}_{n-1}] = Z_{n-1} - \mathbb{E}[Z_n|\mathscr{F}_{n-1}]$. Since from Proposition 3.5, $Z_{n-1} > 0$ on $\{1\le n\le\tau\}$, we conclude that, on $\{n\le\tau\}$, one has $\Delta K_n = \lambda_n$ where $\lambda_n = \frac{\Delta H_n^{p,\mathbb{F}}}{Z_{n-1}}\mathbb{1}_{\{Z_{n-1}>0\}}$ is $\mathbb{F}$-predictable.

Remark 3.10 If Z is predictable, then the martingale part M in its Doob decomposition is constant equal to Z_0. Indeed, the martingale M would be predictable as the difference of two predictable processes, hence, a constant, thanks to Lemma 1.2. In that case, Z is decreasing.

Proposition 3.11 *Suppose Z positive. The multiplicative predictable decomposition of Z is given by $Z_n = N_n^Z\mathscr{E}\big(-\Lambda\big)_n$, $n\ge 0$ where N^Z is a positive $\mathbb{F}$-martingale and Λ is defined in Lemma 3.9.*

Proof We have seen that there exist an $\mathbb{F}$-martingale N^Z and an $\mathbb{F}$-predictable process K^Z such that $Z = N^Z K^Z$ with

$$K_n^Z = \prod_{k=1}^n \frac{\mathbb{E}[Z_k|\mathscr{F}_{k-1}]}{Z_{k-1}} = \prod_{k=1}^n\Big[1-\frac{Z_{k-1}-\mathbb{E}[Z_k|\mathscr{F}_{k-1}]}{Z_{k-1}}\Big],\quad \forall n\ge 1\,.$$

From Lemma 3.9 and the positivity of Z, we have

$$\Delta\Lambda_n = \frac{Z_{n-1}-\mathbb{E}[Z_n|\mathscr{F}_{n-1}]}{Z_{n-1}},\quad \forall n\ge 1\,,$$

then by definition of the exponential process, we get that $K^Z = \mathscr{E}\big(-\Lambda\big)$, which is a predictable decreasing process.

Lemma 3.12 *If $\widetilde{Z}$ is predictable and Z is positive, then $\mathbb{E}[\Delta N_n|\mathscr{F}_n] = -\Delta M_n$ for all $n\ge 1$, where M is the martingale part in the Doob decomposition of Z defined in (19) and N is defined in (27).*

Proof By definition of N, we have that, for $n\ge 1$,

$$\begin{aligned}\mathbb{E}[\Delta N_n|\mathscr{F}_n] &= \mathbb{E}[\mathbb{1}_{\{\tau\le n\}} - \mathbb{1}_{\{\tau\le n-1\}} - \lambda_n\mathbb{1}_{\{\tau\ge n\}}|\mathscr{F}_n]\\ &= \mathbb{E}[-\mathbb{1}_{\{\tau>n\}} + \mathbb{1}_{\{\tau\ge n\}}|\mathscr{F}_n] - \lambda_n\mathbb{E}[\mathbb{1}_{\{\tau\ge n\}}|\mathscr{F}_n]\\ &= -Z_n + \widetilde{Z}_n - \lambda_n\widetilde{Z}_n = -\Delta Z_n - \lambda_n Z_{n-1}\,,\end{aligned}$$

where we have used that, since $\widetilde{Z}$ is predictable, $\widetilde{Z}_n = Z_{n-1}$ (see Comment 3.4). Finally, using the fact that $\Delta Z_n + \Delta H_n^{p,\mathbb{F}} = \Delta M_n$ and, Z being positive, one has $\lambda_n Z_{n-1} = \Delta H_n^{p,\mathbb{F}}$, we get $\mathbb{E}[\Delta N_n|\mathscr{F}_n] = -\Delta M_n$.

The following proposition provides another decomposition of H as the sum of a martingale and an increasing process Γ^τ, this process being only optional. This decomposition will be important in the following.

Proposition 3.13 *Let* $\Pi := H - (\widetilde{Z})^{-1}\mathbb{1}_{[\![0,\tau]\!]} \centerdot H^{o,\mathbb{F}} = H - \Gamma^\tau$, *with*

$$\Delta\Gamma_n = (\widetilde{Z}_n)^{-1}\mathbb{1}_{\{\widetilde{Z}_n>0\}}\Delta H_n^{o,\mathbb{F}}, \ \Gamma_0 = 0\,. \tag{28}$$

Then, for any integrable $\mathbb{F}$*-adapted process* Y*, the process* $Y \centerdot \Pi$ *is a* $\mathbb{G}$*-martingale. In particular,* Π *is a* $\mathbb{G}$*-martingale.*

Proof Since $\Delta\Pi_n = \mathbb{1}_{\{\tau=n\}} - \frac{1}{\widetilde{Z}_n}\mathbb{1}_{\{\tau\geq n\}}\mathbb{P}(\tau = n|\mathscr{F}_n)$, we see that $\Delta\Pi_n\mathbb{1}_{\{\tau<n\}} = 0$ and, from Lemma 3.6 (26), for any $n \geq 1$,

$$\mathbb{E}[Y_n\Delta\Pi_n|\mathscr{G}_{n-1}] = \frac{\mathbb{1}_{\{\tau>n-1\}}}{Z_{n-1}}\mathbb{E}[Y_n\Delta\Pi_n\mathbb{1}_{\{\tau\geq n\}}|\mathscr{F}_{n-1}] + \mathbb{E}[Y_n\Delta\Pi_n\mathbb{1}_{\{\tau<n\}}|\mathscr{G}_{n-1}]$$

$$= \frac{\mathbb{1}_{\{\tau>n-1\}}}{Z_{n-1}}\mathbb{E}\left[Y_n\left(\mathbb{P}(\tau = n|\mathscr{F}_n) - \frac{1}{\widetilde{Z}_n}\mathbb{P}(\tau = n|\mathscr{F}_n)\mathbb{P}(\tau \geq n|\mathscr{F}_n)\mathbb{1}_{\{\widetilde{Z}_n>0\}}\right)|\mathscr{F}_{n-1}\right]$$

$$= \mathbb{1}_{\{\tau>n-1\}}\frac{1}{Z_{n-1}}\mathbb{E}\left[Y_n\mathbb{P}(\tau = n|\mathscr{F}_n)\left(1 - \mathbb{1}_{\{\widetilde{Z}_n>0\}}\right)|\mathscr{F}_{n-1}\right]$$

$$= \mathbb{1}_{\{\tau>n-1\}}\frac{1}{Z_{n-1}}\mathbb{E}\left[Y_n\mathbb{P}(\tau = n|\mathscr{F}_n)\mathbb{1}_{\{\widetilde{Z}_n=0\}}|\mathscr{F}_{n-1}\right],$$

where the fact that Y is $\mathbb{F}$-adapted has been used in the second equality. It remains to note that, on $\{\widetilde{Z}_n = 0\}$ one has $\mathbb{P}(\tau = n|\mathscr{F}_n) = 0$, to obtain

$$\mathbb{E}[\Delta(Y \centerdot \Pi)_n] = \mathbb{E}[Y_n\Delta\Pi_n|\mathscr{G}_{n-1}] = 0\,.$$

The martingale property follows.

Note that, if Y is an $\mathbb{F}$-martingale, then, from Lemma 1.9, the $\mathbb{G}$-martingale part of Y is orthogonal to Π. This result is similar to the one obtained in continuous time by Choulli and Deng [11].

Comment 3.14 There are obviously infinitely many nondecreasing $\mathbb{G}$-adapted processes Θ such that $\mu := H - \Theta$ is a martingale stopped at time τ, e.g. $\Theta = H$ or any convex combination between Λ^τ and Γ^τ, where Λ is the $\mathbb{F}$-predictable process defined in Lemma 3.9 and Γ the $\mathbb{F}$-optional process defined in (28). Assume that μ is a $\mathbb{G}$-martingale stopped at τ (so that $\Delta\mu_n\mathbb{1}_{\{\tau<n\}} = 0$) of the form $\mu = H - K\mathbb{1}_{[0,\tau]} \centerdot J$ where K, J are $\mathbb{F}$-adapted process and $Y \centerdot \mu$ is a $\mathbb{G}$-martingale for any integrable $\mathbb{F}$-adapted process Y. As we show below, the property that $Y \centerdot \mu$ is a martingale for any $\mathbb{F}$-adapted process Y characterizes the pair of processes (K, J) and implies that $\mu = \Pi$.

One can write

$$
\begin{aligned}
0 = \mathbb{E}[Y_n \Delta\mu_n | \mathscr{G}_{n-1}] &= \mathbb{1}_{\{\tau>n-1\}} \frac{1}{Z_{n-1}} \mathbb{E}[\mathbb{1}_{\{\tau>n-1\}} Y_n \Delta\mu_n | \mathscr{F}_{n-1}] \\
&= \mathbb{1}_{\{\tau>n-1\}} \frac{1}{Z_{n-1}} \Big(\mathbb{E}[\mathbb{1}_{\{\tau\geq n\}} Y_n \Delta H_n | \mathscr{F}_{n-1}] - \mathbb{E}[\mathbb{1}_{\{\tau\geq n\}} Y_n K_n \Delta J_n | \mathscr{F}_{n-1}] \Big) \\
&= \frac{\mathbb{1}_{\{\tau\geq n\}}}{Z_{n-1}} \Big(\mathbb{E}[\mathbb{E}[\mathbb{1}_{\{\tau\geq n\}} \Delta H_n | \mathscr{F}_n]\, Y_n | \mathscr{F}_{n-1}] - \mathbb{E}[\mathbb{E}[\mathbb{1}_{\{\tau\geq n\}} | \mathscr{F}_n]\, Y_n K_n \Delta J_n | \mathscr{F}_{n-1}] \Big) \\
&= \mathbb{1}_{\{\tau\geq n\}} \frac{1}{Z_{n-1}} \Big(\mathbb{E}[\mathbb{E}[\Delta H_n | \mathscr{F}_n]\, Y_n | \mathscr{F}_{n-1}] - \mathbb{E}[\widetilde{Z}_n Y_n K_n \Delta J_n | \mathscr{F}_{n-1}] \Big) \\
&= \mathbb{1}_{\{\tau\geq n\}} \frac{1}{Z_{n-1}} \mathbb{E}[Y_n(\Delta H_n^{o,\mathbb{F}} - \widetilde{Z}_n K_n \Delta J_n) | \mathscr{F}_{n-1}]
\end{aligned}
$$

where we have used the fact that $\mathbb{1}_{\{\tau\geq n\}} \Delta H_n = \Delta H_n$. Then, taking conditional expectation w.r.t. $\mathscr{F}_{n-1}$, one obtains, for any $Y_n \in \mathscr{F}_n$

$$\mathbb{1}_{\{Z_{n-1}>0\}} \mathbb{E}[Y_n(\Delta H_n^{o,\mathbb{F}} - \widetilde{Z}_n K_n \Delta J_n) | \mathscr{F}_{n-1}] = 0\,,$$

hence

$$\mathbb{1}_{\{Z_{n-1}>0\}} (\Delta H_n^{o,\mathbb{F}} - \widetilde{Z}_n K_n \Delta J_n) = 0\,.$$

This equality implies that $\Delta\Gamma_n = K_n \Delta J_n$, hence $\mu = \Pi$.

Proposition 3.15 *Assume that Z is positive. Then, Z admits an optional multiplicative decomposition*

$$Z = \widetilde{N} \mathscr{E}(-\Gamma)$$

where $\widetilde{N}$ is an $\mathbb{F}$-martingale and Γ the increasing adapted process defined in (28).

Proof In the case $Z > 0$, we have

$$1 - \Delta\Gamma_n = 1 - (\widetilde{Z}_n)^{-1} \Delta H_n^{o,\mathbb{F}} = 1 - (\widetilde{Z}_n)^{-1} \mathbb{P}(\tau = n | \mathscr{F}_n) > 0\,, \tag{29}$$

and the stochastic exponential $\mathscr{E}(-\Gamma)$ is positive and decreasing.
We check that $Z/\mathscr{E}(-\Gamma)$ is a martingale.

$$\mathbb{E}[\frac{Z_n}{\mathscr{E}(-\Gamma)_n} - \frac{Z_{n-1}}{\mathscr{E}(-\Gamma)_{n-1}} | \mathscr{F}_{n-1}] = \frac{1}{\mathscr{E}(-\Gamma)_{n-1}} \mathbb{E}[\frac{Z_n}{1 - \Delta\Gamma_n} - Z_{n-1} | \mathscr{F}_{n-1}]$$

We obtain

$$\mathbb{E}[\frac{Z_n}{1 - \Delta\Gamma_n} - Z_{n-1} | \mathscr{F}_{n-1}] = \mathbb{E}[\frac{Z_n \widetilde{Z}_n}{\widetilde{Z}_n - \Delta H_n^{o,\mathbb{F}}} - Z_{n-1} | \mathscr{F}_{n-1}] = \mathbb{E}[\widetilde{Z}_n - Z_{n-1} | \mathscr{F}_{n-1}] = 0$$

where we have used (29) in the first equality and $\widetilde{Z}_n - \Delta H_n^{o,\mathbb{F}} = Z_n$ in the second equality. The martingale property follows.

3.1.2 Arbitrages

We now establish that, if $\widetilde{Z} > 0$ (in particular if $Z > 0$), there are no arbitrages before τ. More precisely, we prove that if S is an $\mathbb{F}$-martingale, then there exists a positive $\mathbb{G}$-martingale L such that $S^\tau L$ is a $\mathbb{G}$-local martingale.

Proposition 3.16 *If $\widetilde{Z} > 0$, the process*

$$L_n = \prod_{k=1}^{n}(1 + \Delta U_k) = L_{n-1}(1 + \Delta U_n) = \mathscr{E}(U)_n$$

where $\Delta U_k = \mathbb{1}_{\{\tau \geq k\}}(\frac{Z_{k-1}}{\widetilde{Z}_k} - 1)$, is a positive $\mathbb{G}$-martingale and the process $S^\tau L$ is a $(\mathbb{P}, \mathbb{G})$-martingale, where $S^\tau_n = S_{\tau\wedge n}$.

Proof In a first step, we show that U is a $\mathbb{G}$-martingale.

$$\begin{aligned}
\mathbb{E}[\Delta U_n|\mathscr{G}_{n-1}] &= \mathbb{E}[\mathbb{1}_{\{\tau \geq n\}}(\frac{Z_{n-1}}{\widetilde{Z}_n} - 1)|\mathscr{G}_{n-1}] \\
&= \mathbb{1}_{\{\tau > n-1\}}\frac{1}{Z_{n-1}}\mathbb{E}[\mathbb{1}_{\{\tau \geq n\}}(\frac{Z_{n-1}}{\widetilde{Z}_n} - 1)|\mathscr{F}_{n-1}] \\
&= \mathbb{1}_{\{\tau > n-1\}}\frac{1}{Z_{n-1}}\left(\mathbb{E}[\widetilde{Z}_n\frac{Z_{n-1}}{\widetilde{Z}_n}|\mathscr{F}_{n-1}] - Z_{n-1}\right) = 0
\end{aligned}$$

where we have used Lemma 3.6 in the second equality and the positivity of $\widetilde{Z}$ in the third equality. It follows that $\mathbb{E}[L_n|\mathscr{G}_{n-1}] = L_{n-1}$. Then,

$$\begin{aligned}
&\mathbb{E}[S^\tau_{(n+1)}(1 + \Delta U_{n+1})|\mathscr{G}_n] = \mathbb{E}[S_{(n+1)\wedge\tau}(1 + \mathbb{1}_{\{\tau \geq n+1\}}(\frac{Z_n}{\widetilde{Z}_{n+1}} - 1))|\mathscr{G}_n] \\
&= \mathbb{E}[S_{n+1}\mathbb{1}_{\{\tau \geq n+1\}}(1 + \frac{Z_n}{\widetilde{Z}_{n+1}} - 1)|\mathscr{G}_n] + \mathbb{E}[S_\tau \mathbb{1}_{\{\tau < n+1\}}|\mathscr{G}_n] \\
&= \mathbb{E}[S_{n+1}\mathbb{1}_{\{\tau \geq n+1\}}\frac{Z_n}{\widetilde{Z}_{n+1}}|\mathscr{G}_n] + S_\tau \mathbb{1}_{\{\tau < n+1\}} \\
&= \mathbb{1}_{\{\tau > n\}}\frac{1}{Z_n}\mathbb{E}[S_{n+1}\widetilde{Z}_{n+1}\frac{Z_n}{\widetilde{Z}_{n+1}}|\mathscr{F}_n] + S_\tau \mathbb{1}_{\{\tau \leq n\}} = S_{n\wedge\tau}\,,
\end{aligned}$$

where the next-to-last equality comes from Lemma 3.6. It follows that

$$\mathbb{E}[L_{n+1}S^\tau_{n+1}|\mathscr{G}_n] = L_n\mathbb{E}[S^\tau_{(n+1)}(1 + \Delta U_{n+1})|\mathscr{G}_n] = L_n S^\tau_n\,.$$

Comment 3.17 It is proved in [11] that S satisfies NA($\mathbb{G}$) for any S satisfying NA($\mathbb{F}$) is equivalent to

$$\forall n, \{\widetilde{Z}_n = 0\} = \{Z_{n-1} = 0\}\,.$$

3.1.3 $\mathbb{G}$-Martingales Versus $\mathbb{F}$-Martingales

In this subsection, we give a characterization for $\mathbb{G}$-martingales in terms of $\mathbb{F}$-martingales. We denote as before $\mathbb{P}(\tau = k|\mathscr{F}_n) = p_n(k)$.

We first give a result on $\mathbb{G}$-conditional expectations which will be used in the proof of the following Proposition.

Lemma 3.18 *Let $k \in \mathbb{N}$ and $n \geq k$. Consider $(V_n(i), i \leq n)$ a family of random variables, where $V_n(i)$ is $\mathscr{F}_n$-measurable, then*

$$\mathbb{1}_{\{\tau \leq k\}}\mathbb{E}[V_n(\tau)|\mathscr{G}_k] = \mathbb{1}_{\{\tau \leq k\}}\frac{1}{p_k(\tau)}\mathbb{E}[V_n(i)p_n(i)|\mathscr{F}_k]_{|_{i=\tau}}.$$

Proof From

$$\mathbb{1}_{\{\tau \leq k\}}\mathbb{E}[V_n(\tau)|\mathscr{G}_k] = \sum_{i=0}^{k} \mathbb{1}_{\{\tau=i\}}\mathbb{E}[V_n(i)|\mathscr{G}_k]$$

and using the fact that there exists an $\mathscr{F}_k$-measurable random variable $v_k(i)$ such that, for $i \leq k$

$$\mathbb{1}_{\{\tau=i\}}\mathbb{E}[V_n(i)|\mathscr{G}_k] = \mathbb{E}[\mathbb{1}_{\{\tau=i\}}V_n(i)|\mathscr{G}_k] = \mathbb{1}_{\{\tau=i\}}v_k(i)\,,$$

taking the conditional expectation w.r.t. $\mathscr{F}_k$, we obtain that

$$\mathbb{E}[V_n(i)\mathbb{1}_{\{\tau=i\}}|\mathscr{F}_k] = \mathbb{E}[V_n(i)p_n(i)|\mathscr{F}_k] = v_k(i)\mathbb{P}(\tau = i|\mathscr{F}_k)$$

where we made use of the tower property to obtain the first equality. It follows that

$$\mathbb{1}_{\{\tau \leq k\}}\mathbb{E}[V_n(\tau)|\mathscr{G}_k] = \sum_{i=0}^{k} \mathbb{1}_{\{\tau=i\}}\frac{\mathbb{E}[V_n(i)p_n(i)|\mathscr{F}_k]}{p_k(i)}.$$

Proposition 3.19 *For a $\mathbb{G}$-adapted process of the form $Y := y\mathbb{1}_{[\![0,\tau[\![} + y(\tau)\mathbb{1}_{[\![\tau,\infty[\![}$ where y and $y(k)$ are $\mathbb{F}$-adapted processes, one has*

$$\mathbb{E}[Y_n|\mathscr{F}_n] = y_n Z_n + \sum_{k=0}^{n} y_n(k)p_n(k)\,. \tag{30}$$

The process Y is a $\mathbb{G}$-martingale if and only if the following two conditions are satisfied
(a) for any k, the process $\big(y_n(k)p_n(k), n \geq k\big)$ is an $\mathbb{F}$-martingale,
(b) the process $Y^{\mathbb{F}}$ is an $\mathbb{F}$-martingale, where $Y^{\mathbb{F}}_n := \mathbb{E}[Y_n|\mathscr{F}_n]$.

Proof We start by noting that for $Y = y\mathbb{1}_{[\![0,\tau[\![} + y(\tau)\mathbb{1}_{[\![\tau,\infty[\![}$, one has

$$Y_n^{\mathbb{F}} = \mathbb{E}[y_n \mathbb{1}_{\{n<\tau\}} + y_n(\tau)\mathbb{1}_{\{\tau\le n\}}|\mathscr{F}_n] = y_n Z_n + \sum_{k=0}^{n} \mathbb{E}[y_n(\tau)\mathbb{1}_{\{\tau=k\}}|\mathscr{F}_n]$$
$$= y_n Z_n + \sum_{k=0}^{n} y_n(k) p_n(k) .$$

For the necessity, in a first step, we show that we can reduce our attention to the case where Y is u.i. Indeed, let Y be a $\mathbb{G}$-martingale and $(T_j)_{j\ge 0}$ be a $\mathbb{G}$-localizing sequence such that, for each j, the associated stopped martingale $(Y_{n\wedge T_j}, n \ge 0)$ is u.i. Assuming that the result is established for u.i. martingales will prove that the processes in (a) and (b) are martingales up to T_j for each j. Since $T_j \to \infty$ as $j \to \infty$, the result follows.

Assume that Y is a u.i. $\mathbb{G}$-martingale. Then, it is closed and its terminal value is a $\mathscr{G}_\infty$-measurable random variable that one can write as $Y(\tau)$ where for any k, the random variable $Y(k)$ is $\mathscr{F}_\infty$-measurable, and $Y_n = \mathbb{E}[Y(\tau)|\mathscr{G}_n]$.

• Assuming that Y is a $\mathbb{G}$-martingale, one has $\mathbb{E}[Y_n|\mathscr{G}_{n-1}] = Y_{n-1}$; hence, we obtain $\mathbb{E}[Y_n|\mathscr{G}_{n-1}]\mathbb{1}_{\{\tau=k\}} = Y_{n-1}\mathbb{1}_{\{\tau=k\}}$ which, writing $Y_n\mathbb{1}_{\{\tau=k\}} = y_n(k)\mathbb{1}_{\{\tau=k\}}$, leads to, for $k \le n-1$ one obtains

$$\mathbb{E}[y_n(k)\mathbb{1}_{\{\tau=k\}}|\mathscr{G}_{n-1}] = y_{n-1}(k)\mathbb{1}_{\{\tau=k\}} ,$$

and taking conditional expectation w.r.t. $\mathscr{F}_{n-1}$ $\mathbb{E}[y_n(k)p_n(k)|\mathscr{F}_{n-1}] = y_{n-1}(k)p_{n-1}(k)$, which proves the martingale property of $y(k)p(k)$.

If Y is a $\mathbb{G}$-martingale, $Y^{\mathbb{F}}$ is an $\mathbb{F}$-martingale.

• Conversely, assuming (a) and (b), we shall verify that $\mathbb{E}[Y_n|\mathscr{G}_k] = Y_k$ for $k \le n$. Let us first note that

$$\mathbb{E}[Y_n|\mathscr{G}_k] = \mathbb{1}_{\{k<\tau\}}\frac{1}{Z_k}\mathbb{E}[Y_n\mathbb{1}_{\{k<\tau\}}|\mathscr{F}_k] + \mathbb{1}_{\{\tau\le k\}}\mathbb{E}[Y_n\mathbb{1}_{\{\tau\le k\}}|\mathscr{G}_k] .$$

We then compute the two conditional expectations in the above equality:

$$\begin{aligned}\mathbb{E}[Y_n\mathbb{1}_{\{k<\tau\}}|\mathscr{F}_k] &= \mathbb{E}[Y_n|\mathscr{F}_k] - \mathbb{E}[Y_n\mathbb{1}_{\{\tau\le k\}}|\mathscr{F}_k]\\ &= \mathbb{E}[Y_n^{\mathbb{F}}|\mathscr{F}_k] - \mathbb{E}\left[\mathbb{E}[y_n(\tau)\mathbb{1}_{\{\tau\le k\}}|\mathscr{F}_n]|\mathscr{F}_k\right]\\ &= Y_k^{\mathbb{F}} - \mathbb{E}\left[\sum_{j=0}^{k} y_n(j)p_n(j)|\mathscr{F}_k\right]\\ &= y_k Z_k + \sum_{j=0}^{k} y_k(j)p_k(j) - \sum_{j=0}^{k} y_k(j)p_k(j) = y_k Z_k\end{aligned}$$

where we have used the condition (a) and the equality (30) given in condition (b) to obtain the next-to-last identity.

Furthermore, using Lemma 3.18, for $n > k$,

$$\mathbb{E}[Y_n \mathbb{1}_{\{\tau \le k\}} | \mathscr{G}_k] = \mathbb{E}[y_n(\tau) \mathbb{1}_{\{\tau \le k\}} | \mathscr{G}_k] = \mathbb{1}_{\{\tau \le k\}} \frac{1}{p_k(\tau)} \mathbb{E}[y_n(i) p_n(i) | \mathscr{F}_k]_{|_{i=\tau}}$$
$$= \mathbb{1}_{\{\tau \le k\}} \frac{1}{p_k(\tau)} y_k(\tau) p_k(\tau) = \mathbb{1}_{\{\tau \le k\}} y_k(\tau)$$

where the next-to-last identity holds in view of the condition (a).

Finally, $\mathbb{E}[Y_n | \mathscr{G}_k] = \mathbb{1}_{\{k<\tau\}} \frac{1}{Z_k} y_k Z_k + \mathbb{1}_{\{\tau \le k\}} y_k(\tau) = Y_k$.

3.2 *Immersion in Progressive Enlargement*

In this section, $\mathbb{G}$ is the progressive enlargement of $\mathbb{F}$ with a random time τ. We recall that $\mathbb{F}$ is immersed in $\mathbb{G}$ (we shall write $\mathbb{F} \hookrightarrow \mathbb{G}$) if any $\mathbb{F}$-martingale is a $\mathbb{G}$-martingale.

3.2.1 General Results

Lemma 3.20 $\mathbb{F}$ *immersed in* $\mathbb{G}$ *is equivalent to* $Z_n = \mathbb{P}(\tau > n | \mathscr{F}_\infty) = \mathbb{P}(\tau > n | \mathscr{F}_k)$ *for any* $k \ge n \ge 0$.

Proof Assume that $\mathbb{F}$ is immersed in $\mathbb{G}$. From (b) in Proposition 1.58, $Z_n = \mathbb{P}(\tau > n | \mathscr{F}_\infty)$ which implies, by tower property, $Z_n = \mathbb{P}(\tau > n | \mathscr{F}_k)$ for $k \ge n$.

Conversely, assuming $\mathbb{P}(\tau > n | \mathscr{F}_k) = \mathbb{P}(\tau > n | \mathscr{F}_n)$ for $k \ge n$, we will prove that any $\mathbb{F}$-martingale X is a $\mathbb{G}$-martingale, i.e. $\mathbb{E}[X_n | \mathscr{G}_k] = X_k$ for $k \le n$, or equivalently for any $\mathscr{F}_k$ measurable r.v. U_k and any for any j,

$$\mathbb{E}[X_n U_k \mathbb{1}_{\{\tau \wedge k = j\}}] = \mathbb{E}[X_k U_k \mathbb{1}_{\{\tau \wedge k = j\}}].$$

This equality is obvious for $k < j$. For $j \le k$,

$$\mathbb{E}[X_n U_k \mathbb{1}_{\{\tau \wedge k = j\}}] = \mathbb{E}[X_n U_k \mathbb{P}(\tau \wedge k = j | \mathscr{F}_n)] = \mathbb{E}[X_n U_k \mathbb{P}(\tau = j | \mathscr{F}_n)]$$
$$= \mathbb{E}[X_n U_k \mathbb{P}(\tau = j | \mathscr{F}_k)]$$

where we have used the hypothesis in the last equality. It follows that, using the $\mathbb{F}$-martingale property of X

$$\mathbb{E}[X_n U_k \mathbb{P}(\tau = j | \mathscr{F}_k)] = \mathbb{E}[X_k U_k \mathbb{P}(\tau = j | \mathscr{F}_k)] = \mathbb{E}[X_k U_k \mathbb{1}_{\{\tau \wedge k = j\}}].$$

Lemma 3.21 $\mathbb{F}$ *is immersed in* $\mathbb{G}$ *if and only if the process* $\widetilde{Z}$ *is predictable and* $\widetilde{Z}_n = \mathbb{P}(\tau \ge n | \mathscr{F}_\infty)$, $n \ge 0$.

Proof Assume that $\mathbb{F}$ is immersed in $\mathbb{G}$. Then, for $n \ge 0$,

$$\widetilde{Z}_n = \mathbb{P}(\tau \geq n|\mathscr{F}_n) = \mathbb{P}(\tau > n-1|\mathscr{F}_n) = \mathbb{P}(\tau > n-1|\mathscr{F}_\infty) = \mathbb{P}(\tau \geq n|\mathscr{F}_\infty)\,,$$

where the third equality follows from immersion assumption. Applying immersion again $\widetilde{Z}_n = \mathbb{P}(\tau > n-1|\mathscr{F}_n) = \mathbb{P}(\tau > n-1|\mathscr{F}_{n-1})$, which implies

$$\widetilde{Z}_n = \mathbb{P}(\tau > n-1|\mathscr{F}_{n-1}) = Z_{n-1}$$

and establishes the predictability of $\widetilde{Z}$.

Assume now that the process $\widetilde{Z}$ is predictable and that $\widetilde{Z}_n = \mathbb{P}(\tau \geq n|\mathscr{F}_\infty),\ \forall n \geq 0$. It follows that $\widetilde{Z}_n = \mathbb{P}(\tau \geq n|\mathscr{F}_{n-1}),\ \forall n \geq 1$ and

$$\mathbb{P}(\tau > n|\mathscr{F}_n) = \mathbb{P}(\tau \geq n+1|\mathscr{F}_n) = \widetilde{Z}_{n+1} = \mathbb{P}(\tau > n|\mathscr{F}_\infty)\,.$$

The immersion property follows by Lemma 3.20.

Proposition 3.22 *Assume that Z is positive and that immersion holds. Then, one has $Z = Z_0\mathscr{E}(-\Gamma)$ where Γ is defined in* (28).

Proof If immersion holds, the same kind of computation as the one in the proof of Proposition 3.15 leads to

$$\frac{Z_n}{\mathscr{E}(-\Gamma)_n} = \frac{1}{\mathscr{E}(-\Gamma)_{n-1}}\frac{Z_n}{1-\Delta\Gamma_n} = \frac{Z_n}{\mathscr{E}(-\Gamma)_{n-1}}\frac{\widetilde{Z}_n}{\widetilde{Z}_n - \Delta H^o_n} = \frac{\widetilde{Z}_n}{\mathscr{E}(-\Gamma)_{n-1}}$$

and using the fact that immersion implies that $\widetilde{Z}_n = Z_{n-1}$, we obtain, by recursion, that $\frac{Z_n}{\mathscr{E}(-\Gamma)_n} = Z_0$.

Lemma 3.23 *Under immersion, $p_n(k) = p_k(k)$ for any $n \geq k \geq 0$.*

Proof The result is obtained using the same arguments as in the first part of Lemma 3.20.

Theorem 3.24 *Suppose $\mathbb{F} \hookrightarrow \mathbb{G}$ and $Z > 0$. Let N be defined in (27). Then the following assertions are equivalent*
(a) *Z is $\mathbb{F}$-predictable.*
(b) *For any $\mathbb{G}$-predictable process U, one has $\mathbb{E}\Big[(U \centerdot N)_n|\mathscr{F}_n\Big] = 0,\ \forall n \geq 1$, in particular, $\mathbb{E}[\Delta N_n\big|\mathscr{F}_n] = 0,\ \forall n \geq 1$.*
(c) *Any $\mathbb{F}$-martingale X is orthogonal to N.*

Proof (a)$\Rightarrow$ (b). Let $k \geq 1$. By uniqueness of Doob's decomposition and the predictability of Z, one has $Z_k = M_0 - H^{p,\mathbb{F}}_k$, hence $\Delta M_k = 0$.

By Lemmas 3.12 and 3.21, we have that

$$\mathbb{E}[\Delta N_k|\mathscr{F}_k] = -\Delta M_k\,,$$

therefore $\mathbb{E}[\Delta N_k\big|\mathscr{F}_k] = 0$.

For a $\mathbb{G}$-predictable process U, we note u its predictable reduction (see Lemma 3.7). Then, using the fact that $\Delta N_k = \Delta H_k - \lambda_k \mathbb{1}_{\{\tau \geq k\}} = \mathbb{1}_{\{\tau \geq k\}} \Delta N_k$,

$$\mathbb{E}[U_k \Delta N_k | \mathscr{F}_k] = u_k \mathbb{E}[\Delta N_k | \mathscr{F}_k] = 0 .$$

We now take the sum of the two sides of the previous equality over all $k \leq n$, and using the fact that, from immersion $\mathbb{E}[\Delta N_k | \mathscr{F}_k] = \mathbb{E}[\Delta N_k | \mathscr{F}_n]$, we obtain $\mathbb{E}[\sum_{k=1}^n U_k \Delta N_k \big| \mathscr{F}_n] = 0$, which is the desired result.

(b) $\Rightarrow$ (c). In the proof, we suppose that X is a square integrable martingale. The general case follows by localization. To prove the orthogonality of X and N, we show that $\mathbb{E}[\Delta X_n \ \Delta N_n | \mathscr{G}_{n-1}] = 0$ for all $n \geq 1$.

From Lemma 3.6, we have that

$$\mathbb{E}[\Delta X_n \ \Delta N_n | \mathscr{G}_{n-1}] \mathbb{1}_{\{\tau \geq n\}} = \frac{1}{Z_{n-1}} \mathbb{E}[\Delta X_n \mathbb{1}_{\{\tau > n-1\}} \Delta N_n | \mathscr{F}_{n-1}] \mathbb{1}_{\{\tau \geq n\}} ,$$

since ΔX_n is $\mathscr{F}_n$-measurable and $\mathbb{1}_{\{\tau > n-1\}}$ is $\mathscr{G}_{n-1}$-measurable we have, from (b) and using the fact that $\Delta N_n = \mathbb{1}_{\{\tau > n-1\}} \Delta N_n$

$$\mathbb{E}[\Delta X_n \mathbb{1}_{\{\tau > n-1\}} \Delta N_n | \mathscr{F}_{n-1}] = \mathbb{E}\big[\Delta X_n \mathbb{E}[\Delta N_n | \mathscr{F}_n] | \mathscr{F}_{n-1}\big] = 0$$

hence

$$\mathbb{E}[\Delta X_n \Delta N_n | \mathscr{G}_{n-1}] \mathbb{1}_{\{\tau \geq n\}} = 0 .$$

On the set $\{\tau < n\}$, using that $\{\tau < n\} = \{\tau \leq n-1\} \in \mathscr{G}_{n-1}$, we obtain

$$\mathbb{E}[\Delta X_n \ \Delta N_n | \mathscr{G}_{n-1}] \mathbb{1}_{\{\tau < n\}} = \mathbb{E}[\Delta X_n \Delta N_n \mathbb{1}_{\{\tau < n\}} | \mathscr{G}_{n-1}] = 0 .$$

Finally, we get $\mathbb{E}[\Delta X_n \Delta N_n | \mathscr{G}_{n-1}] = 0$.

(c) $\Rightarrow$ (a). By (c), we have in the one hand, for $n \geq 1$, $\mathbb{E}[\Delta X_n \Delta N_n | \mathscr{G}_{n-1}] = 0$, then $\mathbb{E}[\Delta X_n \Delta N_n] = 0$. On the other hand, $\mathbb{E}[\Delta X_n \Delta N_n] = \mathbb{E}[\Delta X_n \mathbb{E}[\Delta N_n | \mathscr{F}_n]]$. In the case $X = M$, applying Lemma 3.12, we obtain $\mathbb{E}[\Delta N_n \Delta M_n] = -\mathbb{E}[|\Delta M_n|^2]$, which implies $\mathbb{E}[|\Delta M_n|^2] = 0$. Therefore $\Delta M_n = 0$, or equivalently $\mathbb{E}[Z_n | \mathscr{F}_{n-1}] = Z_n$, which is equivalent to the predictability of Z.

Example 3.25 We now present the basic model of credit risk, where the bankruptcy time (called the default time) is modelled as follows. Assume that $\tau = \inf\{n : V_n \geq \Theta\}$ where V is an increasing $\mathbb{F}$-adapted process with $V_0 = 0$ and Θ is a random variable independent from $\mathbb{F}$, with an exponential law. Then, using the fact that $\{\tau > n\} = \{V_n < \Theta\}$, we deduce $Z_n = \mathbb{P}(V_n < \Theta | \mathscr{F}_n) = e^{-V_n} = \mathbb{P}(V_n < \Theta | \mathscr{F}_\infty)$ and immersion property holds (and $\widetilde{Z}_n = \mathbb{P}(\tau > n-1 | \mathscr{F}_n) = \mathbb{P}(V_{n-1} < \Theta | \mathscr{F}_n) = e^{-V_{n-1}} = Z_{n-1}$).

If V is predictable, the Doob decomposition of Z is $Z_n = 1 - H_n^{p,\mathbb{F}} = 1 - (1 - e^{-V_n})$, and $Z = \mathscr{E}(-\Lambda)$ where Λ is defined in Lemma 3.9. Note that, from Proposition 3.22, $\Lambda = \Gamma$. Moreover, Z is predictable and assertions of Theorem 3.24 hold.

If V is not predictable,

$$\Delta\Lambda_n = \frac{\Delta H_n^{p,\mathbb{F}}}{Z_{n-1}} = \frac{\mathbb{E}[-\Delta Z_n|\mathscr{F}_{n-1}]}{Z_{n-1}} = e^{-V_{n-1}}\frac{1-\mathbb{E}[e^{-\Delta V_n}|\mathscr{F}_{n-1}]}{Z_{n-1}} = 1 - \mathbb{E}[e^{-\Delta V_n}|\mathscr{F}_{n-1}]$$

and $\Delta\Gamma_n = 1 - e^{-\Delta V_n}$.

Example 3.26 We can extend the previous example to $\tau = \inf\{n : V_n \geq \Psi\}$ where Ψ is a positive random variable such that $\mathbb{P}(\Psi > x|\mathscr{F}_n) = \int_x^\infty \psi_n(u)du$ where $\psi_n(u)$, $n \geq 0$ is a family of $\mathbb{F}$-adapted processes and V is strictly increasing.

Then, for $m \leq n$

$$\mathbb{P}(\tau > m|\mathscr{F}_n) = \mathbb{P}(\Psi > V_m|\mathscr{F}_n) = \int_{V_m}^\infty \psi_n(u)du$$

therefore, for $j \leq n$, $\mathbb{P}(\tau = j|\mathscr{F}_n) = \int_{V_{j-1}}^{V_j} \psi_n(u)du$.

For $j > n$, one has

$$\mathbb{P}(\tau = j|\mathscr{F}_n) = \mathbb{E}[\int_{V_{j-1}}^{V_j} \psi_n(u)du|\mathscr{F}_n]\,.$$

In that case, since $\mathbb{P}(\tau > m|\mathscr{F}_n)$ is not equal to $\mathbb{P}(\tau > m|\mathscr{F}_\infty)$, the filtration $\mathbb{F}$ is not immersed in $\mathbb{G}$.

3.2.2 Equivalent Probability Measures

Proposition 3.27 *Suppose* $\mathbb{F} \overset{\mathbb{P}}{\hookrightarrow} \mathbb{G}$. *Let* $\mathbb{Q}$ *be a probability measure which is locally equivalent to* $\mathbb{P}$ *on* $\mathbb{G}$ *and let* L *be its Radon–Nikodym density. If* L *is* $\mathbb{F}$*-adapted, then*

$$\mathbb{Q}(\tau > n|\mathscr{F}_n) = \mathbb{P}(\tau > n|\mathscr{F}_n) = Z_n\ ,\ \forall n \geq 0$$

and $\mathbb{F} \overset{\mathbb{Q}}{\hookrightarrow} \mathbb{G}$. *Consequently, the predictable compensator of* H *is unchanged under such equivalent changes of probability measures, i.e.* N*, defined in (27), is a* $\mathbb{G}$*-martingale under* $\mathbb{P}$ *and* $\mathbb{Q}$.

Proof Let X be a $(\mathbb{Q}, \mathbb{F})$-martingale, then, L being $\mathbb{F}$-adapted, Girsanov's theorem implies that XL is a $(\mathbb{P}, \mathbb{F})$-martingale, and since $\mathbb{F}$ is immersed in $\mathbb{G}$ under $\mathbb{P}$ we have that XL is a $(\mathbb{P}, \mathbb{G})$-martingale which implies by Girsanov's theorem that X is a $(\mathbb{Q}, \mathbb{G})$-martingale, therefore $\mathbb{F} \overset{\mathbb{Q}}{\hookrightarrow} \mathbb{G}$. We have for each $n \leq k$, using Bayes' formula and the fact that L is $\mathbb{F}$-adapted

$$\mathbb{Q}(\tau \leq n|\mathscr{F}_k) = \frac{\mathbb{E}_{\mathbb{P}}[L_k \mathbb{1}_{\{\tau \leq n\}}|\mathscr{F}_k]}{\mathbb{E}_{\mathbb{P}}[L_k|\mathscr{F}_k]} = \mathbb{P}(\tau \leq n|\mathscr{F}_k) ,$$

in particular, $\mathbb{Q}(\tau \leq n|\mathscr{F}_n) = \mathbb{P}(\tau \leq n|\mathscr{F}_n)$, then using the fact that $\mathbb{F} \overset{\mathbb{P}}{\hookrightarrow} \mathbb{G}$, one has $\mathbb{Q}(\tau \leq n|\mathscr{F}_n) = \mathbb{Q}(\tau \leq n|\mathscr{F}_k)$ and the assertion follows.

The following result shows that immersion property is preserved under change of probability if the Radon–Nikodym is an exponential martingale built with the martingale N and if some assumptions on Z are satisfied.

Proposition 3.28 *Assume that $\mathbb{F} \overset{\mathbb{P}}{\hookrightarrow} \mathbb{G}$ and that Z is $\mathbb{F}$-predictable and positive. Let φ be an integrable $\mathbb{G}$-predictable process such that $1 + \varphi \Delta N > 0$ where N is defined in (27). Define the positive $\mathbb{G}$-martingale*

$$L := \mathscr{E}(\varphi \centerdot N) ,$$

and

$$d\mathbb{Q}_n = L_n d\mathbb{P}_n \quad on \ \mathscr{G}_n , \quad \forall n \geq 0$$

Then, the $\mathbb{Q}$-Azéma supermartingale associated with τ has the following multiplicative decomposition:

$$Z_n^{\mathbb{Q}} = \mathscr{E}\Big(- \bar{\varphi} \centerdot \Lambda \Big)_n Z_n , \quad \forall n \geq 0 , \tag{31}$$

where

- *Λ is defined in Lemma 3.9*
- *$\bar{\varphi}$ is the $\mathbb{F}$-predictable reduction of φ.*

Moreover, immersion property holds under $\mathbb{Q}$ and the process

$$H_n - \sum_{k=1}^{n \wedge \tau} \Big(1 - \frac{Z_k}{Z_{k-1}} \bar{\varphi}_k \Big) \Delta \Lambda_k , \quad \forall n \geq 0,$$

is the compensated $(\mathbb{Q}, \mathbb{G})$-martingale associated with H.

Proof The process L is a positive $\mathbb{G}$-martingale as it is a local martingale (exponential of a local martingale) and is positive by assumption on its jumps. In a first step, we compute the $\mathbb{Q}$-Azéma supermartingale associated with τ

$$Z_n^{\mathbb{Q}} = \mathbb{Q}(\tau > n|\mathscr{F}_n) = \frac{\mathbb{E}_{\mathbb{P}}[\mathbb{1}_{\{\tau > n\}} L_n|\mathscr{F}_n]}{\ell_n}$$

where $\ell_n = \mathbb{E}_{\mathbb{P}}[L_n|\mathscr{F}_n] = \mathbb{E}_{\mathbb{P}}[1 + (L_{-}\varphi \centerdot N)_n|\mathscr{F}_n] = 1$, thanks to Theorem 3.24(b). If $\bar{\varphi}$ is the $\mathbb{F}$-predictable reduction of φ, we get, using the definition of N

$$Z_n^{\mathbb{Q}} = \mathbb{E}_{\mathbb{P}}[\mathbb{1}_{\{\tau>n\}} L_n|\mathscr{F}_n] = \mathbb{E}_{\mathbb{P}}[\mathbb{1}_{\{\tau>n\}} \prod_{k=1}^{n}(1 + \bar{\varphi}_k \Delta N_k)|\mathscr{F}_n]$$

$$= \mathbb{E}_{\mathbb{P}}[\mathbb{1}_{\{\tau>n\}} \prod_{k=1}^{n}(1 - \bar{\varphi}_k \Delta \Lambda_k)|\mathscr{F}_n] = Z_n \prod_{k=1}^{n}(1 - \bar{\varphi}_k \Delta \Lambda_k) = \mathscr{E}\Big(-\bar{\varphi} \centerdot \Lambda\Big)_n Z_n.$$

where the third equality is obtained using the fact that, for $k \leq n$, one has $\mathbb{1}_{\{\tau>n\}}\Delta N_k = -\mathbb{1}_{\{\tau>n\}}\Delta\Lambda_k$. Then, (31) follows. We can remark that as Z is predictable and positive, then $Z^{\mathbb{Q}}$ is also predictable and positive.

Let X be a $(\mathbb{Q}, \mathbb{F})$-martingale. Since $\ell \equiv 1$, $\mathbb{P}$ and $\mathbb{Q}$ coincide on $\mathbb{F}$ and the process X is a $(\mathbb{P}, \mathbb{F})$-martingale, hence, by immersion a $(\mathbb{P}, \mathbb{G})$ martingale. From Girsanov's theorem the process $X^{\mathbb{G}} = X - \frac{1}{L_-} \centerdot \langle X, L\rangle$ is a $\mathbb{G}$-martingale. From the fact that any $\mathbb{F}$-martingale is orthogonal to N, the bracket $\langle X, N\rangle$ is null and $\langle X, L\rangle = \varphi L_- \centerdot \langle X, N\rangle = 0$, which implies that $X = X^{\mathbb{G}}$ is a $(\mathbb{Q}, \mathbb{G})$-martingale.

Since $Z^{\mathbb{Q}}$ is predictable, then

$$\Delta\Lambda_n^{\mathbb{Q}} = \frac{Z_{n-1}^{\mathbb{Q}} - Z_n^{\mathbb{Q}}}{Z_{n-1}^{\mathbb{Q}}} \mathbb{1}_{\{Z_{n-1}^{\mathbb{Q}}>0\}} = (1 - \frac{Z_n}{Z_{n-1}}(1 + \bar{\varphi}_n \Delta\Lambda_n))\mathbb{1}_{\{Z_{n-1}^{\mathbb{Q}}>0\}}$$

Since Z is predictable and positive, $\Delta\Lambda_k = \frac{Z_{k-1} - Z_k}{Z_{k-1}}$ and, on $\{n \leq \tau\}$, we obtain $\Delta\Lambda_n^{\mathbb{Q}} = (1 - \frac{Z_n}{Z_{n-1}}\bar{\varphi}_n)\Delta\Lambda_n$.

In the general case, immersion property is not stable under change of probability.

Proposition 3.29 *Assume that* $\mathbb{F} \overset{\mathbb{P}}{\hookrightarrow} \mathbb{G}$ *. Let* X *be a* $(\mathbb{P}, \mathbb{F})$*-martingale and let* ψ *be an integrable* $\mathbb{G}$*-predictable process such that* $1 + \psi\Delta X > 0$.

For $L := \mathscr{E}(\psi \centerdot X)$ *define*

$$d\mathbb{Q}_n = L_n d\mathbb{P}_n \quad on\ \mathscr{G}_n\,, \quad \forall n \geq 0\,.$$

Then, the $\mathbb{Q}$*-Azéma supermartingale associated with* τ *has the following multiplicative decomposition:*

$$Z_n^{\mathbb{Q}} = \mathscr{E}\Big((\bar{\psi} - {}^p\psi) \centerdot X^{\mathbb{Q}}\Big)_n Z_n\,, \quad \forall n \geq 0\,,$$

where

- ${}^p\psi$ *is the* $\mathbb{F}$*-predictable projection of the process* ψ *under the probability* $\mathbb{Q}$*, i.e.* ${}^p\psi := \psi_0$ *and* ${}^p\psi_n := \mathbb{E}_{\mathbb{Q}}[\psi_n|\mathscr{F}_{n-1}]$*, for all* $n \geq 1$,
- $\bar{\psi}$ *is the* $\mathbb{F}$*-predictable reduction of* ψ,

- $X^{\mathbb{Q}}$ *is the* $(\mathbb{Q}, \mathbb{F})$*-martingale defined by* $X_0^{\mathbb{Q}} := X_0$ *and* $X_n^{\mathbb{Q}} := X_n - \sum_{k=1}^{n} \frac{\Delta[X, \ell]_k}{\ell_k}$ *with* $\ell_k = \mathbb{E}_{\mathbb{P}}[L_k|\mathscr{F}_k]$, *for all* $k \geq 0$.

Furthermore, the process N *defined in (27) as the compensated* $\mathbb{P}$*-martingale associated with* H *is equal to the compensated* $\mathbb{Q}$*-martingale associated with* H.

If the process ψ *is* $\mathbb{F}$*-predictable,* $Z^{\mathbb{Q}} = Z$ *and the immersion property holds under* $\mathbb{Q}$.

Proof In a first step, we compute $\mathbb{E}_{\mathbb{P}}[\mathbb{1}_{\{\tau>n\}}L_n|\mathscr{F}_n]$, for $n \geq 0$.

$$\begin{aligned}\mathbb{E}_{\mathbb{P}}[\mathbb{1}_{\{\tau>n\}}L_n|\mathscr{F}_n] &= \mathbb{E}_{\mathbb{P}}[\mathbb{1}_{\{n<\tau\}}\prod_{k=1}^{n}(1+\bar{\psi}_k\Delta X_k)|\mathscr{F}_n] \\ &= \mathbb{E}_{\mathbb{P}}[\mathbb{1}_{\{n<\tau\}}|\mathscr{F}_n]\prod_{k=1}^{n}(1+\bar{\psi}_k\Delta X_k) = Z_n\mathscr{E}\big(\bar{\psi}\centerdot X\big)_n . \quad (32)\end{aligned}$$

Moreover,

$$\ell_n := \mathbb{E}_{\mathbb{P}}[L_n|\mathscr{F}_n] = \mathbb{E}_{\mathbb{P}}[L_{n-1}(1+\psi_n\Delta X_n)|\mathscr{F}_n] = \ell_{n-1} + \mathbb{E}_{\mathbb{P}}[L_{n-1}\psi_n|\mathscr{F}_n]\Delta X_n .$$

As $L_{n-1}\psi_n$ is $\mathscr{G}_{n-1}$-measurable, immersion property 1.58(b) implies that

$$\mathbb{E}_{\mathbb{P}}[L_{n-1}\psi_n|\mathscr{F}_n] = \mathbb{E}_{\mathbb{P}}[L_{n-1}\psi_n|\mathscr{F}_{n-1}] .$$

From Bayes' formula, $\mathbb{E}_{\mathbb{P}}[L_{n-1}\psi_n|\mathscr{F}_{n-1}] = \mathbb{E}_{\mathbb{Q}}[\psi_n|\mathscr{F}_{n-1}]\ell_{n-1} = {}^p\psi_n\ell_{n-1}$, hence

$$\ell_n = \ell_{n-1}(1+{}^p\psi_n\Delta X_n) = \mathscr{E}\big({}^p\psi\centerdot X\big)_n . \tag{33}$$

Then combining (32) and (33), we obtain that

$$Z^{\mathbb{Q}} = \frac{\mathscr{E}(\bar{\psi}\centerdot X)}{\mathscr{E}({}^p\psi\centerdot X)} Z .$$

To end the proof, we now show that $Z^{\mathbb{Q}} = \mathscr{E}\Big((\bar{\psi} - {}^p\psi)\centerdot X^{\mathbb{Q}}\Big)Z$, or that

$$\mathscr{E}(\bar{\psi}\centerdot X) = \mathscr{E}({}^p\psi\centerdot X)\mathscr{E}\Big((\bar{\psi} - {}^p\psi)\centerdot X^{\mathbb{Q}}\Big) .$$

From Lemma 1.27, one notes that the right-hand side is

$$\mathscr{E}({}^p\psi\centerdot X + (\bar{\psi} - {}^p\psi)\centerdot X^{\mathbb{Q}} + {}^p\psi(\bar{\psi} - {}^p\psi)\centerdot[X, X^{\mathbb{Q}}]\Big) := \mathscr{E}(K) .$$

Some algebra, based on the definition of $X^{\mathbb{Q}}$ and (33), which implies that $1 - \frac{\Delta \ell}{\ell} = \frac{1}{1+{}^p\psi \Delta X}$, leads to $\Delta K = \bar{\psi} \centerdot \Delta X$. The martingale N remains a $\mathbb{Q}$-martingale as N and X are orthogonal (see Theorem 3.24).

We just apply Girsanov's theorem with the predictable bracket.

If the process ψ is $\mathbb{F}$-predictable, the process L is $\mathbb{F}$-adapted and Proposition 3.27 implies immersion property.

We are now able to give a result in our discrete time setting analogous to the one obtained in a continuous time setting in [12, Th. 6.32].

Theorem 3.30 *Assume that* $\mathbb{F} \overset{\mathbb{P}}{\hookrightarrow} \mathbb{G}$ *and that* Z *is* $\mathbb{F}$*-predictable and positive. Let* X *be a* $(\mathbb{P}, \mathbb{F})$*-martingale,* ψ *and* φ *be two integrable* $\mathbb{G}$*-predictable processes satisfying* $1 + \psi \Delta X > 0$ *and* $1 + \varphi \Delta N > 0$ *and*

$$L := \mathcal{E}(\psi \centerdot X)\mathcal{E}(\varphi \centerdot N) ,$$

where N *is defined in (27).*

Define

$$d\mathbb{Q}_n = L_n d\mathbb{P}_n \quad on\ \mathcal{G}_n , \quad \forall n \geq 0 .$$

Then, the $\mathbb{Q}$*-Azéma supermartingale associated with* τ *has the following multiplicative decomposition:*

$$Z_n^{\mathbb{Q}} = \mathcal{E}\Big((\bar{\psi} - {}^p\psi) \centerdot X^{\mathbb{Q}}\Big)_n \mathcal{E}\Big(-\varphi \centerdot \Lambda\Big)_n Z_n , \quad \forall n \geq 0 ,$$

where

- Λ *is defined in Lemma 3.9*
- ${}^p\psi$ *is the* $\mathbb{F}$*-predictable projection of the process* ψ *under the probability* $\mathbb{Q}$*, i.e.* ${}^p\psi := \psi_0$ *and* ${}^p\psi_n := \mathbb{E}^{\mathbb{Q}}[\psi_n | \mathcal{F}_{n-1}]$*, for all* $n \geq 1$*,*
- $\bar{\psi}$ *is the* $\mathbb{F}$*-predictable reduction of* ψ*, and*
- $X^{\mathbb{Q}}$*, defined by* $X_0^{\mathbb{Q}} := X_0$ *and* $X_n^{\mathbb{Q}} := X_n - \sum_{k=1}^{n} \frac{\Delta[X, \ell]_k}{\ell_k}$ *for all* $n \geq 0$*, is a* $(\mathbb{Q}, \mathbb{F})$*-martingale with* $\ell_k = \mathbb{E}[L_k | \mathcal{F}_k]$*, for all* $k \geq 0$*.*

Furthermore, the process

$$H_n - \sum_{k=1}^{n \wedge \tau} \Big(1 - \frac{Z_k}{Z_{k-1}} \varphi_k\Big) \Delta \Lambda_k , \quad \forall n \geq 0,$$

is the compensated $\mathbb{Q}$*-martingale associated with* H*. In particular, if the process* ψ *is* $\mathbb{F}$*-predictable, then*

$$Z_n^{\mathbb{Q}} = \mathcal{E}(-\varphi \centerdot \Lambda)_n Z_n \ , \quad \forall n \geq 0$$

and the immersion property holds under $\mathbb{Q}$.

Proof Let $n \geq 0$ be fixed. Note that L is a martingale: this is a local martingale by orthogonality of X and N (due to Theorem 3.24) and a martingale by Proposition 1.37.

The result is obtained combining Propositions 3.29 and 3.28.

Corollary 3.31 *Suppose that* $\mathbb{F} \overset{\mathbb{P}}{\hookrightarrow} \mathbb{G}$ *and that* Z *is positive and* $\mathbb{F}$*-predictable. Define* $\mathbb{Q}_n$ *on* $\mathcal{G}_n$ *by*

$$d\mathbb{Q}_n = \mathcal{E}(\psi \centerdot X)_n d\mathbb{P}_n \ , \quad \forall n \geq 0 \ ,$$

with X *an* $\mathbb{F}$*-martingale and* ψ *a* $\mathbb{G}$*-predictable process such that* $1 + \psi \Delta X > 0$. *Then, the process* $N = H - \Lambda^\tau$ *is a* $(\mathbb{Q}, \mathbb{G})$*-martingale.*

Proof It suffices to take $\varphi = 0$ in Theorem 3.30.

Comment 3.32 (a) Theorem 3.24, Theorem 3.30, and Corollary 3.31 are the discrete version of Lemma 5.1, Theorem 6.4, and Corollary 6.5 in [12]. In continuous time the results holds under Assumption (A): the random time τ avoids every $\mathbb{F}$-stopping time T, i.e. $\mathbb{P}(\tau = T) = 0$, but in discrete time Assumption (A) is never true and it is replaced by Z positive and predictable.
(b) If $\mathcal{E}\Big((\bar{\psi} - {}^p\psi) \centerdot X^{\mathbb{Q}}\Big)$ is not equal to 1, immersion does not hold. This is the case when $(\bar{\psi} - {}^p\psi) \neq 0$.

4 Progressive Enlargement Before τ

We continue the study of progressive enlargement. We are now interested with the behaviour of $\mathbb{F}$-martingales, when they are considered as $\mathbb{G}$-semimartingales. We split the study into two cases: before τ and after τ. In continuous time, the case before τ is easy and there is a general answer: any $\mathbb{F}$-martingale stopped at time τ is a $\mathbb{G}$-semimartingale, and the decomposition is known. After τ, only partial answers are known to insure that $\mathbb{F}$-martingales remain $\mathbb{G}$-semimartingales.

4.1 Semimartingale Decomposition

We give in this section several $\mathbb{G}$-semimartingale decompositions for an $\mathbb{F}$-martingale X stopped at τ.

Proposition 4.1 *Any square integrable $\mathbb{F}$-martingale X stopped at τ is a $\mathbb{G}$-semimartingale with decomposition*

$$X_n^\tau = X_n^{\mathbb{G}} + \sum_{k=1}^{n\wedge\tau} \frac{1}{Z_{k-1}} \Delta\langle \widetilde{M}, X\rangle_k^{\mathbb{F}},$$

where $X^{\mathbb{G}}$ is a $\mathbb{G}$-martingale (stopped at τ). Here, $\widetilde{M}$ is the martingale part of the Doob decomposition of the supermartingale $\widetilde{Z}$ defined in (18).

Proof We compute, on the set $\{0 \le n < \tau\}$, the predictable part of the $\mathbb{G}$-semimartingale X using Lemma 3.6:

$$\mathbb{1}_{\{\tau>n\}}\mathbb{E}[\Delta X_{n+1}|\mathscr{G}_n] = \mathbb{1}_{\{\tau>n\}}\frac{1}{Z_n}\mathbb{E}[\widetilde{Z}_{n+1}\Delta X_{n+1}|\mathscr{F}_n].$$

Using now the Doob decomposition of $\widetilde{Z}$, and the martingale property of X, we obtain

$$\begin{aligned}\mathbb{E}[\widetilde{Z}_{n+1}\Delta X_{n+1}|\mathscr{F}_n] &= \mathbb{E}[(\widetilde{M}_{n+1} - \widetilde{A}_{n+1})\Delta X_{n+1}|\mathscr{F}_n] \\ &= \mathbb{E}[\widetilde{M}_{n+1}\Delta X_{n+1}|\mathscr{F}_n] = \Delta\langle \widetilde{M}, X\rangle_{n+1}^{\mathbb{F}}\end{aligned}$$

and finally

$$\mathbb{1}_{\{\tau>n\}}\mathbb{E}[\Delta X_{n+1}|\mathscr{G}_n] = \mathbb{1}_{\{\tau>n\}}\frac{1}{Z_n}\Delta\langle \widetilde{M}, X\rangle_{n+1}^{\mathbb{F}}.$$

The result follows.

Remark 4.2 It seems important to note that the Doob decomposition of Z is not needed. Indeed, Eq. (23) implies that Z admits the optional decomposition $Z = \widetilde{M} - H^{o,\mathbb{F}}$ and hence, $\widetilde{M}$ can be viewed as the martingale part of this optional decomposition. This "explains" why, in continuous time, such an optional decomposition of Z is required. However, since optional decompositions are not unique, we prefer, in discrete time, to refer to $\widetilde{M}$ as the martingale part of the (unique) Doob decomposition of $\widetilde{Z}$.

Remark 4.3 If a filtration $\mathbb{K}$ is such that $\mathbb{F} \subset \mathbb{G} \subset \mathbb{K}$ and, for any $\mathbb{K}$ predictable process U, there exists an $\mathbb{F}$-predictable process u such that $U_n\mathbb{1}_{\{n\le\tau\}} = u_n\mathbb{1}_{\{n\le\tau\}}$, an immediate extension of Lemma 3.6 shows that any $\mathbb{F}$-martingale stopped at time τ is a $\mathbb{K}$-semimartingale with the same decomposition as the one presented in Proposition 4.1.

Comment 4.4 From the result of Proposition 4.1, we can hope that, in continuous time the $\mathbb{G}$-semimartingale decomposition formula of an $\mathbb{F}$-martingale X stopped at time τ will be

$$X_t^\tau = \widehat{X}_t + \int_0^{t\wedge\tau} \frac{1}{Z_{s-}} d\langle \widetilde{M}, X\rangle_s^{\mathbb{F}},$$

with $Z_t = \mathbb{P}(\tau > t|\mathscr{F}_t)$, $\widetilde{Z}_t = \mathbb{P}(\tau \geq t|\mathscr{F}_t) = \widetilde{M}_t - H^{o,\mathbb{F}}_{t-}$ where $\widetilde{M}$ is an $\mathbb{F}$-martingale and the process $\widehat{X}$ is a $\mathbb{G}$-martingale. This is indeed the case and known as the Jeulin formula (see Comment 4.6). Note that, as $\widetilde{Z}$ is not càdlàg, the decomposition $\widetilde{Z} = \widetilde{M} - H^{o,\mathbb{F}}_{-}$ is not the standard Doob–Meyer decomposition established only for càdlàg supermartingales.

In the two following propositions, we present the discrete time version of well-known decomposition in continuous time. The first result (Proposition 4.5) is a predictable decomposition, the second one (Proposition 4.7) an "optional" decomposition. This optional decomposition is important to solve arbitrages problem and optional decomposition of $\mathbb{G}$-martingales.

Proposition 4.5 *Let X be an $\mathbb{F}$-martingale and M be the martingale part of the Doob decomposition of Z. Then*

$$X_n^\tau = X_n^{\mathbb{G}} + \sum_{k=1}^{n\wedge\tau} \frac{1}{Z_{k-1}} \Big(\Delta\langle M, X\rangle_k^{\mathbb{F}} + (\Delta X_\tau \, \mathbb{1}_{[\![\tau,\infty[\![})_k^{p,\mathbb{F}} \Big),$$

where $X^{\mathbb{G}}$ is a $\mathbb{G}$-martingale.

Proof The process $(\Delta X_\tau \, \mathbb{1}_{[\![\tau,\infty[\![})^{p,\mathbb{F}}$ is the dual $\mathbb{F}$-predictable projection of the process $\Delta X_\tau \, \mathbb{1}_{[\![\tau,\infty[\![} = (\mathbb{1}_{\{k\geq\tau\}}\Delta X_\tau,\ k \geq 0)$, which is null strictly before τ and equal to ΔX_τ after τ. In the proof of Proposition 4.1, we can note that

$$\mathbb{E}[\widetilde{Z}_{n+1}\Delta X_{n+1}|\mathscr{F}_n] = \mathbb{E}[Z_{n+1}\Delta X_{n+1}|\mathscr{F}_n] + \mathbb{E}[\mathbb{1}_{\{\tau=n+1\}}\Delta X_{n+1}|\mathscr{F}_n]$$

and that the decomposition can be written as stated, following the proof of Proposition 4.1.

Note that, by uniqueness of the Doob decomposition, the process $X^{\mathbb{G}}$ in Proposition 4.5 is the same as the one of Proposition 4.1.

Comment 4.6 In continuous time (see Jeulin [19, Proposition 4.16] and Jeulin and Yor [20, Théorème 1, pp. 87–88]), one has that every càdlàg $\mathbb{F}$-local martingale X stopped at time τ is a $\mathbb{G}$-special semimartingale with canonical decomposition

$$X_t^\tau = X_t^{\mathbb{G}} + \int_0^{t\wedge\tau} \frac{d\langle X, M\rangle_s^{\mathbb{F}} + dJ_s}{Z_{s-}}$$

where $X^{\mathbb{G}}$ is a $\mathbb{G}$-local martingale, M is the martingale part of the $\mathbb{F}$-Doob Meyer decomposition of Z and J is the $\mathbb{F}$-dual predictable projection of the process $(\Delta X_\tau)\mathbb{1}_{]\!]\tau,\infty]\!]}$.

We introduce two $\mathbb{F}$-stopping times which play an important role in the following optional decomposition. Let $R := \inf\{n \geq 0,\ Z_n = 0\}$ and $\widetilde{R} := R\mathbb{1}_{\{\widetilde{Z}_R=0<Z_{R-}\}} + \infty\mathbb{1}_{\{\widetilde{Z}_R=0<Z_{R-}\}^c}$.

Proposition 4.7 *Any $\mathbb{F}$-local martingale X stopped at τ admits the following optional decomposition:*

$$X_n^\tau = \widehat{X}_n^{\mathbb{G}} + \sum_{k=1}^{n\wedge\tau} \frac{1}{\widetilde{Z}_k} \Delta[\widetilde{M}, X]_k + \sum_{k=1}^{n\wedge\tau} \left(\Delta X_{\widetilde{R}} \, \mathbb{1}_{[\![\widetilde{R},\infty[\![}\right)_k^{p,\mathbb{F}}, \tag{34}$$

where $\widehat{X}^{\mathbb{G}}$ is a $\mathbb{G}$-martingale and $\widetilde{M}$ is the martingale part of the Doob decomposition of the supermartingale $\widetilde{Z}$ defined in (18).

Proof The proof is inspired of the one in continuous time given in [4]. We give it for the ease of the reader. We first remark that, for any $\mathbb{F}$-martingale Y, one has $\mathbb{E}[Y_\tau \mathbb{1}_{\{\tau<\infty\}}] = \mathbb{E}[[Y, \widetilde{M}]_\infty]$. Indeed

$$\begin{aligned}\mathbb{E}[Y_\tau] &= \sum_{k\geq 0} \mathbb{E}[Y_k \mathbb{1}_{\{\tau=k\}}] = \sum_{k\geq 0} \mathbb{E}[Y_k \Delta\widetilde{Z}_k] \\ &= \sum_{k\geq 0} \mathbb{E}[\Delta Y_k \Delta\widetilde{Z}_k] = \mathbb{E}[\,[Y, \widetilde{M}]_\infty]\,.\end{aligned}$$

The second equality is a consequence of the tower property, and the two last equalities are obtained using that Y is an $\mathbb{F}$-martingale.

We now prove that $\widehat{X}^{\mathbb{G}}$ defined by (34) is a $\mathbb{G}$-martingale, or, equivalently that for all $\mathbb{G}$-predictable process K, one has $\mathbb{E}[(K \centerdot \widehat{X}^{\mathbb{G}})_\infty] = 0$ (cf. Proposition 1.16). For any $\mathbb{G}$-predictable process K, we note $\widetilde{K}$ its $\mathbb{F}$-predictable reduction which satisfies $\widetilde{K}\mathbb{1}_{\{Z_-=0\}} = 0$. Then

$$\begin{aligned}\mathbb{E}[(K \centerdot X^\tau)_\infty] &= \mathbb{E}[(\widetilde{K} \centerdot X)_\tau] = \mathbb{E}[[\widetilde{K} \centerdot X, \widetilde{M}]_\infty] = \mathbb{E}\left[\sum_{k\geq 1} \widetilde{K}_k \Delta[X, \widetilde{M}]_k\right] \\ &= \mathbb{E}\left[\sum_{k\geq 1} \widetilde{K}_k \widetilde{Z}_k (\widetilde{Z}_k)^{-1} \mathbb{1}_{\{\widetilde{Z}_k>0\}} \Delta[X, \widetilde{M}]_k\right] + \mathbb{E}\left[\sum_{k\geq 1} \widetilde{K}_k \mathbb{1}_{\{\widetilde{Z}_k=0\}} \Delta[X, \widetilde{M}]_k\right].\end{aligned} \tag{35}$$

As $\widetilde{K}$, $\Delta[X, \widetilde{M}]$, $\widetilde{Z}$ and $\widetilde{M}$ are $\mathbb{F}$-adapted, one has, by tower property

$$\mathbb{E}\left[\sum_{k\geq 1} \widetilde{K}_k \widetilde{Z}_k \,(\widetilde{Z}_k)^{-1} \mathbb{1}_{\{\widetilde{Z}_k>0\}} \Delta[X, \widetilde{M}]_k\right] = \mathbb{E}\left[\sum_{k\geq 1} \widetilde{K}_k \mathbb{1}_{\{k\leq\tau\}} (\widetilde{Z}_k)^{-1} \mathbb{1}_{\{\widetilde{Z}_k>0\}} \Delta[X, \widetilde{M}]_k\right].$$

From Proposition 3.5, we obtain that the first term of the right-hand side of (35) is equal to $\mathbb{E}\left[\sum_{k\geq 1} K_k \mathbb{1}_{\{k\leq\tau\}} (\widetilde{Z}_k)^{-1} \Delta[X, \widetilde{M}]_k\right]$.

For the second term of the right-hand side of (35), we remark that, due to the choice of $\widetilde{K}$, one has

$$\mathbb{E}\left[\sum_{k\geq 1}\widetilde{K}_k\mathbb{1}_{\{\widetilde{Z}_k=0\}}\Delta[X,\widetilde{M}]_k\right]=\mathbb{E}\left[\sum_{k\geq 1}\widetilde{K}_k\mathbb{1}_{\{\widetilde{Z}_k=0<Z_{k-1}\}}\Delta[X,\widetilde{M}]_k\right]$$

and that $\{\widetilde{Z}=0<Z_-\}=[\![\widetilde{R}]\!]$ and $\Delta\widetilde{M}_{\widetilde{R}}=-Z_{\widetilde{R}}$ on $\{\widetilde{R}<\infty\}$. Then

$$\mathbb{1}_{\{\widetilde{Z}=0<Z_-\}}\Delta[X,\widetilde{M}]=-Z_{\widetilde{R}_-}\,\Delta X_{\widetilde{R}}\mathbb{1}_{[\![\widetilde{R},\infty[\![}.$$

Hence

$$\mathbb{E}\left[\sum_{k\geq 1}\widetilde{K}_k\mathbb{1}_{\{\widetilde{Z}_k=0<Z_{k-1}\}}\Delta[X,\widetilde{M}]_k\right]=-\mathbb{E}\left[\sum_{k\geq 1}\widetilde{K}_kZ_{k-1}(\Delta X_{\widetilde{R}}\,\mathbb{1}_{[\![\widetilde{R},\infty[\![})_k\right].$$

Using that $\widetilde{K}Z_-$ is predictable, we obtain

$$\mathbb{E}\left[\sum_{k\geq 1}\widetilde{K}_k\mathbb{1}_{\{\widetilde{Z}_k=0\}}\Delta[X,\widetilde{M}]_k\right]=-\mathbb{E}\left[\sum_{k\geq 1}\widetilde{K}_kZ_{k-1}\left(\Delta X_{\widetilde{R}}\mathbb{1}_{[\![\widetilde{R},\infty[\![}\right)_k^{p,\mathbb{F}}\right]$$

$$=-\mathbb{E}\left[\sum_{k\geq 1}K_k\mathbb{1}_{\{k-1<\tau\}}\left(\Delta X_{\widetilde{R}}\,\mathbb{1}_{[\![\widetilde{R},\infty[\![}\right)_k^{p,\mathbb{F}}\right]=-\mathbb{E}\left[\sum_{k=1}^{\tau}K_k\left(\Delta X_{\widetilde{R}}\,\mathbb{1}_{[\![\widetilde{R},\infty[\![}\right)_k^{p,\mathbb{F}}\right].$$

It now follows that $\mathbb{E}[(K\centerdot\widehat{X}^{\mathbb{G}})_\infty]=0$.

4.2 Viability

In the case of progressive enlargement, we distinguish arbitrages which can occur before τ and those which can occur after τ. In this subsection, we study the situation before τ.

Definition 4.8 We denote by $\mathbb{G}^\tau$ the filtration $\mathcal{G}^\tau_n=\mathcal{G}_{\tau\wedge n}$, $n\geq 0$. The enlargement $(\mathbb{F},\mathbb{G}^\tau)$ is viable if there exists a positive $\mathbb{G}$-martingale L such that, for any $\mathbb{F}$-martingale X, the process LX^τ is a $\mathbb{G}^\tau$-martingale.

Lemma 4.9 *Let $\mathbb{G}^{\tau-}$ be the filtration $\mathbb{G}$ "strictly before τ", i.e. $\mathcal{G}^{\tau-}_n=\mathcal{G}_{n\wedge(\tau-1)}=\mathcal{G}^{\tau-1}_n$. There exists a positive $\mathbb{G}$-martingale L such that, for any $\mathbb{F}$-martingale X, the process $LX^{\tau-}$ is a $\mathbb{G}^{\tau-}$-martingale, where $X^{\tau-}_n=X_{n\wedge(\tau-1)}$.*

Proof The proof is based on Lemma 1.60. For any $\mathbb{F}$-martingale X, we are looking for a positive $\mathbb{F}$-adapted process ψ, such that on the set $\{1\leq n<\tau\}$ (strictly before τ)

$$\mathbb{1}_{\{n-1\leq\tau\}}\mathbb{E}[\psi_nX_n|\mathcal{G}_{n-1}]=\mathbb{1}_{\{n-1\leq\tau\}}X_{n-1}\mathbb{E}[\psi_n|\mathcal{G}_{n-1}]$$

that is

$$\mathbb{1}_{\{n-1\le\tau\}}\frac{1}{\widetilde{Z}_{n-1}}\mathbb{E}[\psi_n X_n Z_n|\mathscr{F}_{n-1}] = \mathbb{1}_{\{n-1\le\tau\}}\frac{X_{n-1}}{\widetilde{Z}_{n-1}}\mathbb{E}[\psi_n Z_n|\mathscr{F}_{n-1}]\,.$$

This equality holds if

$$\mathbb{E}[\psi_n X_n Z_n|\mathscr{F}_{n-1}] = X_{n-1}\mathbb{E}[\psi_n Z_n|\mathscr{F}_{n-1}]\,.$$

The choice $\psi = (1/Z)\mathbb{1}_{\{Z>0\}} + \mathbb{1}_{\{Z=0\}}$ provides a solution, valid for any martingale X.

Theorem 4.10 *Assume that τ is not an $\mathbb{F}$-stopping time. Then, the enlargement $(\mathbb{F}, \mathbb{G}^\tau)$ is viable if and only if, for any n, the set $\{0 = \widetilde{Z}_n < Z_{n-1}\}$ is negligible.*

This result was established in Choulli and Deng [11] and is a particular case of the general results obtained in Aksamit et al. [5]. We give here a slightly different proof, in the two following propositions.

Proposition 4.11 *Assume that for any n, the set $\{\widetilde{Z}_n = 0 < Z_{n-1}\}$ is negligible. Then the process Y defined by $\Delta Y_k = \mathbb{1}_{\{\tau\ge k\}}(\frac{Z_{k-1}}{\widetilde{Z}_k} - 1)$ for $k \ge 1$ and $Y_0 = 0$ is a $\mathbb{G}$-martingale. If X is an $\mathbb{F}$-martingale and $L = \mathscr{E}(Y)$, the process $X^\tau L$ is a $(\mathbb{P}, \mathbb{G})$-martingale.*

Proof Note that the fact that $\{Z_{n-1} = 0\} \subset \{\widetilde{Z}_n = 0\}$ implies that the inclusion $\{\widetilde{Z}_n = 0\} \subset \{Z_{n-1} = 0\}$ is equivalent to $\{\widetilde{Z}_n = 0\} = \{Z_{n-1} = 0\}$ or to $\{\widetilde{Z}_n = 0 < Z_{n-1}\}$ is negligible. We now prove that the process Y is a $\mathbb{G}$-martingale. Indeed, for $n \ge 1$,

$$\begin{aligned}
\mathbb{E}[\Delta Y_n|\mathscr{G}_{n-1}] &= \mathbb{E}[\mathbb{1}_{\{\tau\ge n\}}\frac{Z_{n-1} - \widetilde{Z}_n}{\widetilde{Z}_n}|\mathscr{G}_{n-1}] \\
&= \mathbb{1}_{\{\tau\ge n\}}\frac{1}{Z_{n-1}}\mathbb{E}[\mathbb{1}_{\{\widetilde{Z}_n>0\}}(Z_{n-1} - \widetilde{Z}_n)|\mathscr{F}_{n-1}] \\
&= \mathbb{1}_{\{\tau\ge n\}}\frac{1}{Z_{n-1}}\mathbb{E}[Z_{n-1} - \widetilde{Z}_n - \mathbb{1}_{\{\widetilde{Z}_n=0\}}(Z_{n-1} - \widetilde{Z}_n)|\mathscr{F}_{n-1}] \\
&= \mathbb{1}_{\{\tau\ge n\}}\frac{1}{Z_{n-1}}\mathbb{E}[Z_{n-1} - \widetilde{Z}_n|\mathscr{F}_{n-1}] = 0\,,
\end{aligned}$$

where we have used (26), the fact that $\mathbb{E}[\widetilde{Z}_n|\mathscr{F}_{n-1}] = Z_{n-1}$ and that, by assumption $\{\widetilde{Z}_n = 0\} \subset \{Z_{n-1} = 0\}$, hence $\mathbb{1}_{\{\widetilde{Z}_n=0\}}(Z_{n-1} - \widetilde{Z}_n) = 0$.

Hence, L is a martingale (see Proposition 1.24). On the set $\{\tau \ge k\}$, one has $Z_{k-1} > 0$ which implies that $\Delta Y_k = \frac{Z_{k-1}}{\widetilde{Z}_k} - 1 > -1$ for any k; therefore, from Proposition 1.24, L is positive. Furthermore, for X an $\mathbb{F}$-martingale, since $L_{n+1} = L_n(1 + \Delta Y_{n+1})$,

$$\mathbb{E}[X^\tau_{n+1}\frac{L_{n+1}}{L_n}|\mathscr{G}_n] = \mathbb{E}[X_{(n+1)\wedge\tau}(1+\mathbb{1}_{\{\tau\geq n+1\}}\frac{Z_n-\widetilde{Z}_{n+1}}{\widetilde{Z}_{n+1}})|\mathscr{G}_n]$$
$$= \mathbb{E}[X_{n+1}\mathbb{1}_{\{\tau\geq n+1\}}\frac{Z_n}{\widetilde{Z}_{n+1}}|\mathscr{G}_n] + \mathbb{E}[X_\tau\mathbb{1}_{\{\tau<n+1\}}|\mathscr{G}_n].$$

From Lemma 3.6

$$\mathbb{E}[X^\tau_{n+1}\frac{L_{n+1}}{L_n}|\mathscr{G}_n] = \mathbb{E}[X_{n+1}\mathbb{1}_{\{\tau\geq n+1\}}\frac{Z_n}{\widetilde{Z}_{n+1}}|\mathscr{G}_n] + X_\tau\mathbb{1}_{\{\tau<n+1\}}$$
$$= \mathbb{1}_{\{\tau>n\}}\frac{1}{Z_n}\mathbb{E}[X_{n+1}Z_n\mathbb{1}_{\{\widetilde{Z}_{n+1}>0\}}|\mathscr{F}_n] + X_\tau\mathbb{1}_{\{\tau\leq n\}}$$
$$= \mathbb{1}_{\{\tau>n\}}\frac{1}{Z_n}\mathbb{E}[X_{n+1}Z_n(1-\mathbb{1}_{\{\widetilde{Z}_{n+1}=0\}})|\mathscr{F}_n] + X_\tau\mathbb{1}_{\{\tau\leq n\}}$$
$$= \mathbb{1}_{\{\tau>n\}}\frac{1}{Z_n}\mathbb{E}[X_{n+1}Z_n|\mathscr{F}_n] + X_\tau\mathbb{1}_{\{\tau\leq n\}} = X_{n\wedge\tau}\,,$$

where we have used that, by assumption, $Z_n\mathbb{1}_{\{\widetilde{Z}_{n+1}=0\}} = 0$. Hence, the universal deflator property of L.

Remark 4.12 In case of immersion, there are no arbitrages (indeed any e.m.m. in $\mathbb{F}$ will be an e.m.m. in $\mathbb{G}$). This can be also obtained using the previous result, since, under immersion hypothesis, one has $Z_{n-1} = \widetilde{Z}_n$, and the universal deflator is equal to 1.

Proposition 4.13 *If there exists $n \geq 1$ such that the set $\{0 = \widetilde{Z}_n < Z_{n-1}\}$ is not negligible, and if τ is not an $\mathbb{F}$-stopping time, there exists an $\mathbb{F}$-martingale X such that X^τ is a $\mathbb{G}$-adapted increasing process with $X^\tau_0 = 1$, $\mathbb{P}(X^\tau_\tau > 1) > 0$. Hence, the enlargement $(\mathbb{F}, \mathbb{G}^\tau)$ is not viable.*

Proof The proof is the discrete time version of the one in Acciaio et al. [1, Th.1.4].

Let $\vartheta = \inf\{n : 0 = \widetilde{Z}_n < Z_{n-1}\}$. The random time ϑ is an $\mathbb{F}$-stopping time satisfying $\tau \leq \vartheta$ and $\mathbb{P}(\tau < \vartheta) > 0$ (as $\widetilde{Z}_\tau$ is positive). Being increasing, the process $I_n = \mathbb{1}_{\{\vartheta\leq n\}}$ is a submartingale, and we denote by D the $\mathbb{F}$-predictable increasing process part of its Doob decomposition. One has $D_0 = 0$ and $\Delta D_n = \mathbb{P}(\vartheta = n|\mathscr{F}_{n-1}) \leq 1$. We introduce an $\mathbb{F}$-predictable increasing process U setting $U_n = \frac{1}{\mathscr{E}(-D)_n}$, where the stochastic exponential $\mathscr{E}(-D)$ is positive on the set $\{\vartheta > n\}$. To prove the positivity of the exponential, we prove that $\Delta D_n < 1$. For $A = \{\Delta D_n = 1\}$, one has $\mathbb{E}[\mathbb{1}_A\mathbb{1}_{\theta>n)}] = \mathbb{E}[\mathbb{1}_A\mathbb{P}(\theta > n|\mathscr{F}_{n-1})] = 0$ since $\mathbb{P}(\theta > n|\mathscr{F}_{n-1}) = 0$ on A. It follows that $A \cap \{\theta > n\}$ is negligible. Then, on $\{\vartheta > n\}$,

$$\Delta U_n = \frac{1}{\mathscr{E}(-D)_{n-1}}\Big(\frac{1}{1-\Delta D_n} - 1\Big) = \frac{1}{\mathscr{E}(-D)_{n-1}}\frac{\Delta D_n}{1-\Delta D_n} = U_n\Delta D_n\,.$$

We consider the process $X = UK$ where $K = 1 - I$. Then,

$$\Delta X_n = -U_n\Delta I_n + K_{n-1}\Delta U_n = -U_n\left(\Delta I_n - K_{n-1}\Delta D_n\right)$$

and

$$\begin{aligned}\mathbb{E}[\Delta X_n|\mathscr{F}_{n-1}] &= -U_n\mathbb{E}[\Delta I_n - K_{n-1}\Delta D_n|\mathscr{F}_{n-1}] \\ &= U_n\left(\mathbb{P}(\vartheta = n|\mathscr{F}_{n-1}) - K_{n-1}\mathbb{P}(\Delta D_n|\mathscr{F}_{n-1})\right) \\ &= U_n K_{n-1}\left(\mathbb{P}(\vartheta = n|\mathscr{F}_{n-1}) - \mathbb{P}(\Delta D_n|\mathscr{F}_{n-1})\right) = 0\,,\end{aligned}$$

where we have used that

$$K_{n-1}\mathbb{P}(\vartheta = n|\mathscr{F}_{n-1}) = \mathbb{E}[K_{n-1}\mathbb{1}_{\{\vartheta=n\}}|\mathscr{F}_{n-1}] = \mathbb{P}(\vartheta = n|\mathscr{F}_{n-1})\,.$$

Hence X is an $\mathbb{F}$-martingale.

We now prove that $X_\tau \geq 1$ and $\mathbb{P}(X_\tau > 1) > 0$, or, equivalently, that $D_\tau \geq 0$ and $\mathbb{P}(D_\tau > 0) > 0$. For that, we compute

$$\begin{aligned}\mathbb{E}[D_\tau\mathbb{1}_{\tau<\infty}] &= \sum_{n=0}^{\infty}\mathbb{E}[D_n\mathbb{1}_{\{\tau=n\}}] = \sum_{n=0}^{\infty}\mathbb{E}[D_n\mathbb{P}(\tau = n|\mathscr{F}_n)] \\ &= -\sum_{n=1}^{\infty}\mathbb{E}[D_n\mathbb{P}(\tau > n|\mathscr{F}_n)] + \sum_{n=1}^{\infty}\mathbb{E}[D_n\mathbb{P}(\tau > n-1|\mathscr{F}_n)] + D_0\mathbb{P}(\tau = 0) \\ &= \mathbb{E}[\sum_{n=1}^{\infty}Z_{n-1}\Delta D_n] = \mathbb{E}[Z_{\vartheta-1}\mathbb{1}_{\{\vartheta<\infty\}}] > 0\,,\end{aligned}$$

where in the third equality, we have used integration by parts and $Z_\infty = 0$. In the last inequality, we used that the definition of ϑ implies $Z_{\vartheta-1} > 0$ and that as the set $\{0 = \widetilde{Z}_n < Z_{n-1}\}$ is not negligible, $\mathbb{1}_{\{\vartheta<\infty\}} > 0$. The process X^τ is then an increasing process and cannot be turned in a martingale by change of probability.

More generally, we have the following results.

Lemma 4.14 *The following assertions hold.*
(a) The process $L^{\mathbb{G}}$ defined by

$$L_n^{\mathbb{G}} := \prod_{k=1}^{n}(1 + \Delta X_k)$$

where

$$\Delta X_n = -\mathbb{P}(\widetilde{Z}_n > 0|\mathscr{F}_{n-1})\mathbb{1}_{\{\tau\geq n\}} + Z_{n-1}(\widetilde{Z}_n)^{-1}\mathbb{1}_{\{\tau\geq n\}}$$

is a positive $\mathbb{G}$-martingale.
(b) The process $L^{\mathbb{F}}$ given by

$$L_n^{\mathbb{F}} := \prod_{k=1}^{n}(1 + \Delta Y_k)$$

and

$$\Delta Y_n = \widetilde{Z}_n \mathbb{P}(\widetilde{Z}_n = 0|\mathscr{F}_{n-1}) - Z_{n-1}\mathbb{1}_{\{\widetilde{Z}_n=0\}}$$

is a positive $\mathbb{F}$*-martingale.*

Proof We leave the proof to the reader, and there is no difficulty.

Proposition 4.15 *Suppose that* S *is an* $\mathbb{F}$*-martingale, and consider the equivalent probability measures* $\mathbb{Q}^{\mathbb{F}}$ *and* $\mathbb{Q}^{\mathbb{G}}$ *given by* $\mathbb{Q}^{\mathbb{F}} := L^{\mathbb{F}}\mathbb{P}$ *and* $\mathbb{Q}^{\mathbb{G}} := L^{\mathbb{G}}\mathbb{P}$*. Then,* S *is an* $(\mathbb{F}, \mathbb{Q}^{\mathbb{F}})$*-martingale if and only if* S^{τ} *is a* $(\mathbb{G}, \mathbb{Q}^{\mathbb{G}})$*-martingale.*

Proof See Choulli and Deng [11].

5 Progressive Enlargement After τ

In continuous time, strong conditions are needed to keep the semimartingale property after τ; here, it is no more the case. As we mentioned at the beginning, in a discrete time setting, any $\mathbb{F}$-martingale is a $\mathbb{G}$-semimartingale. In a progressive enlargement of filtration with a random time valued in $\mathbb{N}$, one can give the decomposition formula. We start with the general case, then we study the particular case where τ is honest, to provide comparison with the classical results known in continuous time. We also study the case of pseudo-stopping times.

5.1 General Case

Mixing the results obtained in initial enlargement and progressive enlargement before τ, we obtain the following decomposition.

Proposition 5.1 *Any* $\mathbb{F}$*-martingale* X *decomposes as*

$$X_n = X_n^{\mathbb{G}} + \sum_{k=1}^{n\wedge\tau} \frac{1}{Z_{k-1}} \Delta\langle \widetilde{M}, X\rangle_k^{\mathbb{F}} + \sum_{k=\tau+1}^{n} \frac{\Delta\langle X, p(j)\rangle_k^{\mathbb{F}}|_{j=\tau}}{p_{k-1}(\tau)}, \tag{36}$$

where $X^{\mathbb{G}}$ *is a* $\mathbb{G}$*-martingale.*

Proof Let $X = M^{\mathbb{G}} + V^{\mathbb{G}}$ be the $\mathbb{G}$-semimartingale decomposition of X. On the set $\{0 \le n < \tau\}$, the predictable part of X can be computed following the proof of Proposition 4.1. On the set $\{\tau \le n\}$, for $j \le n$

$$\begin{aligned}
\mathbb{1}_{\{\tau\le n\}}\mathbb{1}_{\{\tau=j\}}(V_{n+1}^{\mathbb{G}} - V_n^{\mathbb{G}}) &= \mathbb{1}_{\{\tau\le n\}}\mathbb{1}_{\{\tau=j\}}\mathbb{E}[X_{n+1} - X_n|\mathscr{G}_n] \\
&= \mathbb{1}_{\{\tau\le n\}}\mathbb{E}[\mathbb{1}_{\{\tau=j\}}(X_{n+1} - X_n)|\mathscr{G}_n] \\
&= \mathbb{1}_{\{\tau\le n\}}\mathbb{E}[\mathbb{1}_{\{\tau=j\}}(X_{n+1} - X_n)|\mathscr{F}_n \vee \sigma(\tau)]\,.
\end{aligned}$$

We conclude using the proof of Proposition 2.3.

Proposition 5.2 *Let X be a process such that $X = \mathbb{1}_{]\!]\tau,\infty[\![} \cdot X$. The process X is an $\mathbb{F}^{\sigma(\tau)}$-local martingale if and only if it is a $\mathbb{G}$-local martingale.*

Proof We prove that, if X is an $\mathbb{F}^{\sigma(\tau)}$-martingale satisfying $X = \mathbb{1}_{]\!]\tau,\infty[\![} \cdot X$, it is a $\mathbb{G}$-martingale. The extension to $\mathbb{F}^{\sigma(\tau)}$-local martingale is standard. Then, as on the set $\{\tau < n+1\}$ one has $\mathscr{G}_n = \mathscr{F}_n^{\sigma(\tau)}$, we deduce

$$
\begin{aligned}
\mathbb{E}[X_{n+1}|\mathscr{G}_n] &= \mathbb{E}[X_{n+1}\mathbb{1}_{\{\tau<n+1\}}|\mathscr{G}_n] = \mathbb{E}[X_{n+1}\mathbb{1}_{\{\tau<n+1\}}|\mathscr{F}_n^{\sigma(\tau)}] \\
&= \mathbb{1}_{\{\tau<n+1\}}\mathbb{E}[X_{n+1}|\mathscr{F}_n^{\sigma(\tau)}] = \mathbb{1}_{\{\tau<n+1\}}X_n \\
&= \mathbb{1}_{\{\tau<n+1\}}\mathbb{1}_{\{\tau<n\}}X_n = \mathbb{1}_{\{\tau<n\}}X_n = X_n
\end{aligned}
$$

The first, fourth and last equality are due to the hypothesis on X which states that $X_k = \mathbb{1}_{\{\tau<k\}}X_k$.

Conversely, the proof is obtained as $\{\tau < n+1\} \in \mathscr{G}_n$.

5.2 *Honest Times*

We now consider the case where τ is honest (and valued in $\mathbb{N}$). We recall the definition (Barlow [7]) and some of the main properties.

Definition 5.3 A random time is honest if, for any $n \geq 0$, there exists an $\mathscr{F}_n$-measurable random variable $\tau(n)$ such that

$$\mathbb{1}_{\{\tau\leq n\}}\tau = \mathbb{1}_{\{\tau\leq n\}}\tau(n)\,. \tag{37}$$

Remark 5.4 Following Jeulin [19, Chapitre V], τ is honest if there exists an $\mathscr{F}_n$-measurable random variable $\widehat{\tau}(n)$ such that

$$\mathbb{1}_{\{\tau< n\}}\tau = \mathbb{1}_{\{\tau< n\}}\widehat{\tau}(n)\,. \tag{38}$$

The two definitions are equivalent. Indeed, starting with the equality (38) and setting $\tau(n) = \widehat{\tau}(n) \wedge n$, we obtain that one has $\tau(n) = n$ and $\mathbb{1}_{\{\tau\leq n\}}\tau = \mathbb{1}_{\{\tau\leq n\}}\tau(n)$, on the set $\{\tau = n\}$.

We now show that, for honest times, the decomposition (16) has a particular form.

Lemma 5.5 *If τ is honest, any $\mathbb{G}$-predictable process V can be written as*

$$V_n = V_n^b\mathbb{1}_{\{n\leq\tau\}} + V_n^a\mathbb{1}_{\{\tau<n\}}$$

*where V^a, V^b are $\mathbb{F}$-predictable processes (the superscript a is for **after** τ and b for **before**).*

Proof Suppose that $V_n = X_n h(\tau \wedge n)$ where $X_n \in \mathscr{F}_{n-1}$ and h a bounded Borel function. In this case, $V_n \mathbb{1}_{\{\tau<n\}} = X_n h(\tau \wedge n)\mathbb{1}_{\{\tau\le n-1\}}$. As τ is honest, there exists an $\mathscr{F}_{n-1}$-measurable random variable $\tau(n-1)$ such that

$$V_n \mathbb{1}_{\{\tau<n\}} = X_n h(\tau(n-1) \wedge n)\mathbb{1}_{\{\tau\le n-1\}}$$

Then, the result is obvious with $V_n^a = X_n h(\tau(n-1) \wedge n)$. The result for general V follows. The process V^b can be chosen as in Lemma 3.7.

Lemma 5.6 *If τ is honest, then $Z_n = \widetilde{Z}_n$ on the set $\{n > \tau\}$ and $\widetilde{Z}_\tau = 1$ on $\{\tau < \infty\}$. If $\widetilde{Z}_\tau = 1$, then τ is honest.*

Proof For any $n \ge 0$,

$$\begin{aligned}\mathbb{P}(\tau = n|\mathscr{F}_n)\mathbb{1}_{\{n>\tau\}} &= \mathbb{P}(\tau = n|\mathscr{F}_n)\mathbb{1}_{\{n>\tau\}}\mathbb{1}_{\{n>\tau(n)\}} = \mathbb{E}[\mathbb{1}_{\{\tau=n\}}\mathbb{1}_{\{n>\tau(n)\}}|\mathscr{F}_n]\mathbb{1}_{\{n>\tau\}}\\ &= \mathbb{E}[\mathbb{1}_{\{\tau=n\}}\mathbb{1}_{\{n>\tau(n)\}}\mathbb{1}_{\{n>\tau\}}|\mathscr{F}_n]\mathbb{1}_{\{n>\tau\}} = 0\,.\end{aligned}$$

It follows that $Z_n \mathbb{1}_{\{\tau<n\}} = \widetilde{Z}_n \mathbb{1}_{\{\tau<n\}}$. Furthermore,

$$\begin{aligned}\widetilde{Z}_n \mathbb{1}_{\{\tau=n\}} &= \mathbb{1}_{\{\tau=n\}}\mathbb{P}(\tau \ge n|\mathscr{F}_n) = \mathbb{1}_{\{\tau=n\}}\mathbb{1}_{\{\tau(n)=n\}}\mathbb{P}(\tau \ge n|\mathscr{F}_n)\\ &= \mathbb{1}_{\{\tau=n\}}\mathbb{E}[\mathbb{1}_{\{\tau(n)=n\}}\mathbb{1}_{\{\tau\ge n\}}|\mathscr{F}_n] = \mathbb{1}_{\{\tau=n\}}\end{aligned}$$

which implies $\widetilde{Z}_\tau = 1$ on the set $\{\tau < \infty\}$.

If $\widetilde{Z}_\tau = 1$, let $\ell(n) = \sup\{k \le n : \widetilde{Z}_k = 1\}$. Then, for any $n \ge 0$, Proposition 3.5 implies $\tau = \ell(n)$ on the set $\{\tau \le n\}$, and τ is honest.

Proposition 5.7 *If the random time τ is honest then $\tau = \sup\{n : \widetilde{Z}_n = 1\}$.*

Proof It follows from the previous lemma and Proposition 3.5.

We give now some examples of honest times.

Example 5.8 (a) Any $\mathbb{F}$-stopping time τ is honest. Let $\tau(n) := \tau \wedge n$, then $\tau(n)$ is an $\mathscr{F}_n$-measurable random variable and on the set $\tau < n$, one has $\tau = \tau(n)$.
(b) Let X be an $\mathbb{F}$-adapted process and $\tau_1 = \sup\{n,\ X_n \le a\}$. Then, τ_1 is honest. Indeed, on the set $\{\tau_1 < n\}$, one has $\tau_1 = \tau(n) := \sup\{k \le n, X_k \le a\} \in \mathscr{F}_n$.
(c) If $\tau_2 = \sup\{n,\ X_n \le X_{n-1}\}$, then τ_2 is honest. On the set $\{\tau_2 < n\}$, one has $\tau_2 = \tau(n) := \sup\{k \le n, X_k \le X_{k-1}\} \in \mathscr{F}_n$.

We now give a specific decomposition of $\mathbb{F}$-martingales, in the case of honest times.

Proposition 5.9 *Let τ be an honest time and X an $\mathbb{F}$-martingale. Then,*

$$X_n = X_n^{\mathbb{G}} + \sum_{k=1}^{n\wedge\tau} \frac{1}{Z_{k-1}}\Delta\langle \widetilde{M}, X\rangle_k^{\mathbb{F}} - \sum_{k=\tau+1}^{n} \frac{1}{1 - Z_{k-1}}\Delta\langle \widetilde{M}, X\rangle_k^{\mathbb{F}} \tag{39}$$

where $X^{\mathbb{G}}$ is a $\mathbb{G}$-martingale.

Proof Let $X = M^{\mathbb{G}} + V^{\mathbb{G}}$ be the $\mathbb{G}$-semimartingale decomposition of X. From the property of honest times, there exists $\widetilde{V}$, an $\mathbb{F}$-predictable process, such that, for any $n \geq 0$

$$V_n^{\mathbb{G}} \mathbb{1}_{\{\tau \leq n\}} = \widetilde{V}_n \mathbb{1}_{\{\tau \leq n\}} .$$

Then,

$$\begin{aligned} \mathbb{1}_{\{\tau \leq n\}}(\widetilde{V}_{n+1} - \widetilde{V}_n) &= \mathbb{1}_{\{\tau \leq n\}}(V_{n+1}^{\mathbb{G}} - V_n^{\mathbb{G}}) = \mathbb{1}_{\{\tau \leq n\}} \mathbb{E}[X_{n+1} - X_n | \mathscr{G}_n] \\ &= \mathbb{E}[\mathbb{1}_{\{\tau \leq n\}}(X_{n+1} - X_n) | \mathscr{G}_n] . \end{aligned}$$

We now take the conditional expectation w.r.t. $\mathscr{F}_n$ in the previous equality. Taking into account that $\widetilde{V}$ is $\mathbb{F}$-predictable, and the fact that $\mathbb{F} \subset \mathbb{G}$, we get

$$\begin{aligned} \mathbb{E}[\mathbb{1}_{\{\tau \leq n\}} | \mathscr{F}_n](\widetilde{V}_{n+1} - \widetilde{V}_n) &= \mathbb{E}[\mathbb{1}_{\{\tau \leq n\}}(X_{n+1} - X_n) | \mathscr{F}_n] \\ &= \mathbb{E}[\mathbb{E}[\mathbb{1}_{\{\tau \leq n\}} | \mathscr{F}_{n+1}](X_{n+1} - X_n) | \mathscr{F}_n] . \end{aligned}$$

Now, using the fact that

$$\begin{aligned} \mathbb{E}[\mathbb{1}_{\{\tau \leq n\}} | \mathscr{F}_n] &= 1 - \mathbb{E}[\mathbb{1}_{\{\tau > n\}} | \mathscr{F}_n] = 1 - Z_n \\ \mathbb{E}[\mathbb{1}_{\{\tau \leq n\}} | \mathscr{F}_{n+1}] &= 1 - \mathbb{E}[\mathbb{1}_{\{\tau > n\}} | \mathscr{F}_{n+1}] = 1 - \mathbb{E}[\mathbb{1}_{\{\tau \geq n+1\}} | \mathscr{F}_{n+1}] = 1 - \widetilde{Z}_{n+1} \end{aligned}$$

and that X is an $\mathbb{F}$-martingale, we obtain

$$(1 - Z_n)(\widetilde{V}_{n+1} - \widetilde{V}_n) = -\mathbb{E}[\widetilde{Z}_{n+1}(X_{n+1} - X_n) | \mathscr{F}_n] = -\Delta \langle \widetilde{M}, X \rangle_{n+1}^{\mathbb{F}} .$$

We deduce that, on the set $\{\tau \leq n\}$, one has

$$\Delta V_n^{\mathbb{G}} = -\frac{1}{1 - Z_{n-1}} \Delta \langle \widetilde{M}, X \rangle_n^{\mathbb{F}} .$$

The form of the decomposition before τ was given in Proposition 4.1.

Comment 5.10 We recall, for the ease of the reader, the Jeulin formula for honest times in continuous time:

$$X_t = X_t^{\mathbb{G}} + \int_0^{t \wedge \tau} \frac{1}{Z_{s-}} \Delta \langle \widetilde{M}, X \rangle_s^{\mathbb{F}} - \int_\tau^t \frac{1}{1 - Z_{s-}} d \langle \widetilde{M}, X \rangle_s^{\mathbb{F}} ,$$

where $\widetilde{M}$ is the martingale in the optional decomposition of Z as $Z = \widetilde{M} - H^{o,\mathbb{F}}$. See [19, Chap. 5] or [6, Pro. 5.15].

Comment 5.11 Let τ be an honest time. We have obtained a formula using Jacod's hypothesis in (36). In continuous time, one can show that an honest time satisfies the equivalence Jacod's hypothesis if and only if it takes countably many values (see [2, lemma 4.11]) and, in that case, the equality between the two decompositions is obtained in [2]. In discrete time, honest times satisfy equivalence Jacod's hypothesis and one can check that the decompositions obtained in (36) and the one for honest times are the same. We proceed as in Aksamit [2]. Let $n \geq 1$ be fixed. On $\tau < n$, we have $\tau = \tau(n-1)$ where $\tau(n-1)$ is an $\mathscr{F}_{n-1}$-measurable r.v. We now restrict our attention to $k < n$. On the one hand,

$$\begin{aligned}\mathbb{1}_{\{\tau=k\}}(1-Z_{n-1}) &= \mathbb{1}_{\{\tau=k=\tau(n-1)\}}\mathbb{P}(\tau \leq n-1|\mathscr{F}_{n-1})\\ &= \mathbb{1}_{\{\tau=k\}}\mathbb{E}[\mathbb{1}_{\{\tau(n-1)=k\}}\mathbb{1}_{\{\tau\leq n-1\}}|\mathscr{F}_{n-1}]\\ &= \mathbb{1}_{\{\tau=k\}}\mathbb{E}[\mathbb{1}_{\{\tau(n-1)=k\}}\mathbb{1}_{\{\tau=k\}}|\mathscr{F}_{n-1}] = \mathbb{1}_{\{\tau=k\}}\mathbb{E}[\mathbb{1}_{\{\tau=k\}}|\mathscr{F}_{n-1}]\\ &= \mathbb{1}_{\{\tau=k\}}p_{n-1}(k)\,.\end{aligned}$$

On the other hand,

$$\begin{aligned}&\mathbb{1}_{\{\tau=k\}}\Delta\langle\widetilde{M},X\rangle^{\mathbb{F}}_{n-1} = \mathbb{1}_{\{\tau=k\}}\mathbb{E}[\widetilde{M}_n\Delta X_n|\mathscr{F}_{n-1}]\\ &= -\mathbb{1}_{\{\tau=k\}}\mathbb{E}[(1-\widetilde{M}_n)\Delta X_n|\mathscr{F}_{n-1}] = -\mathbb{1}_{\{\tau=k=\tau(n-1)\}}\mathbb{E}[(1-\widetilde{Z}_n)\Delta X_n|\mathscr{F}_{n-1}]\\ &= -\mathbb{1}_{\{\tau=k\}}\mathbb{E}[\mathbb{1}_{\{k=\tau(n-1)\}}\mathbb{1}_{\{\tau<n\}}\Delta X_n|\mathscr{F}_{n-1}] = -\mathbb{1}_{\{\tau=k\}}\mathbb{E}[\mathbb{E}[\mathbb{1}_{\{k=\tau\}}|\mathscr{F}_n]\Delta X_n|\mathscr{F}_{n-1}]\\ &= -\mathbb{1}_{\{\tau=k\}}\mathbb{E}[p_n(k)\Delta X_n|\mathscr{F}_{n-1}]\,,\end{aligned}$$

and the equality between (36) and (39) is obtained.

5.3 Viability

Let τ be a bounded honest time which is not an $\mathbb{F}$-stopping time. The enlargement $(\mathbb{F}, \mathbb{G})$ is not viable. Indeed, assuming the existence of a universal deflator L implies that $\widetilde{M}L$ is a $\mathbb{G}$-martingale. Since $\widetilde{Z}_\tau = 1$, one has $\widetilde{M}_\tau \geq 1$, and $\mathbb{P}(\widetilde{M}_\tau > 1) > 0$. Therefore, the fact that τ is bounded allows us to use the optional sampling theorem which yields to $1 = \mathbb{E}[\widetilde{M}_\tau L_\tau] > \mathbb{E}[L_\tau] = 1$, hence a contradiction.

We refer to Choulli and Deng [11] for a necessary and sufficient condition to avoid arbitrages after τ and Fontana et al. [14] for the continuous time case.

6 Pseudo-Stopping Times

We end the study of progressive enlargement with a specific class of random times. We assume that $\mathscr{F}_0$ is trivial.

Definition 6.1 The random time τ is an $\mathbb{F}$-pseudo-stopping time if $\mathbb{E}[X_\tau] = \mathbb{E}[X_0]$ for any bounded $\mathbb{F}$-martingale X.

This definition comes from Nikeghbali and Yor [24].

Theorem 6.2 *The following statements are equivalent:*

(a) τ *is an* $\mathbb{F}$*-pseudo-stopping time.*
(b) $H^{o,\mathbb{F}}_{\infty} = \mathbb{P}(\tau < \infty|\mathscr{F}_\infty)$.
(c) $\widetilde{M} \equiv 1$.
(d) $\widetilde{Z}$ *is predictable and decreasing.*
(e) Every $\mathbb{F}$*-martingale stopped at* τ *is a* $\mathbb{G}$*-martingale.*

Proof We omit the superscript $\mathbb{F}$ in the proof.

(a) $\Rightarrow$ (b) Using that the bounded martingale X is closed, $\lim_{n\to\infty} X_n =: X_\infty$ exists and one can write the integration by parts formula

$$H^o_\infty X_\infty = \sum_{n=1}^{\infty} H^o_{n-1}\Delta X_n + \sum_{n=0}^{\infty} X_n \Delta H^o_n \,.$$

Taking expectation, and using the fact that X is an $\mathbb{F}$-martingale, we obtain

$$\mathbb{E}[H^o_\infty X_\infty] = \mathbb{E}[\sum_{n=0}^{\infty} X_n \Delta H^o_n]$$

and, from property (25) of H^o, one has

$$\mathbb{E}[H^o_\infty X_\infty] = \mathbb{E}[X_\tau \mathbb{1}_{\{\tau<\infty\}}]\,.$$

It follows that

$$\begin{aligned} X_0 &= \mathbb{E}[X_\tau] = \mathbb{E}[X_\tau \mathbb{1}_{\{\tau<\infty\}}] + \mathbb{E}[X_\infty \mathbb{1}_{\{\tau=\infty\}}] \\ &= \mathbb{E}[X_\tau \mathbb{1}_{\{\tau<\infty\}}] + \mathbb{E}[X_\infty] - \mathbb{E}[X_\infty \mathbb{1}_{\{\tau<\infty\}}] \\ &= \mathbb{E}[H^o_\infty X_\infty] + X_0 - \mathbb{E}[X_\infty \mathbb{1}_{\{\tau<\infty\}}]\,, \end{aligned}$$

hence $\mathbb{E}[H^o_\infty X_\infty] = \mathbb{E}[X_\infty \mathbb{1}_{\{\tau<\infty\}}] = \mathbb{E}[X_\infty \mathbb{P}(\tau < \infty|\mathscr{F}_\infty)]$, and we obtain that $H^o_\infty = \mathbb{P}(\tau < \infty|\mathscr{F}_\infty)$.

(b) $\Rightarrow$ (c) Obvious

(c) $\Rightarrow$ (d) By definition of H^o and (23), we have that

$$\widetilde{M}_n = H^o_n + Z_n = H^o_{n-1} + \widetilde{Z}_n\,, \quad \forall n \geq 1\,, \tag{40}$$

therefore, by (c), we deduce that $\widetilde{Z}_n = 1 - H^o_{n-1}$ which, since H^o is $\mathbb{F}$-adapted, implies that the random variable $\widetilde{Z}_n$ is $\mathscr{F}_{n-1}$-measurable for all $n \geq 1$, i.e. $\widetilde{Z}$ is $\mathbb{F}$-predictable. The decreasing property follows by the same argument that the one in Remark 3.10.

(d) $\Rightarrow$ (e) If $\widetilde{Z}$ is predictable, $\widetilde{M}$ is a predictable martingale, hence a constant (see Lemma 1.2) and for any $\mathbb{F}$-martingale X, $\Delta\langle X, \widetilde{M}\rangle_n = 0$ for all $n \geq 1$. The result follows from Proposition 4.1.

(e) $\Rightarrow$ (a) For any bounded $\mathbb{F}$-martingale X, the stopped process X^τ is a bounded (hence a uniformly integrable) $\mathbb{G}$-martingale. Then, as a consequence of the optional stopping theorem applied in $\mathbb{G}$ at time τ, we get $\mathbb{E}[X_\tau] = \mathbb{E}[X_0]$; hence, τ is an $\mathbb{F}$-pseudo-stopping time.

Any random time such that immersion holds is a pseudo-stopping time. This is in particular the case for the random times introduced in Example 3.25 where $\widetilde{Z}$ is indeed predictable.

The following example gives a pseudo-stopping time for which we don't have immersion property.

Example 6.3 Let $Y_n = \prod_{k=0}^n X_k$, where $(X_k)_{k\geq 0}$ are i.i.d. and positive random variables, and $\tau = \sup\{n \leq N, Y_n - Y_N \leq 0\}$. Then τ is a pseudo-stopping time.

Indeed one has $\{\tau < n\} = \{\inf_{n<k\leq N} Y_k > Y_N\} = \{\inf_{n<k\leq N} Y_n^k > Y_n^N\}$ where $Y_n^k = \prod_{j=n+1}^k X_j$. Due to the independence of the variables $(X_k)_{k\geq 0}$, the random variables $(\inf_{n<k\leq N} Y_n^k)$ and Y_n^N are independent from $\mathscr{F}_n$; therefore, $\mathbb{P}(\tau < n|\mathscr{F}_n)$ is deterministic, and hence predictable. One can see that $\mathbb{P}(\tau \geq n|\mathscr{F}_n)$ is not equal to $\mathbb{P}(\tau \geq n|\mathscr{F}_\infty)$, and hence immersion does not hold.

Proposition 6.4 *If a pseudo-stopping time is honest, it is a stopping time.*

Proof From Proposition 5.7, $\tau = \sup\{n : \widetilde{Z}_n = 1\}$ on $\{\tau \leq \infty\}$. The pseudo-stopping time property of τ implies $Z = 1 - H^{o,\mathbb{F}}$. Moreover, $\widetilde{Z} - Z = \Delta H^{o,\mathbb{F}}$. Then, on $\{\tau < \infty\}$ we obtain

$$\begin{aligned}\tau &= \sup\{n : \widetilde{Z}_n = 1\} = \sup\{n : Z_n + \Delta H_n^{o,\mathbb{F}} = 1\}\\ &= \sup\{n : 1 - H_n^{o,\mathbb{F}} + \Delta H_n^{o,\mathbb{F}} = 1\} = \sup\{n : H_{n-1}^{o,\mathbb{F}} = 0\} = \inf\{n : H_n^{o,\mathbb{F}} > 0\}.\end{aligned}$$

So, τ is equal to the $\mathbb{F}$-stopping time $\inf\{n : H_n^{o,\mathbb{F}} > 0\}$ on $\{\tau < \infty\}$.

Comment 6.5 (a) Obviously, pseudo-stopping times do not create arbitrages before τ.

(b) Pseudo-stopping time have been introduced in [24]. In continuous time, the links between pseudo-stopping times and immersion property are presented in [3], and it is proved that τ is a pseudo-stopping time if and only if $\widetilde{Z}$ is a càglàd decreasing process. In discrete time, we have obtained a similar result, τ is a pseudo-stopping time if and only if $\widetilde{Z}$ is a predictable process (note that we do not require the decreasing assumption, since if τ is a pseudo-stopping time, then $\widetilde{Z}$ is decreasing, since $\widetilde{M}$ is a constant).

7 An Optional Representation for Martingales in a Progressive Enlargement

In [10], Choulli et al. present an important optional martingale representation for a particular class of $\mathbb{G}$-martingales in continuous time. We give here the analogue in discrete time, but with a slightly different proof. In the following, if X is an $\mathbb{F}$-martingale, we denote by $\widehat{X}$ the $\mathbb{G}$-martingale which appears in the optional decomposition given in Proposition 4.7.

Theorem 7.1 *For any $\mathbb{F}$-adapted integrable process φ such that $\varphi_\tau \mathbb{1}_{\{\tau<\infty\}}$ is integrable, the martingale $\Phi_n = \mathbb{E}[\varphi_\tau \mathbb{1}_{\{\tau<\infty\}}|\mathscr{G}_n]$ can be decomposed as*

$$\Phi = \Phi_0 + \frac{1}{Z_-}\mathbb{1}_{[\![0,\tau]\!]} \centerdot \widehat{M^\varphi} - \frac{\Psi_-}{Z_-}\mathbb{1}_{[\![0,\tau]\!]} \centerdot \widetilde{M}^{\mathbb{G}} + (\varphi - \Psi) \centerdot \Pi \tag{41}$$

where

- *M^φ is the $\mathbb{F}$-martingale $M^\varphi_n = \mathbb{E}[\sum_{j\geq 0}\varphi_j \Delta H^{o,\mathbb{F}}_j|\mathscr{F}_n]$ and $\widehat{M^\varphi}$ is its $\mathbb{G}$-martingale part.*
- *$\widetilde{M}^{\mathbb{G}}$ is the $\mathbb{G}$-martingale part of the $\mathbb{G}$-semimartingale $\widetilde{M}$.*
- *$\Psi = \dfrac{M^\varphi - (\varphi \centerdot H^{o,\mathbb{F}})}{Z + Z_{R-}\mathbb{1}_{[\![R,+\infty]\!]}}$, where $R := \inf\{n \geq 0, Z_n = 0\}$.*
- *Π is defined in Proposition 3.13.*

The proof of the theorem is based on the following lemma.

Lemma 7.2 *Let $Y_n = \mathbb{E}[\varphi_\tau \mathbb{1}_{\{n<\tau<\infty\}}|\mathscr{F}_n]$, then, using the notation of the previous theorem,*
(a) $Y_n \mathbb{1}_{\{n=R\}} = 0, \; n \geq 0.$
(b) $Y_n = M^\varphi_n - (\varphi \centerdot H^{o,\mathbb{F}})_n, \; n \geq 0.$

In the two following proofs, we omit the superscript $\mathbb{F}$.

Proof *(of the lemma).*
(a) If φ is positive, then Y is non-negative. As $\tau \leq R$, we have

$$\mathbb{E}[Y_n \mathbb{1}_{\{n=R\}}] = \mathbb{E}[\mathbb{E}[\varphi_\tau \mathbb{1}_{\{n<\tau<\infty\}}\mathbb{1}_{\{n=R\}}|\mathscr{F}_n]] = 0\,,$$

therefore $Y_n \mathbb{1}_{\{n=R\}} = 0$ *a.s.* The general case follows using $\varphi = \varphi^+ - \varphi^-$.
(b) We remark that

$$\begin{aligned} Y_n = \mathbb{E}[\varphi_\tau \mathbb{1}_{\{n<\tau<\infty\}}|\mathscr{F}_n] &= \mathbb{E}[\sum_{j\geq n+1} \varphi_j \Delta H_j|\mathscr{F}_n] \\ &= \sum_{j\geq n+1} \mathbb{E}[\varphi_j \Delta H^o_j|\mathscr{F}_n] = \mathbb{E}[\sum_{j\geq 0}\varphi_j \Delta H^o_j|\mathscr{F}_n] - \sum_{j=0}^{n}\varphi_j \Delta H^o_j, \end{aligned}$$

where the third equality is obtained by the tower property. The result follows.

Proof *(of theorem)* The proof is done in three steps.

First step: We set $B := Z + Z_{R-}\mathbb{1}_{[\![R,+\infty[\![}$, and we note that $B > 0$ and, since $\tau \leq R$, that $B_n \mathbb{1}_{\{n<\tau\}} = Z_n \mathbb{1}_{\{n<\tau\}}$ and $B_\tau = Z_\tau + Z_{R-}\mathbb{1}_{\{\tau=R\}} > 0$.

Using the key lemma, we see that

$$\begin{aligned}\Phi_n &= \mathbb{E}[\varphi_\tau \mathbb{1}_{\{\tau<\infty\}}|\mathcal{G}_n] = \varphi_\tau \mathbb{1}_{\{\tau\le n\}} + \mathbb{1}_{\{n<\tau\}} \frac{\mathbb{E}[\varphi_\tau \mathbb{1}_{\{n<\tau<\infty\}}|\mathcal{F}_n]}{Z_n} \\ &= \varphi_\tau \mathbb{1}_{\{\tau\le n\}} + \mathbb{1}_{\{n<\tau\}} \frac{Y_n}{Z_n}\,.\end{aligned}$$

Using the previous lemma, the equality $B_n \mathbb{1}_{\{n<\tau\}} = Z_n \mathbb{1}_{\{n<\tau\}}$ and the definition of Ψ, we obtain

$$\Phi_n = \varphi_\tau \mathbb{1}_{\{\tau\le n\}} + \mathbb{1}_{\{n<\tau\}} \Psi_n\,.$$

Then

$$\Delta\Phi_n = (\varphi_n - \Psi_n)\mathbb{1}_{\{n=\tau\}} + \mathbb{1}_{\{n-1<\tau\}}\Delta\Psi_n.$$

From $\Psi = Y/B$, we compute $\Delta\Psi_n$ taking into account the decomposition (34) applied to M^φ

$$\begin{aligned}\Delta\Psi_n &= -\frac{Y_n \Delta B_n}{B_n B_{n-1}} + \frac{\Delta Y_n}{B_{n-1}} \\ &= \frac{1}{B_{n-1}}\Big(- \frac{Y_n \Delta B_n}{B_n} + \widehat{\Delta M_n^\varphi} + \frac{\Delta \widetilde{M}_n \Delta M_n^\varphi}{\widetilde{Z}_n} + \Big(\Delta M_{\widetilde{R}}^\varphi \mathbb{1}_{[\![\widetilde{R},\infty[\![}\Big)_n^p - \varphi_n \Delta H_n^o\Big).\end{aligned}$$

Using the fact that $B_{n-1} = Z_{n-1}$ on the set $\{n \le \tau\}$, and recalling that $\Pi = H - (\widetilde{Z})^{-1}\mathbb{1}_{[\![0,\tau]\!]} \centerdot H^o$

$$\begin{aligned}\Delta\Phi_n &= (\varphi_n - \Psi_n)\Delta\Pi_n + (\varphi_n - \Psi_n)(\widetilde{Z}_n)^{-1}\mathbb{1}_{\{n\le\tau\}}\Delta H_n^o \\ &+ \frac{\mathbb{1}_{\{n\le\tau\}}}{Z_{n-1}}\left(- \frac{Y_n \Delta B_n}{B_n} + \widehat{\Delta M_n^\varphi} + \frac{\Delta \widetilde{M}_n \Delta M_n^\varphi}{\widetilde{Z}_n} + \Big(\Delta M_{\widetilde{R}}^\varphi \mathbb{1}_{[\![\widetilde{R},\infty[\![}\Big)_n^p - \varphi_n \Delta H_n^o\right) \\ &= (\varphi_n - \Psi_n)\Delta\Pi_n + \mathbb{1}_{\{n\le\tau\}}(Z_{n-1})^{-1}\widehat{\Delta M_n^\varphi} + \mathbb{1}_{\{n\le\tau\}} K_n \end{aligned} \tag{42}$$

where

$$K_n = -\frac{Y_n \Delta B_n}{B_n Z_{n-1}} + \frac{1}{Z_{n-1}\widetilde{Z}_n}\Delta\widetilde{M}_n \Delta M_n^\varphi + \frac{\Big(\Delta M_{\widetilde{R}}^\varphi \mathbb{1}_{[\![\widetilde{R},\infty[\![}\Big)_n^p}{Z_{n-1}} - \frac{\varphi_n}{Z_{n-1}}\Delta H_n^o + (\varphi_n - \Psi_n)\frac{\Delta H_n^o}{\widetilde{Z}_n}\,.$$

Step two: Now, we compute K_n in a more convenient form. In this step, We are working on the set $\{n \le \tau\}$ and all the equalities are considered only on that set. Since $\Psi = \frac{Y}{B}$, we obtain

$$K_n = -\frac{Y_n \Delta B_n}{B_n Z_{n-1}} + \frac{\Delta\widetilde{M}_n \Delta M_n^\varphi}{Z_{n-1}\widetilde{Z}_n} + \frac{\Big(\Delta M_{\widetilde{R}}^\varphi \mathbb{1}_{[\![\widetilde{R},\infty[\![}\Big)_n^p}{Z_{n-1}} + \varphi_n \Delta H_n^o \left(\frac{1}{\widetilde{Z}_n} - \frac{1}{Z_{n-1}}\right) - \frac{Y_n}{B_n \widetilde{Z}_n}\Delta H_n^o.$$

Form equality (24), we have $\widetilde{Z}_n - Z_{n-1} = \Delta\widetilde{M}_n$, then

$$K_n = -\varphi_n \Delta H_n^o \frac{\Delta\widetilde{M}_n}{\widetilde{Z}_n Z_{n-1}} - \frac{Y_n}{B_n}\left(\frac{\Delta B_n}{Z_{n-1}} + \frac{\Delta H_n^o}{\widetilde{Z}_n}\right) + \frac{\Delta\widetilde{M}_n \Delta M_n^\varphi}{Z_{n-1}\widetilde{Z}_n} + \frac{\left(\Delta M_{\widetilde{R}}^\varphi \mathbb{1}_{[\![\widetilde{R},\infty[\![}\right)_n^p}{Z_{n-1}}.$$

Noting that

$$-\varphi_n \Delta H_n^o \frac{\Delta\widetilde{M}_n}{\widetilde{Z}_n Z_{n-1}} + \frac{\Delta\widetilde{M}_n \Delta M_n^\varphi}{Z_{n-1}\widetilde{Z}_n} = \frac{\Delta\widetilde{M}_n}{\widetilde{Z}_n Z_{n-1}}(\Delta M_n^\varphi - \varphi_n \Delta H_n^o) = \frac{\Delta\widetilde{M}_n}{\widetilde{Z}_n Z_{n-1}}\Delta Y_n$$

we deduce that

$$K_n = \frac{\Delta\widetilde{M}_n}{\widetilde{Z}_n Z_{n-1}}\Delta Y_n - \frac{Y_n}{\widetilde{Z}_n Z_{n-1}} \frac{\widetilde{Z}_n \Delta B_n + Z_{n-1}\Delta H_n^o}{B_n} + \frac{\left(\Delta M_{\widetilde{R}}^\varphi \mathbb{1}_{[\![\widetilde{R},\infty[\![}\right)_n^p}{Z_{n-1}}.$$

Since $B_{n-1} = Z_{n-1}$ and from Eq. (24), we get

$$\widetilde{Z}_n \Delta B_n + Z_{n-1}\Delta H_n^o = B_n \widetilde{Z}_n - Z_{n-1}(\widetilde{Z}_n - \Delta H_n^o) = B_n \widetilde{Z}_n - Z_{n-1} Z_n\,,$$

therefore, always on the set $\{n \leq \tau\}$

$$K_n = \frac{\Delta\widetilde{M}_n}{\widetilde{Z}_n Z_{n-1}}\Delta Y_n - \frac{Y_n}{\widetilde{Z}_n Z_{n-1}} \frac{(B_n \widetilde{Z}_n - Z_{n-1} Z_n)}{B_n} + \frac{\left(\Delta M_{\widetilde{R}}^\varphi \mathbb{1}_{[\![\widetilde{R},\infty[\![}(n)\right)^p}{Z_{n-1}}.$$

Using the definition of B, it follows

$$\begin{aligned}\frac{B_n \widetilde{Z}_n - Z_{n-1} Z_n}{B_n} &= \mathbb{1}_{\{n<\tau\leq R\}}\Delta\widetilde{M}_n + \widetilde{Z}_R \mathbb{1}_{\{n=\tau=R\}} \\ &= \mathbb{1}_{\{n<\tau\}}\Delta\widetilde{M}_n + (\widetilde{Z}_R - Z_{R-1} + Z_{R-1})\mathbb{1}_{\{n=\tau=R\}} \\ &= \mathbb{1}_{\{n\leq\tau\}}\Delta\widetilde{M}_n + Z_{R-1}\mathbb{1}_{\{n=\tau=R\}}\end{aligned}$$

where the second equality follows from $\tau \leq R$ a.s. and the third one from (24).
Then

$$\begin{aligned}\frac{Y_n}{\widetilde{Z}_n Z_{n-1}} \frac{(B_n \widetilde{Z}_n - Z_{n-1} Z_n)}{B_n} &= \frac{Y_n}{\widetilde{Z}_n Z_{n-1}}\Delta\widetilde{M}_n \mathbb{1}_{\{n\leq\tau\}} + \frac{Y_n}{\widetilde{Z}_n}\mathbb{1}_{\{n=\tau=R\}} \\ &= \frac{Y_n}{\widetilde{Z}_n Z_{n-1}}\Delta\widetilde{M}_n \mathbb{1}_{\{n\leq\tau\}}\,.\end{aligned}$$

From Lemma 7.2, $\frac{Y_n}{\widetilde{Z}_n}\mathbb{1}_{\{n=\tau=R\}} = 0$, so that, using again $\widetilde{Z}_n - Z_{n-1} = \Delta\widetilde{M}_n$

$$K_n = -\frac{Y_{n-1}}{\widetilde{Z}_n Z_{n-1}} \Delta \widetilde{M}_n + \frac{\left(\Delta M^{\varphi}_{\widetilde{R}} \mathbb{1}_{[\![\widetilde{R},\infty[\![}\right)^p_n}{Z_{n-1}}$$

$$= -\frac{Y_{n-1}}{Z^2_{n-1}} \Delta \widetilde{M}_n + \frac{Y_{n-1}}{\widetilde{Z}_n Z^2_{n-1}} (\Delta \widetilde{M}_n)^2 + \frac{\left(\Delta M^{\varphi}_{\widetilde{R}} \mathbb{1}_{[\![\widetilde{R},\infty[\![}\right)^p}{Z_{n-1}}.$$

Using the optional decomposition of $\widetilde{M}$ given by (34), we obtain

$$K_n = -\frac{Y_{n-1}}{Z^2_{n-1}} \Delta \widetilde{M}^{\mathbb{G}}_n - \frac{Y_{n-1}}{\widetilde{Z}_n Z^2_{n-1}} (\Delta \widetilde{M}_n)^2 - \frac{\Psi_{n-1}}{Z_{n-1}} \left(\Delta \widetilde{M}_{\widetilde{R}} \mathbb{1}_{[\![\widetilde{R},\infty]\!]}\right)^p$$

$$+ \frac{Y_{n-1}}{\widetilde{Z}_n Z^2_{n-1}} (\Delta \widetilde{M}_n)^2 + \frac{\left(\Delta M^{\varphi}_{\widetilde{R}} \mathbb{1}_{[\![\widetilde{R},\infty[\![}\right)^p_n}{Z_{n-1}}.$$

As $Z_R = Y_R = 0$ and $\widetilde{R} \geq \tau$, $\Delta M^{\varphi}_{\widetilde{R}} = \Delta Y_{\widetilde{R}} + \varphi_{\widetilde{R}} \Delta H^{o}_{\widetilde{R}} = -\Psi_{\widetilde{R}_-} Z_{\widetilde{R}_-}$ and $\Delta \widetilde{M}_{\widetilde{R}} = -Z_{\widetilde{R}_-}$, we obtain

$$\left(\Delta M^{\varphi}_{\widetilde{R}} \mathbb{1}_{[\![\widetilde{R},\infty]\!]}\right)^p_n = \mathbb{E}[\Delta M^{\varphi}_{\widetilde{R}} \mathbb{1}_{\{\widetilde{R}=n\}} | \mathscr{F}_{n-1}] = -\Psi_{n-1} Z_{n-1} \mathbb{E}[\mathbb{1}_{\{\widetilde{R}=n\}} | \mathscr{F}_{n-1}]$$

and

$$\left(\Delta \widetilde{M}_{\widetilde{R}} \mathbb{1}_{[\![\widetilde{R},\infty]\!]}\right)^p_n = \mathbb{E}[\Delta \widetilde{M}_{\widetilde{R}} \mathbb{1}_{\{\widetilde{R}=n\}} | \mathscr{F}_{n-1}] = -Z_{n-1} \mathbb{E}[\mathbb{1}_{\{\widetilde{R}=n\}} | \mathscr{F}_{n-1}].$$

Then, on the set $\{n \leq \tau\}$

$$\frac{1}{Z_-} \mathbb{1}_{[0,\tau]} \left(\Delta M^{\varphi}_{\widetilde{R}} \mathbb{1}_{[\![\widetilde{R},\infty]\!]}\right)^p - \frac{\Psi_-}{Z_-} \mathbb{1}_{[\![0,\tau]\!]} \left(\Delta \widetilde{M}_{\widetilde{R}} \mathbb{1}_{[\![\widetilde{R},\infty]\!]}\right)^p = 0\,.$$

It remains $K_n = -\frac{Y_{n-1}}{Z^2_{n-1}} \Delta \widetilde{M}^{\mathbb{G}}_n$.

Step Three: It suffices to plug the value of K, obtained in step two on the set $\{n \leq \tau\}$ into (42) and the result follows.

Remark 7.3 In the definition of the process B in the previous theorem, we remark what we can add to Z any process provided that it is not null at time τ and that B coincide with Z strictly before τ.

8 Other Enlargements

8.1 Enlargement with a Pair (ζ, τ)

A random variable, ζ, valued in $\mathbb{Z}$ and a random time, τ, valued in $\mathbb{N}$ are given. We define the filtration $\mathbb{K}$ as the smallest filtration with contains $\mathbb{F}$, makes τ a stopping time and ζ belonging to $\mathcal{K}_\tau$. We denote by $p_n(j,k) = \mathbb{P}(\tau = j, \zeta = k|\mathcal{F}_n)$. Let X be an $\mathbb{F}$-martingale, and hence a $\mathbb{K}$-semimartingale with decomposition $X = V^{\mathbb{K}} + M^{\mathbb{K}}$ and

$$\Delta V_n^{\mathbb{K}} = \mathbb{E}[\Delta X_n|\mathcal{K}_{n-1}] = \mathbb{1}_{\{\tau\leq n-1\}}\mathbb{E}[\Delta X_n|\mathcal{K}_{n-1}] + \mathbb{1}_{\{n-1<\tau\}}\frac{1}{Z_{n-1}}\mathbb{E}[\Delta X_n\widetilde{Z}_n|\mathcal{F}_{n-1}]\,.$$

Before τ we can apply the results for progressive enlargement. We now compute the after τ part.

$$\mathbb{1}_{\{\tau\leq n-1\}}\mathbb{E}[\Delta X_n|\mathcal{K}_{n-1}] = \sum_{j=1,k=-\infty}^{j=n-1,k=\infty}\frac{\mathbb{1}_{\{\tau=j\}}\mathbb{1}_{\{\zeta=k\}}}{p_{n-1}(j,k)}\mathbb{E}\left[\mathbb{1}_{\{\tau=j\}}\mathbb{1}_{\{\zeta=k\}}\Delta X_n|\mathcal{F}_{n-1}\right].$$

It follows that

$$X = X^{\mathbb{K}} + \mathbb{1}_{[0,\tau[}(Z_-)^{-1}\centerdot\langle X,\widetilde{M}\rangle^{\mathbb{F}} + \sum_{k=-\infty}^{k=\infty}\mathbb{1}_{[\tau,\infty]}\frac{1}{p_-(j,k)}\centerdot\langle X, p(j,k)\rangle^{\mathbb{F}}|_{k=\zeta,j=\tau}\,,$$

where $X^{\mathbb{K}}$ is a $\mathbb{K}$-martingale. This decomposition is a mix of Equation (11) and (36). The case of (11) corresponds to the particular case $\tau \equiv 0$ whereas (36) corresponds to $\zeta \equiv 0$.

8.2 Enlargement with a Process

Let Y be a process and consider the enlargement of $\mathbb{F}$ of the form $\mathcal{G}_n = \mathcal{F}_n \vee \sigma(Y_k, k = 0, \dots, n)$.[3]

We assume that X is a given $\mathbb{F}$-martingale. For $n \geq 0$ let $U_n(dy)$ be the regular conditional distribution of the n-dimensional vector random vector $\mathbf{Y_{n-1}} = (Y_0, \dots, Y_{n-1})$ with respect to $\mathcal{F}_n$ and let $V_n(dy)$ be the regular conditional distribution of $\mathbf{Y_{n-1}}$ with respect to $\mathcal{F}_{n-1}$. In the following we will make the following assumption.

(AC) $U_n(dy)$ is absolutely continuous w.r.t. $V_n(dy)$ for all $n \geq 1$.

[3]The results of this section are based on some notes written by Stefan Ankirchner during his stay in Evry in September 2014.

If Condition (AC) is satisfied, then we can define the density $\ell_n(y) := \frac{U_n(dy)}{V_n(dy)}$. We next show that we can express the information drift in terms of ℓ.

Proposition 8.1 *Suppose that (AC) holds true. Then the information drift of X w.r.t. to $\mathbb{G}$ is given by*

$$A_n = \sum_{k=1}^{n} \langle X, \ell(y)\rangle_k \Big|_{y=\mathbf{Y_{k-1}}}. \tag{43}$$

For the proof of Proposition 8.1 we need the following auxiliary result.

Lemma 8.2 *Let $n \geq 1$ and $f : \mathbb{R}^n \times \Omega \to \mathbb{R}$ be a $\mathscr{B}(\mathbb{R}^n) \otimes \mathscr{F}_{n-1}$-measurable non-negative map. Then*

$$f(\mathbf{Y_{n-1}}, \cdot) = \int f(y, \cdot) V_n(dy).$$

Proof The lemma follows from a monotone class theorem.

Moreover, we have the following.

Lemma 8.3 *Let $n \geq 1$ and $\psi : \mathbb{R}^n \times \Omega \to \mathbb{R}$ be a $\mathscr{B}(\mathbb{R}^n) \otimes \mathscr{F}_n$-measurable non-negative map. Then*

$$\mathbb{E}\left[\int \psi(y, \cdot)\, V_n(dy) \middle| \mathscr{F}_{n-1}\right] = \int \mathbb{E}\left[\psi(y, \cdot) | \mathscr{F}_{n-1}\right]\, V_n(dy). \tag{44}$$

Proof We use a monotone class argument. Suppose first that $\psi(y, \omega) = \mathbb{1}_B(y)\mathbb{1}_C(\omega)$ with $B \in \mathscr{B}(\mathbb{R}^n)$ and $C \in \mathscr{F}_n$. Then we have

$$\begin{aligned}
&\mathbb{E}\left[\int \psi(y, \cdot)\, V_n(dy) \middle| \mathscr{F}_{n-1}\right] = \mathbb{E}\left[\int \mathbb{1}_B(y)\mathbb{1}_C(\omega)\, V_n(dy) \middle| \mathscr{F}_{n-1}\right] \\
=&\mathbb{E}\left[\mathbb{1}_C(\omega) V_n(B) | \mathscr{F}_{n-1}\right] = V_n(B)\mathbb{P}(C|\mathscr{F}_{n-1}) \\
=&\int \mathbb{1}_B(y)\mathbb{P}(C|\mathscr{F}_{n-1})\, V_n(dy) = \int \mathbb{E}\left[\psi(y, \cdot)|\mathscr{F}_{n-1}\right]\, V_n(dy).
\end{aligned}$$

The claim follows for arbitrary non-negative $\mathscr{B}(\mathbb{R}^n) \otimes \mathscr{F}_n$-measurable functions ψ via a monotone class theorem.

Proof *(of Proposition* 8.1*).* Let $A \in \mathscr{F}_{n-1}$ and $C \in \mathscr{B}(\mathbb{R}^n)$. Moreover, let

$$\psi(y, \omega) = \mathbb{1}_A(\omega)\mathbb{1}_C(y)\Delta X_n(\omega)\ell_n(y, \omega)\,.$$

Then Lemma 8.2 implies, with $f(y, \cdot) = \mathbb{E}[\psi(y, \cdot)|\mathscr{F}_{n-1}]$,

$$\begin{aligned}
&\mathbb{E}[\mathbb{1}_A \mathbb{1}_{\{\mathbf{Y_{n-1}} \in C\}} \mathbb{E}[\Delta X_n \ell_n(y)|\mathscr{F}_{n-1}]|_{y=\mathbf{Y_{n-1}}}] = \mathbb{E}[\mathbb{E}[\mathbb{1}_A \mathbb{1}_C(y)\Delta X_n \ell_n(y)|\mathscr{F}_{n-1}]|_{y=\mathbf{Y_{n-1}}}] \\
=&\mathbb{E}[\mathbb{E}[\psi(y, \cdot)|\mathscr{F}_{n-1}]|_{y=\mathbf{Y_{n-1}}}] = \mathbb{E}\Big[\int \mathbb{E}[\psi(y, \cdot)|\mathscr{F}_{n-1}] V_n(dy)\Big].
\end{aligned}$$

Now Lemma 8.3 further yields

$$\mathbb{E}[\mathbb{1}_A \mathbb{1}_{\{\mathbf{Y_{n-1}} \in C\}} \mathbb{E}[\Delta X_n \ell_n(y)|\mathscr{F}_{n-1}]|_{y=\mathbf{Y_{n-1}}}] = \mathbb{E}[\int \mathbb{1}_A \mathbb{1}_C(y) \Delta X_n U_n(dy)]$$
$$= \mathbb{E}[\mathbb{1}_A \Delta X_n \mathbb{P}(\{\mathbf{Y_{n-1}} \in C\}|\mathscr{F}_n)] = \mathbb{E}[\mathbb{1}_A \mathbb{1}_{\{\mathbf{Y_{n-1}} \in C\}} \Delta X_n].$$

This shows $\mathbb{E}[\Delta X_n|\mathscr{G}_{n-1}] = \mathbb{E}[\Delta X_n \ell_n(y)|\mathscr{F}_{n-1}]|_{y=\mathbf{Y_{n-1}}}$, and hence the result.

9 Credit Risk

In credit risk, one classical way to model the bankruptcy time of a firm is to use a default time. The default time τ is a random time and $\mathbb{F}$ is a reference filtration, for example, the filtration of default-free prices. In the literature, the default time is supposed to be not a stopping time in the filtration $\mathbb{F}$. The information of an agent who observes $\mathbb{F}$ and the time when the bankruptcy is appeared is modelled by the progressive enlargement of $\mathbb{F}$ with τ, denoted by $\mathbb{G}$. One can refer to [8] for a deep presentation in continuous time and to [16] for a study in discrete time. We introduce $p_n(k) = \mathbb{P}(\tau = k|\mathscr{F}_n)$ and H, the indicator of the default where $H_n = \mathbb{1}_{\{\tau \le n\}}$. The process $H - \Lambda$ is a $\mathbb{G}$-martingale where Λ is defined in Lemma 3.9. We assume that $\mathbb{P}$ is the pricing measure. The price of a defaultable zero coupon, that is, a claim with payment at N equal to $\mathbb{1}_{\{N<\tau\}}$ is

$$\mathbb{P}(N < \tau|\mathscr{G}_n) = \mathbb{1}_{\{n<\tau\}} \frac{\mathbb{P}(N < \tau|\mathscr{F}_n)}{\mathbb{P}(n < \tau|\mathscr{F}_n)}$$

where, for any k,

$$\mathbb{P}(\tau > k|\mathscr{F}_n) = \sum_{j=k+1}^{\infty} p_n(j)$$

The price of a recovery $h(\tau)$ paid at time τ, where h is a function, is

$$\mathbb{P}(h(\tau)\mathbb{1}_{n<\tau\le N}|\mathscr{G}_n) = \mathbb{1}_{\{n<\tau\}} \frac{\mathbb{E}[h(\tau)\mathbb{1}_{n<\tau\le N}|\mathscr{F}_n]}{\mathbb{P}(\tau > n|\mathscr{F}_n)}$$

where

$$\mathbb{E}[h(\tau)\mathbb{1}_{n<\tau\le N}|\mathscr{F}_n] = \sum_{k=n+1}^{N} h(k)\mathbb{P}(\tau = k|\mathscr{F}_n)$$

We present now the case of two defaults, for which we shall enter in enlargement of filtration methodology. This simple example underlines the role of filtration in a pricing framework.

Let τ_1, τ_2 be two default times, and $H^i_n = \mathbb{1}_{\{\tau_i \le n\}}$. The filtration generated by H^i is denoted $\mathbb{H}^i$. We consider $\mathbb{G}$ the progressive enlargement of $\mathbb{F}$ with the pair τ_1, τ_2, i.e. $\mathscr{G}_n = \mathscr{F}_n \vee \sigma(\tau_1 \wedge n) \vee \sigma(\tau_2 \wedge n)$. For $i = 1, 2$, we denote by $\mathbb{G}^i$ the progressive enlargement of $\mathbb{F}$ with τ_i. Note that $\mathbb{G} = \mathbb{G}^1 \vee \mathbb{H}^2$.

In a first step, we compute the $\mathbb{G}$ predictable process Λ^1 such that $H^1 - \Lambda^1$ is a $\mathbb{G}$ martingale. From Doob's decomposition $\Delta\Lambda^1_n = \mathbb{P}(\tau_1 = n|\mathscr{G}_{n-1})$.

$$\mathbb{P}(\tau_1 = n|\mathscr{G}_{n-1}) = \mathbb{1}_{\{\tau_1 > n-1\}} \frac{\mathbb{P}(\tau_1 = n|\mathscr{G}^2_{n-1})}{\mathbb{P}(\tau_1 > n-1|\mathscr{G}^2_{n-1})}$$

where

$$\begin{aligned}\mathbb{P}(\tau_1 = n|\mathscr{G}^2_{n-1}) &= \mathbb{1}_{\{\tau_2 > n-1\}} \frac{\mathbb{P}(\tau_1 = n,\ \tau_2 > n-1|\mathscr{F}_{n-1})}{\mathbb{P}(\ \tau_2 > n-1|\mathscr{F}_{n-1})} \\ &\quad + \sum_{k=0}^{n-1} \mathbb{1}_{\{\tau_2 = k\}} \frac{\mathbb{P}(\tau_1 = n,\ \tau_2 = k|\mathscr{F}_{n-1})}{\mathbb{P}(\tau_2 = k|\mathscr{F}_{n-1})}\,.\end{aligned}$$

We set $p_n(i, j) = \mathbb{P}(\tau_1 = i, \tau_2 = j|\mathscr{F}_n)$ and $p^{(2)}_n(k) = \mathbb{P}(\tau_2 = k|\mathscr{F}_n)$

$$\mathbb{P}(\tau_1 = n|\mathscr{G}^2_{n-1}) = \mathbb{1}_{\{\tau_2 > n-1\}} \frac{\sum_{k=n}^{\infty} p_{n-1}(n, k)}{\sum_{k=n}^{\infty} p^{(2)}_{n-1}(k)} + \sum_{k=0}^{n-1} \mathbb{1}_{\{\tau_2 = k\}} \frac{p_{n-1}(n, k)}{p^{(2)}_{n-1}(k)}\,.$$

The same computations yield to

$$\mathbb{P}(\tau_1 > n-1|\mathscr{G}^2_{n-1}) = \mathbb{1}_{\{\tau_2 > n-1\}} \frac{\sum_{k=n}^{\infty} \sum_{i=n}^{\infty} p_{n-1}(i, k)}{\sum_{k=n}^{\infty} p^{(2)}_{n-1}(k)} + \sum_{k=0}^{n-1} \mathbb{1}_{\{\tau_2 = k\}} \frac{\sum_{i=n-1}^{\infty} p_{n-1}(i, k)}{p^{(2)}_{n-1}(k)}\,.$$

Finally

$$\mathbb{P}(\tau_1 = n|\mathscr{G}_{n-1}) = \mathbb{1}_{\{\tau_2 > n-1\}} \frac{\sum_{k=n}^{\infty} p_{n-1}(n, k)}{\sum_{k=n}^{\infty} \sum_{i=n}^{\infty} p_{n-1}(i, k)} + \sum_{k=0}^{n-1} \mathbb{1}_{\{\tau_2 = k\}} \frac{p_{n-1}(n, k)}{\sum_{i=n-1}^{\infty} p_{n-1}(i, k)}\,.$$

References

1. Acciaio, B., Fontana, C., Kardaras, C.: Arbitrage of the first kind and filtration enlargements in semimartingale financial models. Stoch. Process. Their Appl. **126**, 1761–1784 (2016)
2. Aksamit, A.: Random times, enlargement of filtration and arbitrages. Thèse de Doctorat. Univ, Evry (2014)

3. Aksamit, A., Li, L.: Projections, pseudo-stopping times and the immersion property. In: Donati-Martin, C., Lejay, A., Rouault, A. (eds.) Séminaire de Probabilités XLVIII. Lecture Notes in Mathematics, vol. 2168. Springer, Berlin (2016)
4. Aksamit, A., Choulli, T., Jeanblanc, M.: On an optional semimartingale decomposition and the existence of a deflator in an enlarged filtration. In: Memoriam Marc Yor - Séminaire de Probabilités XLVII, pp. 187–218. Springer, Berlin (2015)
5. Aksamit, A., Choulli, T., Deng, J., Jeanblanc, M.: Non-arbitrage up to random horizon for semimartingale models. Financ. Stoch. **21**, 1103–1139 (2017)
6. Aksamit, A., Jeanblanc, M.: Enlargement of Filtrations with Finance in View. Springer, Berlin (2017)
7. Barlow, M.T.: Study of filtration expanded to include an honest time Z. Wahr. Verw. Gebiete **44**, 307–323 (1978)
8. Bielecki, T.R., Rutkowski, M.: Credit Risk: Modelling Valuation and Hedging. Springer, Berlin (2001)
9. Brémaud, P., Yor, M.: Changes of filtration and of probability measures. Z. Wahr. Verw. Gebiete **45**, 269–295 (1978)
10. Choulli, T., Daveloose, C., Vanmaele, M.: Hedging mortality risk and optional martingale representation theorem for enlarged filtration (2015). Preprint, arXiv:1510.05858
11. Choulli, T., Deng, J.: Non-arbitrage for informational discrete time market models. Stoch.S Int. J. Probab. Stoch. Process. **89**(3–4), 628–653 (2017)
12. Coculescu, D., Jeanblanc, M., Nikeghbali, A.: Default times, non arbitrage conditions and change of probability measure. Financ. Stoch. **16**(3) 513–535 (2012)
13. Dellacherie, C.: Capacités et processus stochastiques. Springer, Berlin (1972)
14. Fontana, C., Jeanblanc, M., Song, S.: On arbitrages arising from honest times. Financ. Stoch. **18**, 515–543 (2014)
15. Föllmer, H., Schied, A.: Stochastic Finance: An Introduction in Discrete Time. de Gruyter, Berlin (2004)
16. Gouriéroux, Ch., Monfort, A., Renne, J.-P.: Pricing default events: surprise, exogeneity and contagion. J. Econ. **182**, 397–411 (2014)
17. Jacod, J.: Grossissement initial, hypothèse H' et théorème de Girsanov, in Grossissements de filtrations: exemples et applications. Lecture Notes in Mathematics, Séminaire de Calcul Stochastique 1982–83, vol. 1118. Springer, Berlin (1987)
18. Jacod, J., Shiryaev, A.N.: Local martingales and the fundamental asset pricing theorems in the discrete-time case. Financ. Stoch. **2**(3), 259–273 (1998)
19. Jeulin, T.: Semi-martingales et grossissement d'une filtration. Lecture Notes in Mathematics, vol. 833. Springer, Berlin (1980)
20. Jeulin, T., Yor, M.: Grossissement d'une filtration et semi-martingales: formules explicites. In: Séminaire de Probabilités XII. Lecture Notes in Mathematics, vol. 649, pp. 78–97. Springer, Berlin (1978)
21. Kabanov, Y.: In discrete time a local martingale is a martingale under an equivalent probability measure. Financ. Stoch. **12**(3), 293–297 (2008)
22. Privault, N.: Discrete time models. http://www.ntu.edu.sg/home/nprivault/MA5182/discrete-time-model.pdf
23. Neveu, J.: Martingales à temps discret. Masson (1972)
24. Nikeghbali, A., Yor, M.: A definition and some properties of pseudo-stopping times. Ann. Probab. **33**, 1804–1824 (2005)
25. Prokaj, V., Ruf, J.: Local martingales in discrete time (2016). Preprint, arXiv:1701.04025.pdf
26. Revuz, D., Yor, M.: Continuous Martingales and Brownian Motion, 3rd edn. Springer, Berlin (1999)
27. Shreve, S.: Stochastic Calculus Models for Finance, Discrete Time. Springer, Berlin (2004)
28. Shiryaev, A.N.: Essentials of Stochastic Finance. World Scientific, Singapore (1999)
29. Spreij, P.: Measure Theoretic Probability. Lecture Notes. Amsterdam University, Amsterdam (2015). https://staff.fnwi.uva.nl/p.j.c.spreij/onderwijs/master/mtp.pdf

Clustering Effects via Hawkes Processes

Guillaume Bernis and Simone Scotti

Abstract Hawkes processes are a class of self-exciting point processes. They are characterized by the presence of clusters of jumps, which can be found in many natural phenomena (from biology to seismology) as well as in finance. First, we study the general properties of point processes. Then, we propose a constructive approach to Hawkes processes. Main properties as well as extensions are also studied. Special attention has been paid to the application of Hawkes processes in finance. We also propose to illustrate theses processes as a special case of branching processes with immigration.

Keywords Hawkes processes · Self-exciting property · Clusters · Branching processes with immigration

Mathematics Subject Classification (2010) 60G55 · 60G18 · 62M05

JEL Classification C60 · G12

1 Introduction

Hawkes processes are a class of self-exciting point processes, named after the mathematician A. G. Hawkes. This class of processes was introduced by Hawkes in [28]. Although his work was mainly theoretical, Hawkes suggested that this type of

Bernis—The analysis and views expressed in this chapter are those of the authors and do not necessarily reflect those of Natixis Assurances.

Scotti—This research is supported by Institute Europlace de Finance, research program "Clusters and Information Flow: Modelling, Analysis and Implications".

G. Bernis
Natixis Assurances, Paris, France
e-mail: guillaume.bernis@natixis.com

S. Scotti (✉)
LPSM - Université Paris Diderot, 8 place Aurélie Nemours, 75013 Paris, France
e-mail: scotti@math.univ-paris-diderot.fr

© Springer Nature Singapore Pte Ltd. 2020
Y. Jiao (ed.), *From Probability to Finance*, Mathematical Lectures
from Peking University, https://doi.org/10.1007/978-981-15-1576-7_3

process may be useful to represent some specific patterns in epidemiology. Indeed, the occurrence of a disease may increase the probability of other infections. Further applications in the study of earthquakes and their aftershocks were proposed in [29]. In both contexts, the spirit is the same: the occurrence of an event increases the probability of another one. Thus, paths of Hawkes processes are characterized by the presence of clusters of jumps. This feature can find applications in many areas, especially in finance. Indeed, these clusters can be found in the arrival of orders in trading books or in the prices of many financial assets. They can also represent the occurrence of defaults among credit issuers, capturing a contagion effect. See, among other references, [2, 14, 22].

From a mathematical point of view, the key feature of the Hawkes processes is a feedback of the past points on the current intensity of the point process. At each time of jump, the intensity of the process jumps from a given size and then goes back to its initial level with an exponential decay. Thus, the intensity, at a given time, depends on the strict past history of the point process. Loosely speaking, the intensity is given by

$$\lambda(t) = \lambda_0 + \sum_{T_i < t} g(t - T_i)$$

where $\lambda_0 > 0$, $g \geq 0$ satisfies $\int_0^{+\infty} g(t)dt < +\infty$ and $(T_i)_{i \geq 1}$ is the sequence of jumps of the point process. The classical case, which will be at the core of our analysis, corresponds to the choice $g(t) := e^{-\beta t}$, with $\beta > 0$.

Originally, the Hawkes processes were treated from a spectral point of view, cf. [28, 30], and extended to the case of mutually exciting processes. The estimation of the parameters of a Hawkes process was carried out by Ozaki [47]. An important generalization, due to Brémaud and Massoulié [8], extends the framework of Hawkes processes to the case where the intensity is given by

$$\lambda(t) = \Phi\left(\sum_{T_i < t} g(t - T_i)\right)$$

where Φ is a positive, possibly nonlinear mapping. According to Brémaud and Massoulié, we will call these processes nonlinear Hawkes processes.

One of the fascinating properties of the Hawkes processes is that they can be studied either by focussing on their intensity or on the structure of their clusters. Following the intensity-based approach, a Hawkes process can be easily constructed with a standard Poisson process, at least by two different methods: a change of time or a change of probability. The first method has several interesting implications. First, it shows, in a very constructive way, that the process exists and is non-explosive. Second, it provides a way to simulate it. The second method—the change of probability—is also important because it makes the link between the Hawkes process and its likelihood ratio. These approaches are studied in details in Sect. 3.

The cluster-based approach finds its roots in the analogy with another class of processes, initially used to study the dynamics of populations: the branching processes with immigration. In the case of Hawkes processes, each jump of a Poisson process (the ancestors) produces jumps (second generation) according to an exponentially decreasing intensity. Each jump of the second generation also produces its own clustering effect, and so on. Modern treatment of branching processes can provide useful results on the Hawkes processes that would be difficult to obtain from the intensity-based vision. This specific question will be the main topic of Sect. 7.

The developments of this chapter are the following. First, we study the general theory of point processes, with a specific focus on the Poisson processes, which are closely related to Hawkes processes. Then, we provide rigorous constructions of Hawkes processes. Section 4 is dedicated to the computation of moments of the Hawkes processes. In Sect. 5, we investigate several generalizations of Hawkes processes, especially nonlinear Hawkes processes. Section 6 summarizes some of the numerous applications of Hawkes processes in finance. To conclude this chapter, we put in perspective Hawkes processes with branching processes with immigration. This approach has some fruitful implications and provides several complex properties of Hawkes processes as particular cases of general results on branching processes.

2 Point Processes: An Overview

This section introduces some basic notions on point processes and marked point processes. It sets out the notations that will be used throughout the chapter and gives the properties of point processes that will be required in our study of Hawkes processes.

This introduction focuses on point processes, but it does not intend to provide a comprehensive treatment of this wider field. The reader can find an in-depth exposition of this topic in [39] or in Chapter II of [31].

2.1 What is a Point Processes?

We now work on a given probability space $(\Omega, \mathscr{F}, \mathbb{P})$, on which all random elements are defined, unless explicitly stated otherwise. The expectation under $\mathbb{P}$ is denoted by $\mathbb{E}^{\mathbb{P}}\{\cdot\}$ or simply $\mathbb{E}\{\cdot\}$ when there is no ambiguity on the probability.

We introduce point processes as follows.

Definition 2.1 (*Point process*) A point process (PP) is a (non-decreasing) sequence of positive random times $\mathscr{T} := (T_n)_{n\geq 1}$, such that

$$\begin{cases} T_n < T_{n+1} \text{ if } T_{n+1} < +\infty \\ T_n = T_{n+1} \text{ if } T_n = +\infty \end{cases}$$

The random times, T_n, $n \geq 0$, denote the times of jumps of the PP. Alternatively, it may be interesting to define $\tau_k := T_k - T_{k-1}$, for $k > 1$, and $\tau_1 := T_1$, which are the inter-jump times. For example, if $(\tau_k)_{k\geq 1}$ are independent random variables distributed according to the exponential law with parameter $\gamma > 0$, then the PP constructed on these inter-jump times is the (homogeneous) Poisson process, with intensity γ, that we will study below. It is convenient to work with the *counting measure* of the PP:

Definition 2.2 (*Counting measure*) The *counting measure* of the PP $\mathscr{T}$ is the integer-valued random measure N on $\big(\mathbb{R}^{*+}, \mathscr{B}(\mathbb{R}^+)\big)$, defined by

$$N(A) := \sum_{n\geq 1} \mathbb{I}_{\{T_n \in A\}}, \quad A \in \mathscr{B}(\mathbb{R}^+) \tag{1}$$

Here, we have denoted by $\mathscr{B}(\mathbb{R}^+)$ the Borelian sigma-field of $\mathbb{R}^+$. Note that the measure N is a random element, and hence a mapping from $\Omega \times \mathscr{B}(\mathbb{R}^+)$ into $\mathbb{N}$. Let us set $N_t := N(]0, t])$, in order to define the process $(N_t)_{t\geq 0}$ which is the *counting process*. Under the convention of Definition 2.1, we have $N_0 = 0$. As also specified in Definition 2.1, the times of jumps may be infinite. It naturally leads to the following definition.

Definition 2.3 (*Time of explosion*) The *time of explosion* of the PP $\mathscr{T}$ is $T^e := \sup\{T_n,\ n \geq 1\}$. If $T^e = +\infty$ $\mathbb{P}$-a.s., the process is said to be *non-explosive*.

Characterization of non-explosive PP can be found in [7] section III.A.4, p. 120 or [39] Exercise 1.5.2, p. 9. We shall go back to this notion later on, in the case of Hawkes processes. In many applications, it will be convenient to add a *mark* to a PP. For example, it can represent the magnitude of an earthquake, the recovery rate of a credit event, the size of a jump on the price of a stock, the size of the orders in a trading book, etc. In a general case, the mark will be observed on some measurable space $(X, \mathscr{X})$.

Definition 2.4 (*Marked point process*) A *marked point process* (MMP) is given by a sequence $\mathscr{T}^{\mathscr{M}} := (T_n, X_n)_{n\geq 1}$, where $(T_n)_{n\geq 1}$ is a PP and $(X_n)_{n\geq 1}$ a sequence of random variables on $(X, \mathscr{X})$.

The counting measure can be defined, in this case, as

$$M(B) := \sum_{n\geq 1} \mathbb{I}_{\{(T_n,\ X_n)\in B\}}, \quad B \in \mathscr{B}(\mathbb{R}^+) \otimes \mathscr{X}$$

Accordingly, for any $A \in \mathscr{B}(\mathbb{R}^+)$, $M(A \times X)$ is the counting measure of the PP defined by Eq. (1).

Now, we precise the measurability concepts that will be dealt with in this chapter. In the following, the filtration $\mathbb{F} := (\mathscr{F}_t)_{t\geq 0}$ represents the *natural filtration* of a MPP $\mathscr{T}^{\mathscr{M}}$. This filtration shows specific properties, stemming from the nature of point processes. First, let us consider, for any $t > 0$, the counting measures *stopped* at time t. They are defined for any $A \in \mathcal{B}(\mathbb{R}^+)$ and $B \in \mathcal{X}$, by

$$M^{(t)}(A \times B) := M(\{A \cap]0, t]\} \times B)$$
$$M^{(t^-)}(A \times B) := M(\{A \cap]0, t[\} \times B)$$

These stopped measures generate the natural filtrations associated to the MPP. More precisely, we state the following.

Proposition 2.5 *For any* $t \geq 0$, $\mathcal{F}_t = \sigma(M^{(t)})$ *and* $\mathcal{F}_{t^-} = \sigma(M^{(t^-)})$.

We recall that $\mathcal{F}_{t^-}$ (resp., $\mathcal{F}_{t^+}$) is the sigma-field generated by all the $\mathcal{F}_s$, $s < t$ (resp. $s > t$), if $t > 0$ and $\mathcal{F}_{0^-} = \mathcal{F}_{0^+} = \mathcal{F}_0$. Now, we introduce two concepts of measurability.

Definition 2.6 The *predictable sigma-field* is the sigma-field $\mathcal{P}$ generated by $\mathbb{F}$-adapted left-continuous (hereafter, càg) processes, seen as mappings on $\mathbb{R}^+ \times \Omega$. The *optional sigma-field* is the sigma-field $\mathcal{O}$ generated by $\mathbb{F}$-adapted right continuous with limit on the left (hereafter, càdlàg) processes.

These two concepts can be declined on processes and random measures.

Definition 2.7 (*Predictability—Optionality*) A process $(Y_t)_{t \geq 0}$ is predictable (resp., optional) if the mapping $(t, \omega) \mapsto Y_t(\omega)$, defined on $\mathbb{R}^+ \times \Omega$, is $\mathcal{P}$-measurable (resp., $\mathcal{O}$-measurable). A random measure μ on $\mathbb{R}^+ \times X$ is predictable (resp., optional) if the process $(Y_t)_{t \geq s}$, $Y_t := \mu(]s, t] \times B)$, is predictable (resp., optional), for any $s \geq 0$ and $B \in \mathcal{X}$.

According to this definition, every adapted, càg process is predictable. If Y is a predictable (resp. optional) process, then for any stopping time T, Y_T is $\mathcal{F}_{T^-}$-measurable (resp., $\mathcal{F}_T$-measurable). The following propositions provide some insight on the peculiarities of PP filtrations. Proofs of these results can be found in [39], Chap. 2.

Proposition 2.8 *The natural filtration of an MPP is right continuous:* $\mathcal{F}_t = \mathcal{F}_{t^+}$. *The predictable (resp., optional) sigma-field is generated by* $M^{(t^-)}$ *(resp.,* $M^{(t)}$*)*

Proposition 2.9 *Let* (T_n, X_n) *be a MPP. Then, we have*

$$\mathcal{F}_{T_n} = \sigma\left(T_1, \ldots, T_n, X_1, \ldots, X_n\right)$$
$$\mathcal{F}_{T_n^-} = \sigma\left(T_1, \ldots, T_n, X_1, \ldots, X_{n-1}\right)$$

The intuition behind this result is that the predictable sigma-field allows to see the jump, but not its effect. We also add this result on the form of the predictable mappings.

Proposition 2.10 *Consider a PP and its natural filtration. Then, every predictable process* h *can be written as* $h(t, \omega) = f\left(t, M^{(t^-)}(\omega)\right)$, *where* f *is a deterministic measurable function.*

2.2 Compensator and Intensity

The notion of predictability, presented in the previous section, is at the core of the notion of compensator, which we are about to introduce. The following result can be found, under a more general framework, in [17], Theorem 65 p. 386 or in [31], Theorem 1.8, p 66.

Theorem 2.11 (Compensator existence) *Let $\mathcal{T}$ be a MPP, M its counting measure and $\mathbb{F}$ its natural filtration. Assume that the measure $M(dt, dx) \otimes \mathbb{P}(d\omega)$ is σ-finite. Then, there exists a (non-negative) predictable measure $\mu(dt, dx)$, unique up to a $\mathbb{P}$-null set, such that, for any predictable mapping $f : \Omega \times \mathbb{R}^+ \times X \to \overline{\mathbb{R}}$, we have the following equality in $\overline{\mathbb{R}}$:*

$$\mathbb{E}\left\{\int_0^{+\infty} \int_X f(t,x)\, M(dt,dx)\right\} = \mathbb{E}\left\{\int_0^{+\infty} \int_X f(t,x)\, \mu(dt,dx)\right\} \tag{2}$$

The signed measure $\widetilde{M}(dt, dx) := M(dt, dx) - \mu(dt, dx)$ is called the *compensated* measure of the MPP. Property (2) is equivalent to the fact that $\widetilde{M}(dt, dx)$ is a *martingale measure*: for any predictable mapping h from $\mathbb{R}^+ \times X$ into $\mathbb{R}$, the process $\widetilde{M}[h]_t := \int_0^t \int_X h(s,x)\widetilde{M}(dt, dx)$, $t \geq 0$, is a local martingale.

Lemma 2.12 *Assume that $\mathbb{E}\left\{\int_0^T \int_X |h(s,x)|\mu(ds,dx)\right\} < +\infty$. Then, $(\widetilde{M}[h]_t)_{0\leq t\leq T}$ is a martingale.*

Proof Set $Y_t := \widetilde{M}[h]$. Theorem 47, p. 36 in [49], states that a sufficient condition for the local martingale $(Y_t)_{0\leq t\leq T}$ to be a martingale is that $\sup_{0\leq t\leq T} |Y_t|$ is integrable. We have

$$|Y_t| \leq \int_0^T \int_X |h(s,x)|\, [M(ds,dx) + \mu(ds,dx)]\,,\ \mathbb{P} - \text{a.s.}$$

By definition of the compensator, we obtain the following inequality:

$$\mathbb{E}\left\{\sup_{0\leq t\leq T} |Y_t|\right\} \leq 2\mathbb{E}\left\{\int_0^T \int_X |h(s,x)|\, \mu(ds,dx)\right\}$$

This yields the inequality needed. □

The compensator does depend on the filtration *and on the probability*. This last point will be emphasized in Sect. 2.4.

Definition 2.13 Let $\mathcal{T}$ be a PP, N its counting measure and $\mathbb{F}$ its natural filtration. The intensity λ is the adapted process defined, when this limit exists, by

$$\lambda(t) := \lim_{h\mapsto 0^+} \mathbb{E}\left\{\frac{N_{t+h} - N_t}{h} \,\middle|\, \mathcal{F}_t\right\} \tag{3}$$

The link between intensity and compensator is given by the following lemma.

Lemma 2.14 *Let $\mathscr{T}$ be a PP, N its counting measure and $\mathbb{F}$ its natural filtration. Assume that the PP admits a compensator, which is absolutely continuous with respect to the Lebesgue measure, i.e. $\mu(dt) := \lambda(t)dt$. Then, λ is the intensity of the PP.*

2.3 The Poisson Process

It is interesting to present in details the Poisson process because Hawkes and Poisson processes are closely related. For example, the Hawkes process can be defined by applying either a time change or a change of probability to a Poisson process. In this subsection, we adopt a definition of the Poisson process through the sequence of its inter-jump times. Generally, in the literature, the Poisson process is introduced through the theory of stationary processes with independent increments. In our context, the definition through the inter-jump times is more convenient and gives more insight on the construction of the Hawkes process.

Definition 2.15 The Poisson process with parameter $\gamma > 0$ is a PP such that its inter-jump times are independent, identically distributed (i.i.d.) exponential random variables, with parameter γ.

In the rest of the subsection, $\mathscr{T}^P = (S_n)_{n\geq 1}$ is a Poisson process, with counting measure N and inter-jump times $(\theta_k)_{k\geq 1}$. The following theorem states the main properties of a Poisson process.

Theorem 2.16 *The Poisson process has independent, identically distributed increments. Moreover, for every $0 \leq s < t$, $N_t - N_s$ follows a Poisson law of parameter $\gamma(t-s)$.*

Proof The idea of the proof is to show that $N_t - N_s$ is independent from N_s and follows a Poisson law of parameter $\gamma\,(t-s)$. But, as a starting point, we show that N_t follows a Poisson law of parameter $\gamma\, t$.

We have $\mathbb{P}\,(N_t = k) = \mathbb{P}\,(\{S_k \leq t\} \cap \{S_k + \theta_{k+1} > t\})$. By the independence between S_k and θ_{k+1}, we have

$$\mathbb{P}\,(N_t = k) = \mathbb{E}\left\{\mathbb{I}_{\{S_k \leq t\}} e^{-\gamma(t-S_k)}\right\}$$

An easy but tedious calculation shows that S_k, as a sum of i.i.d. Exponential random variables, follows a Gamma law of parameters (k, γ). Thus,

$$\mathbb{P}\,(N_t = k) = e^{-\gamma t} \int_0^{+\infty} \frac{\gamma^k}{(k-1)!} e^{-\gamma u} u^{k-1} e^{\gamma u} du = e^{-\gamma t} \frac{(\gamma t)^k}{k!}$$

This leads that N_t follows a Poisson law with parameter $\gamma\, t$. Now, let us prove that $N_t - N_s, 0 \le s < t$, is independent from $\{N_s = k\}$, for some $k \ge 0$, and follows a Poisson law of parameter $\gamma\, (t - s)$. Let $d \ge 0$ be an integer.

$$\begin{aligned}
&\mathbb{P}\left(\{N_s = k\} \cup \{N_t - N_s = d\}\right) \\
&= \mathbb{E}\left\{\mathbb{I}_{\{S_k \le s\}}\, \mathbb{I}_{\{S_k > s - \theta_{k+1}\}} \prod_{p=1}^{d} \mathbb{I}_{\left\{\sum_{j=1}^{p} \theta_{k+j} \le t - S_k\right\}} e^{-\gamma\, (t - S_k - \sum_{j=1}^{d} \theta_{k+j})}\right\}
\end{aligned} \tag{4}$$

Let us notice that $S_k + \sum_{j=1}^{d} \theta_{k+j} = S_{k+d-1} + \theta_{k+d}$. Conditionally with respect to S_{k+d-1}, and after integration according to the law of θ_{k+d}, we get

$$\begin{aligned}
&\mathbb{E}\left\{\mathbb{I}_{\{\theta_{k+d} \le t - S_{k+d-1}\}}\, e^{-\gamma\, (t - S_{k+d-1} - \theta_{k+d})} \mid S_{k+d-1}\right\} \\
&= \int_0^{t - S_{k+d-1}} \gamma e^{-\gamma\, u} e^{-\gamma (t - S_{k+d-1} - u)} du \\
&= \gamma\, (t - S_{k+d-1})\, e^{-\gamma\, (t - S_{k+d-1})}
\end{aligned}$$

Let us plug in Eq. (4). It yields

$$\begin{aligned}
&\mathbb{P}\left(\{N_s = k\} \cup \{N_t - N_s = d\}\right) \\
&= \mathbb{E}\left\{\mathbb{I}_{\{S_k \le s\}}\, \mathbb{I}_{\{S_k > s - \theta_{k+1}\}} \prod_{p=1}^{d-1} \mathbb{I}_{\left\{\sum_{j=1}^{p} \theta_{k+j} \le t - S_k\right\}}\, e^{-\gamma\, (t - S_{k+d-1})} \gamma\, (t - S_{k+d-1})\right\}
\end{aligned}$$

After iterating $d - 1$ times the previous step, we get

$$\begin{aligned}
&\mathbb{P}\left(\{N_s = k\} \cup \{N_t - N_s = d\}\right) \\
&= \mathbb{E}\left\{\mathbb{I}_{\{S_k \le s\}}\, \mathbb{I}_{\{S_k > s - \theta_{k+1}\}}\, \mathbb{I}_{\{\theta_{k+1} \le t - S_k\}}\, e^{-\gamma\, (t - S_{k+1})} \frac{[\gamma\, (t - S_{k+1})]^{d-1}}{(d-1)!}\right\} \\
&= \mathbb{E}\left\{\mathbb{I}_{\{S_k \le s\}} \int_{s - S_k}^{t - S_k} \gamma\, e^{-\gamma\, u} e^{-\gamma\, (t - S_k - u)} \frac{\gamma^{d-1}\, (t - S_k - u)^{d-1}}{(d-1)!} du\right\} \\
&= \mathbb{E}\left\{\mathbb{I}_{\{S_k \le s\}} \left[-\frac{(t - S_k - u)^d}{(d-1)!}\right]_{s - S_k}^{t - S_k} \gamma^d e^{-\gamma\, (t - S_k)}\right\} = \mathbb{E}\left\{\mathbb{I}_{\{S_k \le s\}} \frac{\gamma^d\, (t - s)^d}{d!} e^{-\gamma\, (t - S_k)}\right\} \\
&= e^{-\gamma (t - s)} \frac{\gamma^d\, (t - s)^d}{d!} e^{-\gamma s} \frac{(\gamma\, s)^k}{k!} \\
&= \mathbb{P}\left(N_{t-s} = d\right) \times \mathbb{P}\left(N_s = k\right)
\end{aligned}$$

It provides, on the one hand, the independence of $N_t - N_s$ and N_s, and, on the other hand (after taking the sum on k), the equality in law between $N_t - N_s$ and N_s. Hence, the result follows. □

2.4 Change of Probability

The theory of change of probability for Brownian motion is well known, since it has fruitful applications in financial mathematics: notion of risk-neutral probability or change of numéraire for interest rate derivatives. In this case the change of probability modifies the drift of the Brownian motion. In our context of PP, analogous results exist. In this case, the change of probability transforms the intensity of the PP. We are about to give such a result, under a very restrictive form. For general results, see Theorem 10.2.6, p. 339 in [39]. The version given here is adapted to the context of Hawkes processes.

Theorem 2.17 *Let $(S_n)_{n\geq 1}$ be a Poisson process with intensity $\gamma > 0$. Consider some predictable mapping $\zeta > -1$, such that, for any $t \geq 0$, $\mathbb{E}\left\{\int_0^t |\zeta(s)|ds\right\} < +\infty$. Define $L(t)$, for any $t \geq 0$, by*

$$L(t) = \exp\left(-\gamma \int_0^t \zeta(s)ds\right) \prod_{S_i \leq t} (1 + \zeta(S_i)) \tag{5}$$

If, for some $T \geq 0$, $\mathbb{E}\{L(T)\} = 1$, then there exists a probability $\mathbb{Q}$, such that $\frac{d\mathbb{Q}}{d\mathbb{P}}(t) = L(t)$ and, under $\mathbb{Q}$ the PP $(S_n)_{n\geq 1}$ admits a compensator equal to $\nu(t) := \gamma\ [1 + \zeta(t)]$, for $t \leq T$.

The process L, given by Eq. (5), is a Doléans-Dade exponential associated to a pure jump process. See, for instance, [49], Theorem 36, p. 77. If we denote by N the counting measure of the Poisson process, L is the Doléans-Dade exponential of the pure jump martingale $M_t := \int_0^t \zeta(s)\widetilde{N}(ds)$, $t \geq 0$. The Doléans-Dade exponential of a martingale is a local martingale (c.f. [31], Theorem 4.61, p. 59). Hence, we only have $\mathbb{E}\{L(t)\} \leq 1$, by Fatou's lemma. Some additional condition is required for the local martingale to be a true martingale, and, thus, define a change of probability. It is, for example, the case when ζ is bounded, because, in this case $\sup_{0\leq s\leq t} L(s)$ is integrable (see [49], Theorem 47 p. 36).

The quantity L can also be considered as a likelihood ratio, and it is extensively studied in the case of the statistical study of Hawkes processes. See, e.g. [46, 47].

3 Hawkes Processes

This section provides the main features of Hawkes processes. First, it develops a simple constructive proof of the existence of Hawkes processes. For this purpose, it harnesses the properties of the Poisson process, studied in Sect. 2.3. This approach has a very practical interest since it defines a way to simulate a Hawkes process.

3.1 A Time Change Approach to Hawkes Process

Let us consider $(S_n)_{n\geq 1}$ a Poisson process, with counting measure N^P and intensity 1. The inter-jump times are denoted by $(\theta_n)_{n\geq 1}$. By Definition 2.15, they are i.i.d exponential random variables of parameter 1. Set out $\lambda_0 > 0$, $\alpha \geq 0$ and $\beta > 0$. We construct the PP $(T_n)_{n\geq 1}$, with counting measure N^H, and inter-jump times $(\tau_n)_{n\geq 1}$ as follows. First, set $T_0 = S_0 = 0$, and define, for every $n \geq 0$,

$$\Lambda_n(t) := \lambda_0 t + \alpha \sum_{i=1}^{n} \frac{1 - e^{-\beta(t-T_i)}}{\beta} \tag{6}$$

$$\theta_{n+1} := \Lambda_n(T_{n+1}) - \Lambda_n(T_n) \tag{7}$$

Remark 1 The mapping Λ_n is continuous and increasing. Therefore, it defines a bijection from $]T_n, +\infty[$ onto $]\Lambda_n(T_n), +\infty[$. Hence, the recurrence defined by the system (6) is valid: T_{n+1} is always defined, whatever is the value of $\theta_{n+1} > 0$.

Definition 3.1 The PP defined by Eqs. (6)–(7) is called a Hawkes process.

It is easy to see that, when $\alpha = 0$, $\theta_n = \lambda_0 \tau_n$, for every $n \geq 1$, which defines a Poisson process with intensity λ_0. Hence, the Poisson process is included in the class of Hawkes processes. We can now state a first property of the Hawkes process.

Lemma 3.2 (Time change) *For every $n \geq 0$, $S_{n+1} = \Lambda(T_{n+1})$, or equivalently, $N^P_{\Lambda(t)} = N^H_t$, where $\Lambda(t) = \Lambda_n(t)$ on the random interval $]T_n, T_{n+1}]$.*

Proof According to Eq. (6), $\theta_{n+1} = \Lambda_n(T_{n+1}) - \Lambda_n(T_n)$. So, we have

$$S_{n+1} = \sum_{i=1}^{n} \theta_i = \sum_{i=1}^{n} \left[\Lambda_i(T_{i+1}) - \Lambda_i(T_i)\right]$$

But we can notice that $\Lambda_{n-1}(T_n) = \Lambda_n(T_n)$. So, we can write

$$S_{n+1} = \sum_{i=1}^{n} \Lambda_i(T_{i+1}) - \sum_{i=1}^{n} \Lambda_{i-1}(T_i) = \Lambda(T_{n+1})$$

Now, let us turn to the second assertion. Using the definition of the counting process and the first part of the lemma, we can write

$$\left\{N^H_t = n\right\} = \{T_n \leq t < T_{n+1}\} = \left\{\Lambda_n^{-1}(S_n) \leq t < \Lambda_n^{-1}(S_{n+1})\right\}$$

It follows that

$$\left\{N^H_t = n\right\} = \{S_n \leq \Lambda(t) < S_{n+1}\} = \left\{N^P_{\Lambda_n(t)} = n\right\}$$

Hence, the end of the proof. □

Now, let us turn to an important property of the Hawkes process.

Proposition 3.3 (Non-explosivity) *The Hawkes process is non-explosive.*

Proof From Eq. (6) and the convexity of the exponential, we derive the straightforward inequalities: $\lambda_0\tau_{k+1} \leq \theta_{k+1} \leq \lambda_0\tau_{k+1} + \alpha k\tau_{k+1}$. This implies the following inequalities on τ_{k+1}:

$$\frac{\theta_{k+1}}{\lambda_0 + \alpha k} \leq \tau_{k+1} \leq \frac{\theta_{k+1}}{\lambda_0}$$

Using these inequalities in the Laplace transform of τ_{k+1}, we obtain, for any $z > 0$,

$$\frac{1}{1+\frac{z}{\lambda_0}} \leq \mathbb{E}\left\{e^{-z\tau_{k+1}} \mid \mathscr{F}_{T_k}\right\} \leq \frac{1}{1+\frac{z}{\lambda_0+\alpha k}}$$

Set, for any $n \geq 0$, $L_n(z) := \mathbb{E}\left\{e^{-zT_n}\right\}$. We obtain, from the previous calculations and the independence of the θ_k, the following recurrence relationship, for any $n > 1$:

$$\frac{1}{1+\frac{z}{\lambda_0}} L_{n-1}(z) \leq L_n(z) \leq \frac{1}{1+\frac{z}{\lambda_0+\alpha k}} L_{n-1}(z)$$

With this recurrence at hand, we take the logarithm to prove that

$$-n \ln\left(1+\frac{z}{\lambda_0}\right) \leq \ln(L_n(z)) \leq -\sum_{k=0}^{n-1} \ln\left(1+\frac{z}{\lambda_0+\alpha k}\right)$$

The lower and upper boundaries go to minus infinity when n diverges. This shows that the Hawkes process is non-explosive. □

The time change approach has a direct consequence on the definition of the intensity.

Proposition 3.4 *The intensity of the Hawkes process is given by*

$$\lambda(t) = \lambda_0 + \alpha \sum_{T_i<t} e^{-\beta(t-T_i)} \tag{8}$$

The compensator of the Hawkes process is $\nu(dt) := \lambda(t)dt$.

Proof Let us set out $t \geq 0$ and consider the mapping $h \mapsto g(h) := \mathbb{E}\left\{N^H_{t+h} \mid \mathscr{F}_t\right\}$. We have

$$\frac{g(h)-g(0)}{h} = \frac{1}{h}\mathbb{E}\left\{N^P_{\Lambda(t+h)} - N^P_{\Lambda(t)} \mid \mathscr{F}_t\right\} = \frac{1}{h}\mathbb{E}\{\Lambda(t+h) - \Lambda(t) \mid \mathscr{F}_t\}$$

We easily see that $\Lambda'(t) = \lambda(t)$. We deduce from Eq. (8) that, for $h \in [0, H]$, $\lambda(t + h) \leq \lambda(t) + \alpha(N^P_{t+H} - N^P_t)$, which provides a uniform boundary on the derivative. Therefore, we can permute the expectation and the derivation to obtain

$$g'(0^+) = \lambda(t)$$

Now let us turn to the second assertion. Let f be some predictable mapping. Lemma 3.2 implies that

$$\mathbb{E}\left\{\int_0^t f(s)\, N^H(ds)\right\} = \mathbb{E}\left\{\int_0^{\Lambda(t)} f(\Lambda^{-1}(s))\, N^P(ds)\right\} = \mathbb{E}\left\{\int_0^{\Lambda(t)} f(\Lambda^{-1}(s))ds\right\}$$

A change of variable $u = \Lambda^{-1}(s)$ yields

$$\mathbb{E}\left\{\int_0^t f(s)N^H(ds)\right\} = \mathbb{E}\left\{\int_0^t f(u)du\right\}$$

This implies that $\lambda(t)dt$ is the compensator of the Hawkes process. □

We can see that Eq. (8) defines a predictable process, since it is adapted and càg. Later on, we will resort to a right continuous version of the intensity, but, for the purpose of these preliminary results, we stick to original formulations and to the framework defined in Sect. 2 concerning the compensator.

The definition of the recurrence system (6)–(7) is similar to the simulation method defined by Ozaki in [47]. It gives rise to a simple method to simulate a Hawkes process. From a practical point of view, it may be interesting to provide the algorithm as given by [47], since it incorporates some useful numerical requirements.

Algorithm 3.5 (*Ozaki (1978)*) The simulation of the n first times of jump of a Hawkes process can be performed by the following steps:

1. Generate a realization U of a uniform random generator.
2. Set $T_1 = -\ln(U)$ and $S_1 := 1$.
3. For $k = 2$ to n.

- Generate a realization U of a uniform random generator.
- Find T_k the solution in x of the following equation:

$$\ln(U) + \lambda_0\,(x - T_{k-1}) + \alpha S_{k-1}\frac{1 - e^{-\beta(x-T_{k-1})}}{\beta} = 0 \tag{9}$$

  This root finding method can be achieved by performing a Newton method with first guess $x_0 := T_{k-1} - \frac{\ln(U)}{\lambda_0}$.
- Set $S_k = 1 + S_{k-1}\frac{e^{-\beta(T_k - T_{k-1})}}{\beta}$.

Remark 2 A by-product of this algorithm, which is also clear from (6), is that the first jump of the Hawkes process is distributed according to an exponential law with parameter λ_0. Therefore, the probability that no jump occurs on $[0, t]$ is simply $\mathbb{P}\left(\left\{N_t^H = 0\right\}\right) = e^{-\lambda_0 t}$.

Another consequence of the previous algorithm is that it can be used to represent the distribution of N_t^H and analyse the influence of the various parameters on this shape. First, let us consider a fixed horizon of $t = 5$ years. Set $\beta = 0.5$ year^{-1}. We consider several distributions of Hawkes processes, with different values of α, so that the mean of the distributions are the same. In our example, the theoretical mean value is $\mathbb{E}\left\{N_t^H\right\} = 7.76$, which means that, in average, the Hawkes process shows between 7 and 8 jumps in 5 years. The average can be computed by the mean of Eq. (14). By inverting this equation, for each λ_0, we can find a value of α such that N_t^H has the same mean. We consider four values of λ_0: 0.8, 1.0, 1.2 and 1.552 year^{-1}. The values of α are, respectively, 0.425, 0.3, 0.186 and 0. This means that the last distribution is a Poisson distribution. The results are displayed in Fig. 1.

As shown in Fig. 1, both skewness and kurtosis increase with α (decreasing λ_0). Thus, the three distributions of Hawkes processes ($\alpha > 0$) can be seen as mean preserving spreads of the Poisson distribution.

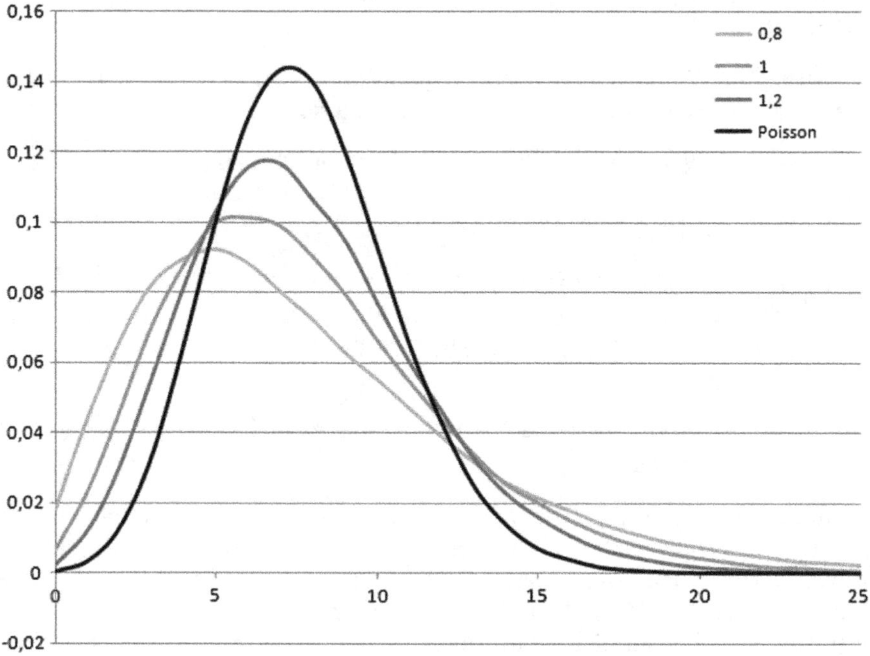

Fig. 1 Empirical distribution of N_5^H, based on 200 000 Monte Carlo simulations, for $\beta = 0.5$, $\lambda_0 \in \{0.8, 1.0, 1.2, 1.552\}$ and $\mathbb{E}\left\{N_5^H\right\} = 7.76$. The case $\lambda_0 = 1.552$ is a Poisson distribution

3.2 A Change of Probability Approach of the Hawkes Process

It is possible to define the Hawkes process as the result of a change of probability, as given in Theorem 2.17. Let us consider $\mathscr{T} := (S_n)_{n\geq 1}$ a Poisson process, with counting measure N^P and intensity $\gamma > 0$.

Theorem 3.6 *Let λ be defined by Eq. (8). Set*

$$L(t) = \exp\left(-\int_0^t (\lambda(s) - \gamma)ds\right) \prod_{S_i \leq t} \frac{\lambda(S_i)}{\gamma}$$

Then, we can define a probability $\mathbb{Q}$, by setting $\frac{d\mathbb{Q}}{d\mathbb{P}}(t) := L(t)$, and, under $\mathbb{Q}$, $\mathscr{T}$ is a Hawkes process.

Proof We can apply Theorem 3.6 with $\zeta(t) := \frac{\lambda(t)}{\gamma} - 1$. It yields the appropriate form for L. Now, we need to show that it actually defines a probability. We can resort to the general result in [36], which provides sufficient conditions for the Doléans-Dade exponential of a martingale to be a true martingale. In our context of PP, this result can be applied because $\lambda(t) \leq \lambda_0 + \alpha N_t^P$. □

The change of probability given by Theorem 3.6 gives rise to a simple Monte Carlo algorithm.

Algorithm 3.7 Consider some mapping f, such that $\mathbb{E}^{\mathbb{Q}}\{f(N_t)\}$. One simulation of $X_t := f(N_t)$ can be performed by the following steps:

1. Generate a realization U of a uniform random generator. Set $k = 1$, $T_1 := \frac{-\ln(U)}{\gamma}$ and $H_1 := \frac{\lambda_0}{\gamma}$.
2. While $T_k \leq t$ do
    - Generate a realization U of a uniform random generator. Set $T_k := T_{k-1} - \frac{\ln(U)}{\gamma}$.
    - Set $H_k := \frac{\lambda(T_k)}{\gamma}$, where λ is given by Eq. (8).
    - Increment k: $k \leftarrow k + 1$.
3. Calculate $\Lambda := \int_0^t \lambda(s)ds$ and $L := exp(-\Lambda + \gamma\, t)\, \prod_{i=1}^{k-1} H_k$.
4. Set $X_t = f(k-1)\, L_t$.

Remark 3 Despite the simplicity of this algorithm, some numerical instability may arise when γ is too small or too large. It is particularly important when simulating extreme events in the distribution of N_t. For example, consider $f(x) = \mathbb{I}_{\{x=n\}}$, for some large integer n. If no Monte Carlo path produces n jumps, then the value will be 0. This event is controlled by the value of γ, which can be seen as an importance sampling parameter.

The density L_t is not only the change of probability but can be seen as a *likelihood ratio*. Therefore, it can be used to estimate the parameters of the Hawkes process (cf.

[47]). This issue will be studied more specifically through examples in Sect. 6. But it may be important to provide directly the fundamental properties of this method. This is the purpose of next subsection.

3.3 Maximum Likelihood Estimation

We focus on the statistical estimation. The results of this subsection are due to Ozaki [47].

Proposition 3.8 *Let $\mathscr{T} = \{T_1, \ldots, T_N\}$ be a realization of a point process, the intensity of which is denoted by λ, then the log-likelihood reads*

$$\mathscr{L}(T_1, \ldots, T_N) = -\int_0^{T_N} \lambda(t)dt + \sum_i \log \lambda(T_i)$$

The next corollary gives the particular result for Hawkes processes.

Corollary 3.9 *Let $\mathscr{T} = (T_n)_{n\geq 1}$ be a Hawkes process, the intensity of which is given by Eq. (8). Then the log-likelihood reads*

$$\begin{aligned}\mathscr{L}(T_1, \ldots, T_N) = &-\lambda_0 T_N + \sum_{i=1}^{N} \frac{\alpha}{\beta}\left[e^{-\beta(T_N - T_i)} - 1\right] \\ &+ \sum_{i=1}^{N} \log\left[\lambda_0 + \alpha \sum_{j=1}^{i-1} e^{-\beta(T_i - T_j)}\right]\end{aligned}$$

Moreover, the first-order conditions that give rise to the maximum likelihood estimator are given by

$$0 = -T_N + \sum_{i=1}^{N}\left\{\lambda_0 + \alpha \sum_{j=1}^{i-1} e^{-\beta(T_i - T_j)}\right\}^{-1}$$

$$0 = \sum_{i=1}^{N} \frac{1}{\beta}\left[e^{-\beta(T_N - T_i)} - 1\right] + \sum_{i=1}^{N} \frac{\sum_{j=1}^{i-1} e^{-\beta(T_i - T_j)}}{\lambda_0 + \alpha \sum_{j=1}^{i-1} e^{-\beta(T_i - T_j)}}$$

$$\begin{aligned}0 = &\frac{\alpha}{\beta^2} \sum_{i=1}^{N}\left\{1 - \left[\beta(T_N - T_i) + 1\right]e^{-\beta(T_N - T_i)}\right\} \\ &- \sum_{i=1}^{N} \frac{\alpha \sum_{j=1}^{i-1}(T_i - T_j)e^{-\beta(T_i - T_j)}}{\lambda_0 + \alpha \sum_{j=1}^{i-1} e^{-\beta(T_i - T_j)}}\end{aligned}$$

Ozaki [47] provides also the Fisher information matrix. Other studies have explored the estimation of Hawkes parameters in various contexts, see Bacry et al. [3], Hardiman et al. [26], Lemonnier and Vayatis [41], etc.

3.4 Limit Behaviour

We now turn to the study of the limit behaviour of Hawkes processes. We first consider the stable case, that is, $\alpha < \beta$. Under this hypothesis, a law of large numbers and a central limit theorem can be proved, see [4].

Proposition 3.10 *Let N be a Hawkes process with intensity λ characterized by a triplet $(\lambda_0, \alpha, \beta)$. Assume that $\alpha < \beta$, then we have the following law of large numbers:*

$$\frac{1}{T}N_T - \frac{\beta}{\beta-\alpha}\lambda_0 \underset{T\to+\infty}{\longrightarrow} 0, \quad \mathbb{P}\text{-a.s.}$$

as T goes to infinity. Moreover, we have the following central limit theorem:

$$\sqrt{T}\left[\frac{1}{T}N_T - \frac{\beta}{\beta-\alpha}\lambda_0\right] \underset{T\to+\infty}{\longrightarrow} \left(\frac{\beta}{\beta-\alpha}\right)^{\frac{3}{2}} \sqrt{\lambda_0}\mathcal{N}$$

where $\mathcal{N}$ is a standard Gaussian random variable and the convergence is in law.

This result is relatively standard and due to the finite variance of Hawkes process if $\alpha < \beta$, as we will see in the next section.

The previous result can be extended to the case of nearly unstable Hawkes process, that is, when $\alpha \to \beta$, see Jaisson and Rosenbaum [32]. In particular, they show that, if the time T goes to infinity as the $\alpha^{(T)}/\beta^{(T)}$ goes to 1, two results can be shown depending on the speed at which $\alpha^{(T)}/\beta^{(T)}$ tends to 1. This is achieved by rescaling the intensity of the Hawkes process by the factor $(1-\alpha^{(T)}/\beta^{(T)})$.

Proposition 3.11 *Assume $T(1-\alpha^{(T)}/\beta^{(T)}) \to \infty$ as T diverges. Then, the sequence of Hawkes processes is asymptotically deterministic and we have the following convergence in L^2 sense:*

$$\sup_{u\in[0,1]} \frac{1-\frac{\alpha^{(T)}}{\beta^{(T)}}}{T}\left|N_{uT}^{(T)} - \mathbb{E}\left[N_{uT}^{(T)}\right]\right| \underset{T\to+\infty}{\longrightarrow} 0$$

Assume $T(1-\alpha^{(T)}/\beta^{(T)}) \to \theta > 0$ as T diverges. Then, the sequence of the renormalized intensities $\lambda_{uT}^{(T)}(1-\alpha^{(T)}/\beta^{(T)})$ of the Hawkes processes converges to the following CIR model for $u \in [0,1]$:

$$X_u = \int_0^u \frac{\theta}{\alpha^{(\infty)}}(\lambda_0 - X_s)ds + \int_0^u \frac{\sqrt{\theta}}{\alpha^{(\infty)}}\sqrt{X_s}dB_s$$

Moreover, the sequence of renormalized Hawkes processes converges in law, for the Skorokhod topology, to the integral of X, that is

$$\frac{1-\alpha^{(T)}/\beta^{(T)}}{T} N_{uT}^{(T)} \underset{T\to+\infty}{\longrightarrow} \int_0^u X_s ds$$

4 Moments

This section investigates the calculation of moments. Except for the first one, the moments are generally hard to obtain under closed-form formula. However, in the present context they can be derived as solutions of ordinary differential equations. This approach has been introduced by [13], for the second-order moments, in a multidimensional framework. In particular, this method is used to perform some fast calibration procedure for the Hawkes process. In this section, we follow the steps of [13] and provide recurrence formulae for moments of every order. This recurrence relation links the moments of the counting process with the moments of its intensity. This is a first reason why we start by the study of the intensity. The second reason is that it provides some interesting results on the limit distribution that we will encounter in Sect. 5.

4.1 Moments of the Intensity

The intensity given by Eq. (8) can be written under the following alternative integral form $\lambda(s) = \lambda_0 + \alpha \int_0^{t^-} e^{-\beta(t-s)} N^H(ds)$. The upper value of the integral is t^-, in order to keep the predictability of the intensity. This new expression can be turned into the following form, which emphasizes the role of β as an mean-reverting coefficient:

$$\lambda(t) = \lambda_0 + \int_0^t \beta(\lambda_0 - \lambda(s))ds + \alpha \int_0^{t^-} N^H(ds) \tag{10}$$

We can also introduce a càdlàg version of λ, which is simply given by

$$\widehat{\lambda}(t) := \lambda_0 + \alpha \int_0^t e^{-\beta(t-s)} N^H(ds) \tag{11}$$

As $\widehat{\lambda}(t) = \lambda(t) + \alpha \sum_{n\geq 1} \mathbb{I}_{\{T_n=t\}}$, the processes λ and $\widehat{\lambda}$ have the same moments: the event $\{T_n = t\}$ is a $\mathbb{P}$-null set.

Lemma 4.1 *Assume that $\alpha \neq \beta$. The first-order moment of the intensity $m_1(t) := \mathbb{E}\{\lambda(t)\}$ is given by*

$$m_1(t) = \frac{\beta\lambda_0}{\beta-\alpha} + \left(\lambda_0 - \frac{\beta\lambda_0}{\beta-\alpha}\right) e^{(\alpha-\beta)t} \tag{12}$$

Proof As noticed above, we also have $\mathbb{E}\left\{\widehat{\lambda}(t)\right\} = m_1(t)$. Using the càdlàg version of Eq. (10), we can apply Fubini theorem to obtain

$$\begin{aligned} m_1(t) &= \lambda_0 + \beta \int_0^t (\lambda_0 - m_1(s))ds + \alpha\,\mathbb{E}\left\{\int_0^t N^H(ds)\right\} \\ &= \lambda_0 + \beta \int_0^t (\lambda_0 - m_1(s))ds + \alpha \int_0^t m_1(s)ds \end{aligned}$$

Or, equivalently,

$$m_1(t) = \lambda_0 + \beta \int_0^t (\lambda_0 - m_1(s))ds + \alpha \int_0^t m_1(s)ds \tag{13}$$

Now, it is sufficient to see that the form given by Eq. (12) is the solution of Eq. (13). □

We see from Lemma 4.1 that the mean-reversion coefficient is, actually, $\beta - \alpha$, and not simply β.

Remark 4 General recurrence equations for the moments of the intensity are given in [6].

As an immediate consequence of Lemma 4.1, we have the following result.

Lemma 4.2 *If $\alpha < \beta$, then m_1 tends to $\frac{\beta\lambda_0}{\beta-\alpha}$ when t tends to infinity.*

This result will be seen under a more general form in Sect. 5.2. With Lemma 4.1 in hand, and using the properties of the compensator, we can state the following results.

Proposition 4.3 *The first-order moment of the Hawkes process is given, for any $t \geq 0$, by*

$$\mathbb{E}\{N(t)\} = \frac{\beta\lambda_0}{\beta-\alpha} t + \left(\lambda_0 - \frac{\beta\lambda_0}{\beta-\alpha}\right) \frac{e^{(\alpha-\beta)t} - 1}{\alpha-\beta} \tag{14}$$

4.2 Recurrence Formulae for Moments

First, set, for any $t \geq 0$, and any integers $p, n \geq 0$: $V_{p,n}(t) := \mathbb{E}\left\{N_t^p \widehat{\lambda}^n(t)\right\}$. This quantity is at the core of the representation of moments. We can observe that $V_{p,0}(t)$ is the moment of order p of N_t and that $V_{0,k}(t) = m_k(t)$, i.e. the moment of order k of the intensity. The following result provides the recurrence formula, which makes the link between the moments of the counting process and the moments of the intensity.

Lemma 4.4 *For any $p \geq 1$, we have*

$$V_{p,0}(t) = \sum_{k=0}^{p-1} \binom{p}{k} \int_0^t V_{k,1}(s)ds$$

For any $p \geq 1$ and $n \geq 1$, we have

$$\begin{aligned} V_{p,n}(t) = & \int_0^t e^{-n(\beta-\alpha)(t-s)} \beta n \lambda_0 V_{p,n-1}(s)ds \\ & + \sum_{k=0}^{p-1} \sum_{i=0}^{n} \binom{p}{k} \binom{n}{i} \alpha^{n-i} \int_0^t e^{-n(\beta-\alpha)(t-s)} V_{k,i+1}(s)ds \\ & + \sum_{i=0}^{n-2} \binom{n}{i} \alpha^{n-i} \int_0^t e^{-n(\beta-\alpha)(t-s)} V_{p,i+1}(s)ds \end{aligned}$$

Proof The first assertion stems from a direct application of the Itō formula for PP (see [49], Theorem 33, p. 74). We have

$$d(N_t^p) = \left[(N_{t^-} + 1)^p - N_{t^-}^p\right] dN_t$$

The definition of the compensator and an application of Fubini theorem yield

$$\mathbb{E}\left\{N_t^p\right\} = \int_0^t \sum_{k=0}^{p-1} \binom{p}{k} \mathbb{E}\left\{N_{s^-}^k \lambda(s)\right\} ds$$

Hence the result needed. For the second one, we use the Itō formula for PP together with the Integration by Parts formula (cf. [49], Corollary 2, p. 60) to deduce

$$\begin{aligned} d\left(N_t^p \widehat{\lambda}^n(t)\right) = & n\beta \left[\lambda_0 - \lambda(t)\right] N_{t^-}^p \lambda^{n-1}(t)dt \\ & + \lambda^n(t) \left[(N_{t^-} + 1)^p - N_{t^-}^p\right] dN_t \\ & + N_{t^-}^p \left[(\lambda(t) + \alpha)^n - \lambda^n(t)\right] dN_t \\ & + \left[(N_{t^-} + 1)^p - N_{t^-}^p\right] \left[(\lambda(t) + \alpha)^n - \lambda^n(t)\right] N_t \end{aligned}$$

Remark that the last term in the previous equation stems from the *bracket* of N and $\widehat{\lambda}$, which are semi-martingales jumping at the same time. This can be reorganized into

$$\begin{aligned} d\left(N_t^p \widehat{\lambda}^n(t)\right) = & n\beta \left[\lambda_0 - \lambda(t)\right] N_{t^-}^p \lambda^{n-1}(t)dt \\ & + \left[(N_{t^-} + 1)^p (\lambda(t) + \alpha)^n - N_{t^-}^p \lambda^n(t)\right] N_t \end{aligned} \tag{15}$$

Using again the compensator and Fubini theorem, we obtain, from Eq. (15),

$$V_{p,n}(t) = -n\beta \int_0^t V_{p,n}(s)ds + n\beta\lambda_0 \int_0^t V_{p,n-1}(s)ds$$
$$+\mathbb{E}\left\{\int_0^t \left[(N_{t^-}+1)^p(\lambda(t)+\alpha)^n - N_{t^-}^p\lambda(t)^n\right]\lambda(t)dt\right\} \tag{16}$$

The last term in Eq. (16) can be developed and simplified in order to obtain

$$V_{p,n}(t) = -n(\beta-\alpha) \int_0^t V_{p,n}(s)ds + n\beta\lambda_0 \int_0^t V_{p,n-1}(s)ds$$
$$+\sum_{k=0}^{p-1}\sum_{i=0}^{n}\binom{p}{k}\binom{n}{i}\alpha^{n-i}\int_0^t V_{k,i+1}(s)ds$$
$$+\sum_{i=0}^{n-2}\binom{n}{i}\alpha^{n-i}\int_0^t V_{p,i+1}(s)ds$$

Solving this ordinary differential equation yields the result. □

5 Generalization of Hawkes Processes

This section displays several generalizations of the standard Hawkes process, introduced in Sect. 3.1. Some of these generalizations are plain and consist in adding marks to the point process. But we also provide a brief overview of the nonlinear Hawkes processes, which require more complex tools to be dealt with. But, as a preliminary remark, we will add a few words on the original formulation, found in [28]. The general form of the intensity proposed by Hawkes was

$$\lambda(t) = \lambda_0 + \int_0^{t^-} g(t-u)dN_u$$

where g is a positive integrable mapping. A stationarity argument imposes that $\int_0^{+\infty} g(s)ds < 1$. The rest of the developments contained in [28] are obtained with a specific form of g, with an exponential decay:

$$g(t) = \sum_{k=1}^{K}\alpha_k e^{-\beta_k t}$$

where $\alpha_k > 0$ and $\beta_k > 0$. In the previous parts of this chapter, we have simplified this framework to a single exponential function. Under this simpler form, the behaviour of the process remains globally the same, but the analysis performed in Sect. 4 becomes tractable.

5.1 Marked Hawkes Processes

It is always possible to add an independent mark to a standard Hawkes process. This is of interest in many applications where both times of jumps and size of jumps are involved. For example, it can be interesting to represent the time of a credit event and its recovery rate, or the arrival of orders in a trading book and the price or quantity of the orders. However, it can be reasonably argued that the size of the jump may also have an influence on the future evolution of the process, as well as the times of jumps themselves. For instance, a credit event with a small recovery rate may have more influence on other debt instruments than a credit event with a large recovery rate. To capture this feature, a second type of feedback can be added to the intensity dynamics. Such an approach has been introduced in [6].

Definition 5.1 (*Marked Hawkes process*) Set out $\lambda_0 > 0, 0 \leq \overline{\alpha} < \beta$. Let $(S_n)_{n\geq 1}$ be a Poisson process, with intensity 1, and $(X_n)_{n\geq 0}$ an independent marking. Consider some measurable function $\widehat{\alpha}(\cdot) \leq \overline{\alpha}$. We construct the MPP $(T_n, X_n)_{n\geq 1}$, with inter-jump times $(\theta_n)_{n\geq 1}$ as follows. First, set $T_0 = S_0 = 0$, and define, for every $n \geq 0$,

$$\Lambda_n(t) := \lambda_0 t + \sum_{i=1}^{n} \widehat{\alpha}(X_i) \frac{1 - e^{-\beta(t-T_i)}}{\beta} \tag{17}$$

$$\theta_{n+1} = \Lambda_n(T_{n+1}) - \Lambda_n(T_n) \tag{18}$$

This process is called Marked Hawkes process.

We can notice that, if we choose $\widehat{\alpha} \equiv \alpha$, we obtain a classic independent marking of the Hawkes process. Let us denote by $\Theta(d\zeta)$ the distribution of the mark on the Borelian space $(X, \mathscr{X})$, as introduced in Definition 2.4. Let us denote by $\mu(dt, d\zeta)$ the counting measure of the MPP. We derive from Eq. (17) the following form for the compensator: $\Theta(d\zeta)\lambda(t)dt$, where

$$\lambda(t) = \lambda_0 + \int_0^{t^-} \int_X \widehat{\alpha}(\zeta) e^{-\beta(t-s)} \mu(ds, d\zeta)$$

This particular form provides results which are very similar to the standard case. Let us give an example with the first-order moment of the intensity:

Lemma 5.2 *The first-order moment of the marked Hawkes process, defined by* $m_1(t) := \mathbb{E}\{\lambda(t)\}$ *satisfies Eq. (12), where* $\alpha := \int_X \widehat{\alpha}(\zeta)\Theta(d\zeta)$.

Proof By property of the compensator, we can write

$$m_1(t) = \lambda_0 + \mathbb{E}\left\{\int_0^t \int_X \widehat{\alpha}(\zeta) e^{-\beta(t-s)} \Theta(d\zeta)\lambda(s)ds\right\}$$

By Fubini Theorem, we have

$$m_1(t) = \lambda_0 + \int_0^t \int_X \widehat{\alpha}(\zeta)\Theta(d\zeta)e^{-\beta(t-s)}m_1(s)ds = \lambda_0 + \int_0^t \alpha e^{-\beta(t-s)}m_1(s)ds$$

The next steps follow those of the proof of Lemma 4.1. □

Remark 5 The Poisson process and the Hawkes process share the same mark. The change of probability result in Theorem 3.6 remains unchanged in the case of the marked Hawkes process.

5.2 *Nonlinear Hawkes Processes*

This subsection deals with an important generalization of Hawkes processes, due to Brémaud and Massoulié [8]. The results in [8] do not only provide existence results for a much more general class of intensity, but give sufficient conditions for the existence of a stationary distribution. In the following, we present one of these results in a slightly less general framework. The class of self-exciting processes studied in [8] was named *nonlinear Hawkes processes*.

Definition 5.3 Let $\Phi : \mathbb{R} \to \mathbb{R}^+$ and $h : \mathbb{R}^+ \to \mathbb{R}$. The nonlinear Hawkes process N is the PP with intensity

$$\lambda(t) = \Phi\left(\int_0^{t^-} h(t-s)dN(s)\right) \tag{19}$$

With $\Phi(x) = \lambda_0 + \alpha x$ and $h(t) = e^{-\beta t}$, we find back the standard Hawkes process, with intensity given by (8). Existence and stability in this setting was first given by Hawkes and Oakes [30]. In the general setting, the existence and stability are given by the following theorem.

Theorem 5.4 (Brémaud and Massoulié) *Assume that Φ is κ-Lipschitz ($\kappa > 0$) and that the following condition holds:*

$$\kappa \int_0^{+\infty} |h(s)|ds < 1 \tag{20}$$

Then, there exists a unique stationary distribution of N, with finite average intensity $\mathbb{E}\{N(]0,1])\} < +\infty$ and dynamics (19).

For the standard Hawkes process, condition (20) simply writes $\alpha < \beta$, which is coherent with Lemma 4.2.

6 Financial Applications of Hawkes Processes

This section investigates some examples of application of Hawkes processes in finance. Clustering effects appear in many areas of finance. An immediate use of these processes consists in modelling the time arrival of orders in a trading book. See, e.g. [2, 13]. Many works emphasize the existence of clustering effects on financial instruments among various markets: foreign exchange [22], interest rates [25] or credit (cf. [43] for a structural model and [14] for a reduced-form model). The presence of clustering effects is also particularly glaring in energy prices (cf. [34]).

First, let us see how we can exhibit the clustering effects in financial data.

6.1 Calibration of Hawkes Processes

As a case study, we show how the clustering effects can be isolated from time series, using the likelihood ratio introduced at Sect. 3.2. We will deal here with daily data of the Eurostoxx 50 index, from February 2007 to August 2017. In this case, the split between volatility part and jumps can be done by taking extreme values (5% and 95%-percentiles in our case) of the time series log returns. For more sophisticated methods to estimate the parameters of diffusions with jumps, see [44]. Once times of jumps $(T_i)_{1\leq i\leq n}$ are isolated, we can perform a maximization of the log-likelihood, introduced in Sect. 3.3. It writes

$$\mathscr{L}(T_1, \ldots, T_n \,|\lambda_0, \alpha, \beta) := \sum_{i=1}^{n} \ln(\lambda(T_i)) - \int_0^t \lambda(s)ds$$

where time t is the total length of the observation sample (in years) and $\theta := (\lambda_0, \alpha, \beta)$. The maximization can be achieved, for instance, by a downhill simplex method, in dimension 3.

However, in [47], the form of the gradient is given in order to implement some gradient descent method. In our example, the maximization of $\mathscr{L}(T_1, \ldots, T_n \,|\theta)$, with respect to the three parameters θ, provides the following result: $\theta^* := (4.679, 11.180, 14.405)$. First, we observe that the limit intensity given by Lemma 4.2, i.e. $\frac{\lambda_0\beta}{\beta-\alpha} = 20.9$, is very close to the value given by direct estimation of the intensity: $\frac{N_t}{t} = 19.9$. Thus, the likelihood method is coherent with the empirical limit value of the process.

Figure 2 displays the evolution of the Eurostoxx 50 (right axis) together with the intensity of the Hawkes process driving the jumps (left axis). This figure sheds light on the clustering effects underlying the index series. In order to measure the accuracy of the fit—and illustrate the change of time property highlighted in Sect. 3.1—we consider the times $S_i := \Lambda(T_i|\theta^*) = \int_0^{T_i} \lambda(s|\theta^*)ds$, $1 \leq i \leq n$. According to Lemma 3.2, the inter-jump times $\tau_1 := S_1$ and $\tau_i := S_i - S_{i-1}$, $2 \leq i \leq n$, should be distributed according to an exponential law with parameter 1. In Fig. 3, we compare the

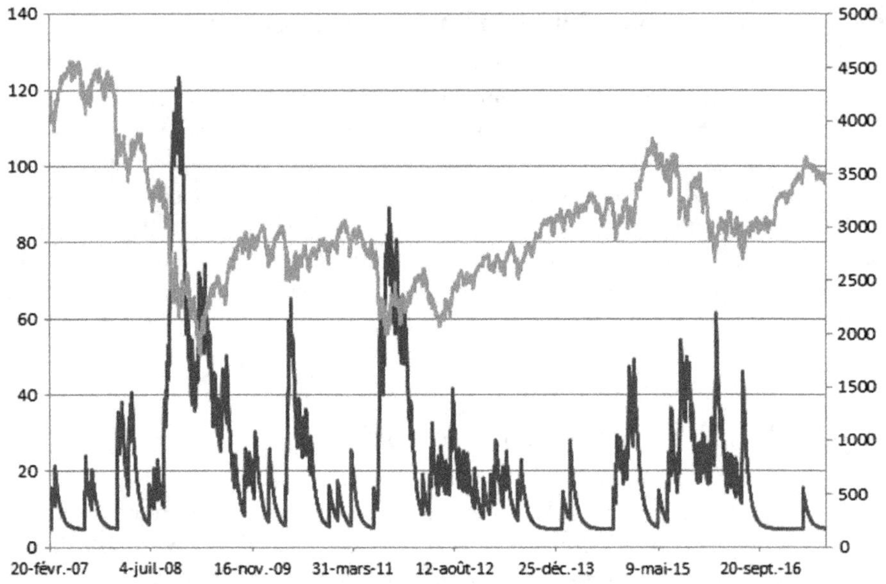

Fig. 2 Historical evolution of the Eurostoxx 50 index (right axis) and the intensity $\lambda(s|\theta^*)$ of the Hawkes process for the optimal parameters of the log-likelihood maximization: $\theta^* = (4.679, 11.180, 14.405)$ (left axis)

empirical distribution of inter-jump times and the theoretical exponential distribution. It shows a very good fit, corresponding to a maximum distance between the two cumulated distribution of 0.07.

The features illustrated in the previous example can also be found in other markets such as foreign exchange, interest rates, credit spreads or implied volatility. Let us give another example in a rather different market: electricity prices. This type of market is characterized by the occurrence of numerous spikes in the prices, due to specific tensions on the supply and demand. The same method as above is applied to a time series representing the price of electricity in Italy, at 7 pm between 2010 and 2014. The calibration of the parameters of the Hawkes process yields $\theta^* := (16.340, 30.317, 93.448)$. The results are represented in Fig. 4. This example is treated in [34], by the means of branching processes, that will be studied in Sect. 7.

As in the previous example, the fit of the model is pretty correct, and the empirical and theoretical long-term means are very close: 24 and 24.2, respectively. However, the parameters α and β are much larger in the case of electricity prices. This is coherent with the observation of these numerous spikes of large amplitude. The intensity soars and it goes back very quickly to its initial value, with a large mean-reverting factor.

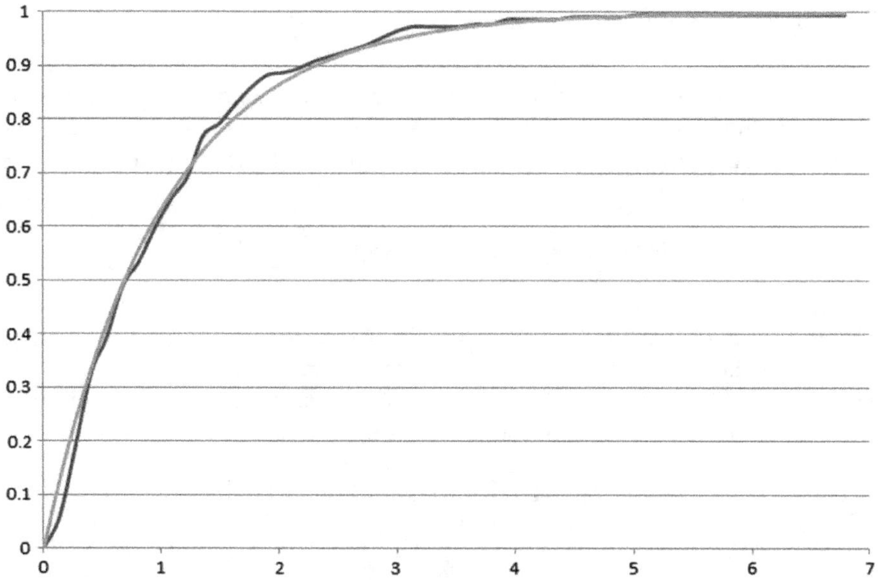

Fig. 3 Empirical cumulated distribution of the inter-jump times $(\tau_i)_{1\leq i\leq n}$ after time change and theoretical cumulated exponential distribution, with parameter 1

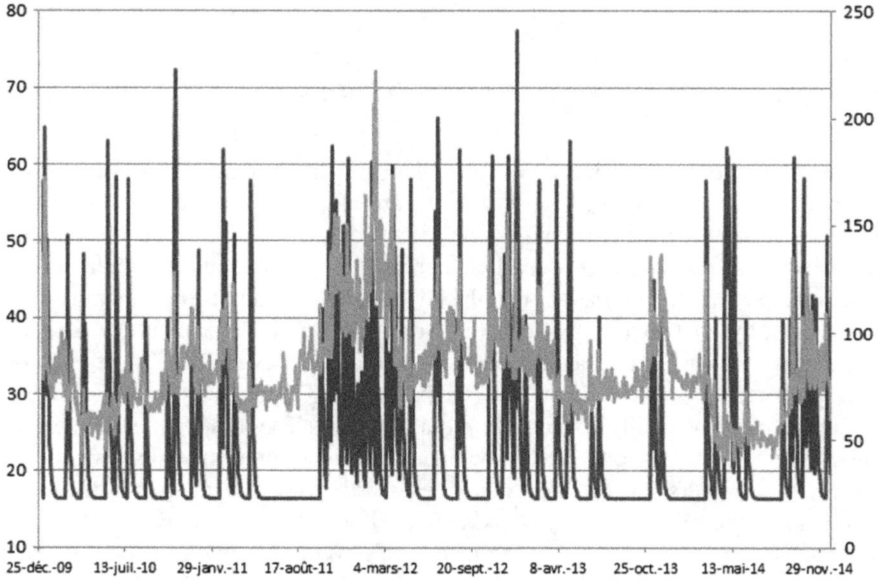

Fig. 4 Price of electricity in Italy, at 7 PM, from 2010 to 2014 (right axis) and the intensity $\lambda(s|\theta^*)$ of the Hawkes process for the optimal parameters of the log-likelihood maximization: $\theta^* = (16.340, 30.317, 93.448)$ (left axis)

6.2 Interest Rate Model with Hawkes Jumps

In the case of interest rates, jumps and clustering effects are particularly important on short-term rates, and especially overnight rates (EONIA for Euro). In [25], the following model is proposed to represent the dynamics of EONIA rates

$$dr_t = (\theta(t) - r_t)\,dt + \sigma dW_t + d\left(\sum_{i=1}^{N_t} J_i\right)$$

where W is a standard Brownian motion, θ a deterministic mapping, $(J_i)_{i\leq 1}$ a sequence of i.i.d random variables independent from W, and N a jump process with clustering effects (independent from the other sources of risk). The intensity of N is very close from the intensity studied above:

$$\begin{aligned} d\lambda_t &= \kappa\,(c - \lambda_t)\,dt + \delta dL_t \\ dL_t &= |J_{N_t}| dN_t \end{aligned}$$

This model is closely related to the work of [21] in the context of credit risk. Exploiting the affine form of the Laplace transform, pricing formulae for bonds are given in [25]. The Laplace transform for Hawkes process will be studied in Sect. 7.

6.3 Hawkes Processes for Credit Modelling

The modelling of credit risk involves several aspects that can be related to clustering effects. Basically, the reduced-form models are based on two elements: First, the credit event itself (roughly speaking, the default of a credit issuer), and second, the intensity of the default. This second element is closely related to the credit spread, which is a part of the remuneration of the bondholder (among other factors: risk-free rates, liquidity…). Clustering effects intervene at these two levels. For instance, the same analysis as in Sect. 6.1 can be applied to credit indices such as iTraxx or CDX. The spread of these synthetic indices (i.e. based on CDS) shows similar features as those shown in the equity-based example of Sect. 6.1. The jumps on credit spreads are triggered either by global tension on financial markets or by specific news inflows on sectors or issuers. Even if they do not necessarily trigger credit events, they have an impact on the value of the credit instruments (bonds, CDS, structured credit notes) and therefore have an impact on the investors. Several references deal with affine processes in credit modelling. See, among other references, [18] for a global survey and [14, 21] for the specific case of clustering effects. The clustering effects are also used in [48] for collateralized debt obligations. Another application in the same context of structured credit products is [6].

An alternative way to deal with credit risk is the structural approach. An application of Hawkes processes in this context can be found in [43].

In order to give a little colour on the influence of clustering effects on the modelling of credit events, let us take a simple example. We consider an equally weighted portfolio of m credit products—loans, bonds or CDS—and assume a constant recovery rate R. In order to represent the expected loss of this portfolio, we should consider the dependence structure of the individual credit events among this portfolio. This may be difficult when the portfolio is large (e.g. in the case of asset-backed securities) or when historical data is poorly available (Leveraged loans, for example). This is why it can be interesting to represent the loss as the number of jumps of a PP, say (N_t). In this case, the expected loss at time t simply writes $(1 - R) \times N_t$ (assuming that the probability that $N_t > m$ is small enough). If we assume that the credit events are identically distributed according to an exponential law with constant intensity λ_0, the distribution of N_t is similar to a Poisson distribution with intensity $m\lambda_0$, for sufficiently large m (by Poisson approximation of the binomial law). Therefore, the case where N is a Poisson process amounts to the case of independent credit issuers. Introducing clustering effects with a Hawkes process can be seen as a way to deal with an implied correlation between credit events. Examples of Hawkes distributions in Fig. 1 show how the parameters of the Hawkes processes control the tail risk (corresponding, here, to the occurrence of a large number of defaults in the portfolio).

6.4 *Clustering for Trading Orders and Optimal Execution*

Looking at high-frequency data in finance, self-exciting behaviour was first observed by Mandelbrot [45]: it is the so-called volatility clustering. A recent literature focuses on tick-by-tick models where the evolution of buy and sell orders, and more general limit order book (LOB) events, are described by coupled Hawkes processes. We highlight in particular the paper of Bacry and Muzy [5], where the authors successfully attempt to describe both price and order flow behaviour. We can mention, although not exhaustive, the papers of Da Fonseca and Zaatour [13], Filimonov and Sornette [23] and Zheng et al. [51].

The basic model is based on the description of the arrival of market orders that are represented by a two-dimensional counting process, the first (resp. the second) one representing the cumulated number of buy (resp. sell) market orders. This representation can be applied in a similar way to model the mid-price as the difference of two increasing pure jump processes.

This representation can be easily generalized to take into account different phenomena and features of high-frequency financial markets. As an example, in order to describe LOB market, we need to model the arrival of both limit and markets orders for both sides of the order books. Moreover, the limit orders are characterized by both the size and the proposed price, that is, the limit. The complexity is further increased by the possibility to cancel a limit order before its execution.

In two recent papers, Alfonsi and Blanc (cf. [1, 2]) make a bridge between high-frequency price models and optimal execution frameworks. They highlight that Hawkes processes seem able to describe satisfactorily the flow of market orders. The main result of their study shows that a price manipulation strategy necessarily appears when the flow of market orders follows a Poisson paradigm. Moreover, these strategies can be implemented without knowing the model parameters. Roughly speaking, it is enough to bite a little bit of all market orders to have a profitable strategy. Finally, they show that Hawkes processes paradigm can avoid these drawbacks.

7 A Branching Processes Point of View

In this section, we propose a link between Hawkes processes and continuous-state branching processes with immigration (CBI). Our objective is twofold. First, we will show that CBI is the natural class to extend Hawkes processes in order to include infinite activity and/or Brownian noise. Second, CBI class has been well studied starting with the seminal works of Lamperti [38] and Kawazu and Watanabe [35], and many theoretical and applied results are now known. Therefore, it can help to understand the behaviour of Hawkes processes and to simplify some proofs, by highlighting the analogies between the two frameworks.

First, we summarize shortly the ideas of CBI setup. Branching processes were mainly used in demography. The first problem was to study the dilemma of Galton about the extinction of distinguished families. Watson refused the usual answer, that is, a diminution in fertility, and started to study the probability that the male line will be extinguished after k generations (see [50]). By studying the probability generating function of the event that a man has i sons, denoted by $f(s) = \sum_i p_i s^i$, and assuming a generational invariance, Watson obtained the probability generating function f_k to have Z_k descendants at generation k, showing that f_k is the k-iterate of the function f. He concluded that the extinction occurs not only if the expected number of son per father is smaller than one but also in the opposite case. The first case clearly describes a fertility problem and the extinction is almost sure. The second case is more surprising and the probability of the extinction is given by the only fixed point of the function $f(s)$. The proof of this nice result constitutes an excellent exercise for a first probability course. We can refer to the bibliographies of Harris [27], Chap. 1, Kyprianou [37], Chap. 10 and Li [42], Chap. 4, for a more general analysis of Galton–Watson model and their generalization to branching processes.

The previous analysis highlights some interesting facts related to branching processes. First of all, branching processes are mathematically tractable and exhibit closed forms for many interesting functions in probability. This is probably the reason why many studies are dedicated to this class and some subclasses, like superprocesses. See, for instance, [40]. Second, branching processes are deeply related to exponential affine processes, that is, stochastic processes having a Laplace transform that can be written as an exponential of an affine function with respect to the initial value of the process itself (cf. [24]). Third, Galton–Watson process has some counter-intuitive features, for instance, the extinction is possible even if the fertility is very high and neglecting overpopulation effects. Branching processes share this ability to give easy but surprising explanations to very different phenomena. Some examples in mathematical finance are presented in [33, 34].

7.1 Branching Property and Integral Representation

We now study explicitly the main properties of branching processes in continuous time. In this section, we shall denote by P_x the law of a process X under the starting condition $X_0 = x$. As well, we will denote by X^x the process X with initial condition x.

Definition 7.1 (*Branching property*) A process X is said to satisfy the branching property if, for any t, and $(x_1, x_2) \in \mathbb{R}_+^2$, $X_t^{x_1+x_2}$ under the law $P_{x_1+x_2}$ is equal in law of the sum of independent process $X_t^{x_1} + X_t^{x_2}$ under the law $P_{x_1} \times P_{x_2}$.

The previous property is equivalent to say that the process X can always be decomposed into two independent processes $X^{(1)}$ and $X^{(2)}$ both satisfying the same stochastic differential equation (hereafter SDE) of X. We can then define the branching processes as the processes infinitely divisible with respect to the starting condition. To avoid possible confusion, we recall that there is another class of infinitely divisible processes but with respect to time: the Lévy processes. There is also an important link between Lévy and branching processes known as Lamperti transform, see [37, 38].

Surprisingly, the main drawback of branching processes, hindering large applications in finance, is the difficulty to exhibit an explicit SDE for a general class of branching processes. The main example often used in literature is the square-root diffusion universally known in finance as CIR model, see [12]. However, no other explicit SDE satisfying the branching property appears in the literature despite the large number of papers related to exponential affine structure, see, for instance, [20, 24].

The keystone papers [15, 16] give rise to a new paradigm able to describe explicitly a SDE for a general class of branching process. In order to introduce their representation, the best way is to come back to the Galton–Watson model. There is a natural relation between generations: the population at generation k is given by the sum of all sons of the generation $k-1$, that is,

$$Z_k = \sum_{i=1}^{Z_{k-1}} \zeta_i^{(k)}$$

where $\zeta_i^{(k)}$ represents the number of sons of the i-father in the previous generation and it is a sequence of iid random variables. Branching processes are the continuous-state and continuous-time extension of Galton–Watson model. The sum will be then replaced by an integral. Dawson and Li finally obtain the following representation:

$$X_t = X_0 + \int_0^t \int_0^{X_{s-}} \left(a\, du\, ds + \sigma W(du,\, ds) + \int_0^\infty \zeta \widetilde{N}(d\zeta,\, du,\, ds) \right).$$

We recognize the integral up to the process itself. Inside the parenthesis, there are three terms: the first term is a constant, the second one, W, is a white noise, and the third one, $\widetilde{N}$, is a spectrally positive compensated jump process with measure $\mu(d\zeta)\, du\, ds$, see Li [42] Chap. 4. We recognize the three terms of the canonical decomposition of an infinitely divisible process, that is, a constant, a continuous martingale and a pure jump martingale, see [17]. Compared with Lévy processes, two main differences appear: the jumps are only positive and, more importantly, each term has an additional dimension u, that is, used to integrate up to the process itself. The linearity of the integral, with the homogeneity of the measures and the positivity of jumps, is the key argument used to check the branching property. See Definition 7.1.

The main drawback of pure branching processes is that only two limit situations exist: extinction or explosion, that is, no ergodic distribution arises. This is a disadvantage in demography applications and more generally in long-term modelling. The solution is to add another term, the immigration, that is, an exogenous infinitely divisible positive term. The general equation becomes

$$\begin{aligned} X_t = X_0 + \int_0^t \int_0^{X_{s-}} \left(a\, duds + \sigma W(du,\, ds) + \int_0^\infty \zeta \widetilde{N}(d\zeta,\, du,\, ds) \right) \\ + \int_0^t \left(b\, ds + \int_0^\infty \zeta M(d\zeta,\, ds) \right). \end{aligned} \tag{21}$$

To force the last term to give a non-negative contribution, we cannot add a continuous martingale, that is, a Brownian motion, nor negative jumps. Accordingly, the measure M is not compensated.

The last two pages may seem disconnected from the main subject of our analysis. The next section, inspired by [33, 34], conciliates the two subjects showing that the intensity of Hawkes processes is a particular case of branching process with immigration. As a consequence, we can extend Hawkes processes using the formalism of CBI. Moreover, these two ways to describe Hawkes processes give birth to different ways to study and understand their properties.

7.2 Hawkes Processes and CBI Processes

We summarize the main result in the next proposition.

Proposition 7.2 *Consider Eq. (21). Let $\sigma = 0$, the measures $M = 0$ and $\mu(d\zeta) = \alpha\delta_1(d\zeta)$, where δ_1 denotes the Dirac delta on the point 1. Then, the (non-compensated) Poisson measure $N(d\zeta,\, du,\, ds)$ reduces to a random measure on $(\mathbb{R}^+)^2$ with intensity ds, du, denoted by $N(ds, du)$. Hence Z can be rewritten as*

$$Z_t = Z_0 + bt - \int_0^t (a+\alpha) Z_s ds + \alpha \int_0^t \int_0^{Z_{s-}} N(ds, du) \tag{22}$$

Moreover, defining

$$J_t := \int_0^t \int_0^{Z_{s-}} N(ds, du), \tag{23}$$

we have that J is a counting process with intensity Z. Finally, J is a Hawkes process and we have the following representation for the intensity:

$$Z_t = Z_0 e^{-(a+\alpha)t} + \frac{b}{a+\alpha}\left(1 - e^{(a+\alpha)t}\right) + \alpha \int_0^t \int_0^{Z_{s-}} e^{-(a+\alpha)(t-s)} N(ds, du) \tag{24}$$

Proof We start rewriting Eq. (21) with the fixed parameters and measures

$$Z_t = Z_0 + bt + \int_0^t \int_0^{X_{s-}} \left(a\, du ds + \alpha \int_0^\infty \zeta \widetilde{N}(d\zeta,\, du,\, ds) \right)$$

Separating the compensator and the pure jump part, we obtain

$$Z_t = Z_0 + bt + \int_0^t \int_0^{X_{s-}} \left(aduds + \alpha \int_0^\infty \zeta N(d\zeta, du, ds) + \alpha \int_0^\infty \zeta \delta_1(d\zeta) duds \right)$$

Integrating with respect to $d\zeta$, we can recombine the first and the third terms

$$Z_t = Z_0 + bt + \int_0^t \int_0^{X_{s-}} (a+\alpha) duds + \alpha \int_0^t \int_0^{X_{s-}} N(du, ds)$$

Finally, integrating with respect to du we obtain (22).

Consider now the process $J_t := \int_0^t \int_0^{Z_{s-}} N(ds, du)$. It is plain that it is a counting process according to Definition 2.2. Moreover, by a direct application of Theorem 2.11 and Lemma 2.14, we have that the intensity of J is the process Z itself. We then can rewrite Z as

$$Z_t = Z_0 + bt - \int_0^t (a+\alpha) Z_s ds + \alpha J_t$$

showing that J is a Hawkes process. Finally, integrating the previous relation for Z, we obtain (24). □

Equation (24) is the equivalent of (10). We can then recognize that the mean-reverting speed β coincides with $a+\alpha$, that is, in CBI representation (21), we have $a = \beta - \alpha$, where the long-term intensity λ_0 is replaced by b/β, i.e.

$$\lambda_t = \lambda_0 + \int_0^t (\beta - \alpha)\left(\frac{\beta}{\beta-\alpha}\lambda_0 - \lambda_s\right) ds + \alpha \int_0^t \int_0^{\lambda_{s-}} \widetilde{N}(du, ds) \tag{25}$$

or, equivalently in representation (11), we have

$$\lambda_t = \lambda_0 e^{-(\beta-\alpha)t} + \frac{\beta}{\beta-\alpha}\lambda_0 \left[1 - e^{-(\beta-\alpha)t}\right] + \alpha \int_0^t \int_0^{\lambda_{s-}} e^{-(\beta-\alpha)s}\, \widetilde{N}(du, ds) \tag{26}$$

Remark 6 (*Main consequences*) Representation (25) for the intensity of a Hawkes process allows to deduce easily many results. First, we remark that the process λ is split into two terms, a deterministic mean-reverting and a martingale term. From representation (25), we easily deduce that the mean-reverting speed is given by $\beta - \alpha$ and that the long-term mean level is given by $\lambda_0\beta/(\beta - \alpha)$.

We also have an interesting result about the Laplace transform.

Proposition 7.3 (Laplace transform) *Let Eq. (25) be the evolution of the intensity of Hawkes process (23). We have the following Laplace transform:*

$$\mathbb{E}\left\{e^{-a\lambda_T - b\int_t^T \lambda_s ds - cJ_T} | \mathscr{F}_t\right\} = exp\left\{-\lambda_t v(T-t) - cJ_t - \beta \int_t^T v(s)ds\right\}$$

where $v(s)$ indicates $v\,(s;\ a, b, c)$ that satisfies

$$\frac{\partial}{\partial t} v\,(s;\ a, b, c) = -\Psi\,(v\,(s;\ a, b, c)\,, b, c)\,,$$

with $v\,(T;\ a, b, c) = a$ and

$$\Psi(q_1, q_2, q_3) := -\beta q_1 + q_2 - \left(e^{-q_1\alpha - q_3} - 1\right)$$

Proof The proof is divided into two steps. The first one is to show that the triplet $\Theta_t := (\lambda_t, \int_0^t \lambda_s ds, J_t)$ is an affine process. This fact stems mainly by the arguments of Chap. 4 in [42]. We recall that the SDE satisfied by Θ_t

$$\begin{aligned}
\Theta_t^{(1)} &:= \lambda_t = \lambda_0 + \int_0^t (\beta-\alpha)\left(\frac{\beta}{\beta-\alpha}\lambda_0 - \lambda_s\right) ds + \alpha \int_0^t \int_0^{\lambda_{s-}} \widetilde{N}(du, ds) \\
\Theta_t^{(2)} &= \int_0^t \lambda_s ds \\
\Theta_t^{(3)} &= \int_0^t \int_0^{\lambda_{s-}} N(du, ds)
\end{aligned}$$

We remark that the first term is autonomous. It is the SDE of a CBI written in integral form. Thanks to Theorem 4.2.1 in [42], the related generator is

$$\mathscr{L} f(x) = (\beta-\alpha)\left(\frac{\beta}{\beta-\alpha}\lambda_0 - x\right) f'(x) + x\left(f(x+\alpha) - f(x) - \alpha f'(x)\right)$$

The two other terms in Θ depend on the first one and using similar arguments, we obtain the whole generator

$$\mathscr{L}f(x) = (\beta-\alpha)\left(\frac{\beta}{\beta-\alpha}\lambda_0 - x_1\right) f_1'(x) + x_1 f_2'(x)$$
$$+x_1 \int_0^{\infty} \Big(f(x_1+\alpha z, x_2, x_3+z) - f(x) - \alpha z f_1'(x)\Big)\delta_1(dz),$$

where $x := (x_1, x_2, x_3)$ and f_i' denotes the first derivative with respect to the i-coordinate.

The second step is to remark that the previous generator is of affine type. The proof is then obtained by a direct application of Proposition 1 in [19]. The computation is quite tedious but straightforward and then it is left to interested readers. □

Proposition 7.3 is very useful for mathematical finance applications since many models are defined as the exponential of a stochastic process in order to fulfil the positivity. In particular, interest rates and credit risk modelling uses exponential model, see [12, 20, 21, 33]. Moreover, fast Fourier transform techniques are known to solve financial problems, see [10]. Finally, martingale representation (25) allows to compute the SDE satisfied by a general function of the intensity process λ via Itō-formula distinguishing automatically martingale and finite variation parts. Interested reader can easily find the recurrence equations for the moments of the intensity in this way.

7.3 Extensions

The main advantage of the CBI point of view is its plasticity. We recall that the intensity of a Hawkes process belongs to the class described by (21). We now summarize some extensions and some related financial applications.

Looking at CBI representation (21), we can first relax the fixed size of the jumps impact. We obtain the marked Hawkes process, see Definition 2.4. That is,

$$\lambda_t = \lambda_0 + \int_0^t (\beta-\alpha)\left(\frac{\beta}{\beta-\alpha}\lambda_0 - \lambda_{s-}\right) ds + \int_0^t \int_0^{\lambda_{s-}} \int_0^{\infty} \zeta \widetilde{N}(d\zeta, du, ds) \tag{27}$$

where the compensated Poisson measure $\widetilde{N}$ has $\mu(d\zeta)duds$ as compensator, with $\mu(d\zeta)$ a finite activity Poisson measure. This situation is useful in finance to describe the proper impact of each jump. An application in credit risk modelling is to consider that $\mu(d\zeta)$ is related to the recovery rate of each default, as seen in Sect. 5.1.

A second possible extension, keeping a finite variation framework, is to add an external source of risk via a second jump process acting on the intensity λ. The couple (λ_t, P_t) reads

$$\begin{cases} \lambda_t = \lambda_0 + \int_0^t (\beta - \alpha) \left(\frac{\beta}{\beta - \alpha} \lambda_0 - \lambda_{s-} \right) ds + \int_0^t \int_0^\infty \zeta M(d\zeta, ds) \\ \qquad + \int_0^t \int_0^{\lambda_{s-}} \int_0^\infty \zeta \widetilde{N}(d\zeta, du, ds) \\ P_t = \int_0^t \int_0^{\lambda_{s-}} \int_0^\infty N(d\zeta, du, ds) \end{cases} \tag{28}$$

where M is an independent pure jump Poisson measure with compensator $\nu(d\zeta)ds$. Note that the counting process P_t is only affected by the jump of the Poisson measure N and is unchanged when the Poisson measure M has a jump. This is in particular the case of the model described in [14].

Third extension is obtained adding a Brownian term

$$\begin{cases} \lambda_t = \lambda_0 + \int_0^t (\beta - \alpha) \left(\frac{\beta}{\beta - \alpha} \lambda_0 - \lambda_{s-} \right) ds + \int_0^t \int_0^\infty \zeta M(d\zeta, ds) \\ \qquad + \sigma \int_0^t \int_0^{\lambda_{s-}} W(du, ds) + \int_0^t \int_0^{\lambda_{s-}} \int_0^\infty \zeta \widetilde{N}(d\zeta, du, ds) \\ P_t = \int_0^t \int_0^{\lambda_{s-}} \int_0^\infty N(d\zeta, du, ds) \end{cases}$$

where W is a two-dimensional white noise. A possible application is on the field of optimal liquidation, and λ can represent the liquidity cost process that exhibit positive jumps due to large trader interventions, see [11]. But small investors act as well and their contribution can be easily represented by a white noise. Finally, self-exciting structure and Hawkes processes are recently introduced in this framework to explain some features, see [2].

Finally, we consider the main extension differentiating Hawkes processes and CBI framework, that is, infinite activity. Levy processes include the possibility to have an infinite number of jumps in a finite interval assuming that there are small. To overcome the explosion, these jumps can be added only compensated, that is, the pure jump part cannot be separated by its compensator. Roughly speaking both the pure jump process and its compensator diverge but their sum stay finite.

The infinite activity case cannot be easily applied in Hawkes framework directly since P is a counting process. As a consequence, in infinite activity case, N diverges in finite time. CBI theory gives a natural framework to overcome this difficulty, paying the cost to neglect the counting process P. We obtain the evolution

$$\begin{aligned} \lambda_t = \lambda_0 &+ \int_0^t (\beta - \alpha) \left(\frac{\beta}{\beta - \alpha} \lambda_0 - \lambda_{s-} \right) ds + \int_0^t \int_0^\infty \zeta M(d\zeta, ds) \\ &+ \sigma \int_0^t \int_0^{\lambda_{s-}} W(du, ds) + \int_0^t \int_0^{\lambda_{s-}} \int_0^\infty \zeta \widetilde{N}(d\zeta, du, ds) \end{aligned} \tag{29}$$

where the measure N has now infinite activity. The crucial reason why there is no difficulty to extend CBI to infinite activity is that the measure N appears directly compensated in (29).

As in Hawkes case, we have access to a quasi-explicit Laplace transform, which is displayed in the following proposition. The proof can be adapted from Proposition 7.3, see also the original proof of Theorem 3.1 in [15].

Proposition 7.4 (Laplace transform) *Let λ satisfies (29), we introduce the two following mechanisms:*

$$\Psi(q) = (\beta - \alpha)q + \frac{1}{2}\sigma^2 q^2 + \int_0^\infty (e^{-qu} - 1 + qu)\mu(du),$$

$$\Phi(q) = \beta\lambda_0 q + \int_0^\infty (1 - e^{-qu})\nu(du),$$

where μ, ν are the two Lévy measures associated, respectively, to N and M such that $\int_0^\infty (u \wedge u^2)\mu(du) < \infty$ and $\int_0^\infty (1 \wedge u)\nu(du) < \infty$. Then, for $p \geq 0$, by

$$\mathbb{E}_x \left\{e^{-p\lambda_t}\right\} = \exp\left(-\lambda_0 v(t, p) - \int_0^t \Phi\big(v(s, p)\big)ds\right),$$

where the function $v : \mathbb{R}_+ \times \mathbb{R}_+ \to \mathbb{R}$ satisfies the following differential equation:

$$\frac{\partial v(t, p)}{\partial t} = -\Psi(v(t, p)), \quad v(0, p) = p.$$

References

1. Alfonsi, A., Blanc, P.: Extension and calibration of a Hawkes-based optimal execution model. Mark. Microstruct. Liq. **2**, 1650005 (2016)
2. Alfonsi, A., Blanc, P.: Dynamic optimal execution in a mixed-market-impact Hawkes price model. Financ. Stoch. **20**(1), 183–218 (2016)
3. Bacry, E., Dayri, K., Muzy, J.: Non-parametric kernel estimation for symmetric hawkes processes. Application to high frequency financial data. Eur. Phys. J. B Condens. Matter Complex Syst. **85.5**, 1–12 (2012)
4. Bacry, E., Delattre, S., Hoffmann, M., Muzy, J.: Some limit theorems for Hawkes processes and application to financial statistics. Stoch. Process. Their Appl. **123–127**, 2475–2499 (2013)
5. Bacry, E., Muzy, J.: Hawkes model for price and trades high-frequency dynamics. Quant. Financ. **14–17**, 1147–1166 (2014)
6. Bernis, G., Salhi, K., Scotti, S.: Sensitivity analysis for marked Hawkes processes—application to CLO pricing (2017). Available via SSRN:https://ssrn.com/abstract=2989193
7. Bouleau, N.: Processus Stochastiques et Applications. Hermann, Paris (1988)
8. Brémaud, P., Massoulié, L.: Stability of Nonlinear Hawkes Processes. Ann. Probab. **24**(3), 563–1588 (1996)
9. Callegaro, G., Gaigi, M.H., Scotti, S., Sgarra, C.: Optimal investment in markets with over and under-reaction to information. Math. Financ. Econ. **11**(3), 299–322 (2017)
10. Carr, P., Madan, D.: Option valuation using the fast Fourier transform. J. Comput. Financ. **2**(4), 61–73 (1999)

11. Chevalier, E., Ly Vath, V., Roch, A., Scotti, S.: Optimal execution cost for liquidation through a limit order market. Int. J. Theor. Appl. Financ. **19**(1), 1650004 (2016)
12. Cox, J., Ingersoll, J., Ross, S.: A theory of the term structure of interest rate. Econometrica **53**, 385–408 (1985)
13. Da Fonseca, J., Zaatour, R.: Hawkes processes: fast calibration, application to trade clustering and diffusive limit. J. Futur. Mark. **34**, 548–579 (2014)
14. Dassios, A., Zhao H.: A generalized contagion process with an application to credit risk. Int. J. Theor. Appl. Financ. **20.01**, 1750003 (2017)
15. Dawson, D.A., Li, Z.: Skew convolution semigroups and affine Markov processes. Ann. Probab. **34**, 1103–1142 (2006)
16. Dawson, A., Li, Z.: Stochastic equations, flows and measure-valued processes. Ann. Probab. **40**(2), 813–857 (2012)
17. Dellacherie, C., Meyer, P.A.: Probabilités et Potentiles—Chapitre I à IV. Hermann, Paris (1978)
18. Duffie, D.: A short course on credit risk modeling with affine processes. J. Bank. Financ. **29**, 2751–2802 (2005)
19. Duffie, D., Pan, J., Singleton, K.: Transform analysis and asset pricing for affine jump-diffusions. Econometrica **68**(6), 1343–1376 (2000)
20. Duffie, D., Filipovic, D., Schachermayer, W.: Affine processes and applications in finance. Ann. Appl. Probab. **13**(3), 984–1053 (2003)
21. Errais, E., Giesecke, K., Goldberg, L.: Affine point processes and portfolio credit risk. SIAM J. Financ. Math. **1**, 642–665 (2010)
22. Fičura, M.: Forecasting jumps in the intraday foreign exchange rate time series with Hawkes processes and logistic regression. In: Procházka, D. (ed.) New Trends in Finance and Accounting, pp. 125–137. Springer, New York (2017)
23. Filimonov, V., Sornette, D.: Quantifying reflexivity in financial markets: toward a prediction of flash crashes. Phys. Rev. E **85**, 056108 (2012)
24. Filipovic, D.: A general characterization of one factor affine term structure models. Financ. Stoch. **5**, 389–412 (2001)
25. Hainault, D.: A model for interest rates with clustering effects. Quant. Financ. **16**, 1203–1218 (2016)
26. Hardiman, S., Bercot, N., Bouchaud, J.: Critical reflexivity in financial markets: a Hawkes process analysis. Eur. Phys. J. B Condens. Matter Complex Syst. **86–10**, 1–9 (2013)
27. Harris, T.E.: The theory of branching processes. Dover Phoenix Editions, Mineola (2002)
28. Hawkes, A.G.: Spectra of some self-exciting and mutually exciting point processes. Biometrika **58**, 83–90 (1971)
29. Hawkes, A.G., Adamopoulos, L.: Cluster models for earthquakes—regional comparisons. Bull. Int. Stat. Inst. **45**, 454–461 (1973)
30. Hawkes, A.G., Oakes, D.: Spectra of some self-exciting and mutually exciting point processes. J. Appl. Probab. **11**, 493–503 (1974)
31. Jacod, J., Shiryaev, N.: Limit Theorems for Stochastic Processes. Springer, Heidelberg (2003)
32. Jaisson, T., Rosenbaum, M.: Limit theorems for nearly unstable Hawkes processes. Ann. Appl. Probab. **25–2**, 600–631 (2015)
33. Jiao, Y., Ma, C., Scotti, S.: Alpha-CIR model with branching processes in sovereign interest rate modeling. Financ. Stoch. **21**(3), 789–813 (2017)
34. Jiao, Y., Ma, C., Scotti, S., Sgarra, C.: A branching process approach to power markets. Energy Econ. **79**, 144–156 (2019)
35. Kawazu, K., Watanabe, S.: Branching processes with immigration and related limit theorems. Theory Probab. Its Appl. **I**(1), 36–54 (1971)
36. Klebaner, F., Liptser, G.: When a stochastic exponential is a true martingale. Extension of the Beneš method. Theory Probab. Appl. **58**(1), 38-DC362 (2012)
37. Kyprianou, A.: Introductory lectures on fluctuations of Lévy processes with applications. Springer Science & Business Media, Berlin (2006)
38. Lamperti, J.: Continuous state branching processes. Bull. Am. Math. Soc. **73**(3), 382–386 (1967)

39. Last, G., Brandt, A.: Marked Point Processes on the Real Line—The Dynamic Approach. Springer, Heidelberg (1995)
40. Le Gall, J.-F.: Spatial Branching Processes. Random Snakes and Partial Differential Equations. Lectures in Mathematics. Birkhauser, ETH Zurich (1999)
41. Lemonnier, R., Vayatis N.: Nonparametric Markovian learning of triggering kernels for mutually exciting and mutually inhibiting multivariate Hawkes processes. In: Machine Learning and Knowledge Discovery in Databases, pp. 161–176. Springer, Berlin (2014)
42. Li, Z. Continuous-state branching processes (2012). arXiv:1202.3223
43. Ma, Y., Xu, W.: Structural credit risk modelling with Hawkes jump diffusion processes. J. Comput. Appl. Math. **303**, 69–80 (2016)
44. Mancini, C.: Non-parametric threshold estimation for models with stochastic diffusion coefficient and jumps. Scand. J. Stat. **36**, 270–296 (2009)
45. Mandelbrot, B.: The variation of certain speculative prices. J. Bus. **36–4**, 394–419 (1963)
46. Ogata, Y.: Statistical models for earthquake occurrences and residual analysis for point processes. J. Am. Stat. Assoc. **83**, 9–27 (1988)
47. Ozaki, T.: Maximum likelihood estimation of Hawkes' self exciting processes. Ann. Inst. Statist. Math. **31**, 145–155 (1979)
48. Peng, X., Kou, S.: Default clustering and valuation of collateralized debt obligations (2009)
49. Protter, Ph: Stochastic Integration and Differential Equations. Springer, Heidelberg (1992)
50. Watson, H.W., Galton, F.: On the probability of the extinction of families. J. Anthropol. Inst. G. B. Irel. **4**, 138–144 (1875)
51. Zheng, B., Roueff, F., Abergel, F.: Modelling bid and ask prices using constrained hawkes processes: ergodicity and scaling limit. SIAM J. Financ. Math. **5–1**, 99–136 (2014)

Bernstein Copulas and Composite Bernstein Copulas

Jingping Yang, Fang Wang and Zongkai Xie

Abstract Copula functions have been widely used in econometrics, finance, statistics, and social science for modeling dependence. Reference [42] presented the Bernstein copulas for approximating copula functions. Inspired by the Bernstein copula put forward by [42], reference [48] introduced a new copula function, named as composite Bernstein copula. The composite Bernstein copulas include Bernstein copulas as its special family. Following [48]'s work, [20] discussed the composite Bernstein copula from its generality, its probability structure, and its application in portfolio credit risk. This paper serves as a summary of main results in the above papers.

Keywords Bernstein copulas · Composite Bernstein copulas

1 Introduction

A copula function is a multivariate distribution function with uniform $[0,1]$ marginal distributions. The important role of the copula function can be stated by Sklar's theorem. Sklar's theorem shows that for each joint distribution function H with marginal distributions $F_1, \cdots, F_k$, there exists a copula function C such that

$$H(x_1, x_2, \cdots, x_k) = C(F_1(x_1), F_2(x_2), \cdots, F_k(x_k)),$$

J. Yang
LMEQF, Department of Financial Mathematics, Peking University,
Beijing 100871, China
e-mail: yangjp@math.pku.edu.cn

F. Wang (✉)
School of Mathematical Sciences, Capital Normal University,
Beijing 100048, China
e-mail: fang72_wang@cnu.edu.cn

Z. Xie
Department of Financial Mathematics, Peking University, Beijing 100871, China
e-mail: xiezongkai@pku.edu.cn

© Springer Nature Singapore Pte Ltd. 2020
Y. Jiao (ed.), *From Probability to Finance*, Mathematical Lectures
from Peking University, https://doi.org/10.1007/978-981-15-1576-7_4

and the copula C is unique when the marginal distributions $F_1, \cdots, F_k$ are continuous. For more detailed introduction about copula theory, see [35]. Now copula functions have been widely used in insurance and finance. Please see [46] for copula methods in non-life insurance and [3] for modeling dependence in insurance claims using Lévy copulas. See also [13] for copula methods in finance and [33] for copula methods in quantitative risk management. Please see [26] for the applications of copula methods in other areas.

Construction of copula functions has become an important research area during the past few years. Reference [42] introduced a new family of copulas called the Bernstein copula (BC), defined in terms of Bernstein polynomials. The BC can be used as an approximation to a known copula. The BC can also be used to approximate an unknown copulas, which is called the empirical Bernstein copula (EBC). Reference [42] proved that any copula function can be approximated by Bernstein copulas. Following [42]'s work, [5, 16, 17, 22, 41, 43] discussed the BC from probabilistic and statistical perspectives, and [15, 47] focused on the applications of the BC in non-life insurance and finance.

Bernstein copulas can also be used in many other areas of science, including medicine, prevention of natural disasters, and engineering physics. Reference [21] applied Bernstein copula to model the dependence structure of petrophysical properties. Reference [34] focused on the two-dimensional fracture network simulations by using Bernstein copula, which is a popular method for modeling fracture systems. Reference [12] used the Bernstein copula-based methods to study the data of wind directions and rainfall.

Inspired by the BC, [48] presented a new family of copulas, called composite Bernstein copula (CBC). The family of CBCs includes the family of BCs. As pointed out in [42], a limitation of the Bernstein copula is that it fails to capture the extreme tail behavior, a challenging modeling problem in insurance and finance. The CBC is able to capture the tail dependence, and it has reproduction property for the three important dependence structures: comonotonicity, countermonotonicity, and independence.

Following [48]'s work, [20] investigated the CBC from its generality, its probability structure and its application in portfolio credit risk. Reference [20] showed that some well-known copulas, such as the Baker's distributions in [5], the Baker's Type II BB distributions in [7], and the mixture copulas in [6], belong to the CBC family. The probabilistic structure for the CBC family was presented in [20], and it was also applied in the modeling of portfolio credit risk.

In the following, we will introduce the main contributions of the BC and the CBC in [20, 42, 48]. In Sect. 2, we will present the main results in [42]. In Sect. 3, we will focus on the main results in [20, 48], and the connection between the CBC and the BC is provided. In Sect. 4, a simulation study and a real data analysis are given to illustrate the main results. The conclusions are drawn in Sect. 5.

2 Bernstein Copulas

2.1 Definition of Bernstein Copulas

The Bernstein copula (BC) was formally introduced by [42] in terms of Bernstein polynomials, with applications to approximation and nonparametric estimation of copulas.

In approximation theory, Bernstein copula approximation has been studied earlier and related works can be traced back to [18, 28–30] to our knowledge. Reference [14] reviewed the connection of Bernstein polynomials to the construction of Bernstein copulas in terms of tensor product Bernstein operators with a discrete skeleton. We now give the definition of the BC, using the notations from [42].

Fix positive integers $m_1, \cdots, m_k$. Let

$$P_{v_j,m_j}(u_j) := \binom{m_j}{v_j} u_j^{v_j}(1-u_j)^{m_j-v_j}, \quad u_j \in [0,1].$$

Let $\alpha(v_1/m_1, \cdots, v_k/m_k)$ be a real-valued constant indexed by $(v_1, \cdots, v_k)$, such that $0 \le v_j \le m_j \in \mathbb{N}$. Given the sequence

$$\{\alpha(v_1/m_1, \cdots, v_k/m_k), v_j = 0, 1, \cdots, m_j \text{ for } j = 1, \cdots, k\},$$

define the following map $C_B : [0,1]^k \to [0,1]$ as

$$C_B(u_1, \cdots, u_k) = \sum_{v_1=0}^{m_1} \cdots \sum_{v_k=0}^{m_k} \alpha\left(\frac{v_1}{m_1}, \cdots, \frac{v_k}{m_k}\right) P_{v_1,m_1}(u_1) \cdots P_{v_k,m_k}(u_k). \quad (1)$$

In the above equation, $C_B(u_1, \cdots, u_k)$ can be regarded as an approximation to the function $\alpha(u_1, \cdots, u_k)$, using independent Binomial random variables.

Remark 2.1 This type of $k-$dimensional Bernstein polynomial can be traced back to the monograph by [32] for approximating the function $\alpha(u_1, \ldots, u_k)$. Here $P_{v_j,m_j}(u_j)$ is the Bernstein basis polynomial, also the Binomial probability, so $P_{v_1,m_1}(u_1) \cdots P_{v_k,m_k}(u_k)$ is the probability law of independent Binomial random variables. The integers $m_1, \cdots, m_k$ play the role of smoothing parameters.

Let $\Delta_{1,\ldots,k}$ denote the k-dimensional forward difference operator, i.e.,

$$\Delta_{1,\ldots,k}\alpha\left(\frac{v_1}{m_1}, \cdots, \frac{v_k}{m_k}\right) := \sum_{\ell_1=0}^{1} \cdots \sum_{\ell_k=0}^{1} (-1)^{k+\ell_1+\cdots+\ell_k} \alpha\left(\frac{v_1+\ell_1}{m_1}, \cdots, \frac{v_k+\ell_k}{m_k}\right).$$

Under some conditions on the array $\alpha(\frac{v_1}{m_1}, \cdots, \frac{v_k}{m_k})$, $C_B(u_1, \cdots, u_k)$ is a copula function. The following statement merges the result of [42] and that of [45] into a new theorem.

Theorem 2.2 ([42, 45]) *The Bernstein polynomial $C_B(u_1, \dots, u_k)$ is a copula function if conditions*

$$\Delta_{1,\dots,k}\alpha\left(\frac{v_1}{m_1}, \cdots, \frac{v_k}{m_k}\right) \geq 0, \tag{2}$$

for all $0 \leq v_j \leq m_j - 1$, $j = 1, \dots, k$,

$$\alpha\left(\frac{v_1}{m_1}, \dots, \frac{v_{j-1}}{m_{j-1}}, 0, \frac{v_{j+1}}{m_{j+1}}, \frac{v_k}{m_k}\right) = 0, \quad \forall j = 1, \dots, k, \tag{3}$$

and

$$\alpha\left(1, \dots, 1, \frac{v_j}{m_j}, 1, \dots, 1\right) = \frac{v_j}{m_j}, \quad \forall j = 1, \dots, k \tag{4}$$

hold. Moreover, (3) *and* (4) *are necessary for* $C_B(u_1, \cdots, u_k)$ *to be a copula.*

Based on the above theorem, the following definition of Bernstein copula is given.

Definition 2.3 If conditions (2), (3), and (4) are satisfied, we call $C_B(u_1, \dots, u_k)$ in (1) the *Bernstein copula* (BC).

Remark 2.4 Actually, [42] gave a little stronger sufficient conditions for the function C_B to be a copula function, that is, the condition

$$\max\left(0, \frac{v_1}{m_1} + \dots + \frac{v_k}{m_k} - (k-1)\right) \leq \alpha\left(\frac{v_1}{m_1}, \dots, \frac{v_k}{m_k}\right) \leq \min\left(\frac{v_1}{m_1}, \dots, \frac{v_k}{m_k}\right) \tag{5}$$

and Eq. (2) hold. Obviously, Eq. (5) implies Eqs. (3) and (4). The conclusion (3) is expressed as $\lim_{v_j \to 0} \alpha(\frac{v_1}{m_1}, \dots, \frac{v_k}{m_k}) = 0$, $\forall j = 1, \dots, k$, in [42].

Remark 2.5 If α is a copula, so is C_B from the above theorem. However, the copula α is usually unknown in practice; [42] proposed to replace α with its empirical version, named as empirical Bernstein copula (EBC). Using Bernstein polynomials, the EBC smooths the empirical copula. The EBC or its variants have found many successful applications recently, such as estimating the copula density nonparametrically ([9, 10, 23]), regression function estimations ([8, 11, 24, 25]), dependence function estimations in the extreme value theory ([2, 27]), etc.

2.2 The BC Generated by a Copula Function

Suppose that C is a k-dimensional copula. If we let

$$\alpha(v_1/m_1, \cdots, v_k/m_k) = C(v_1/m_1, \cdots, v_k/m_k), \tag{6}$$

the conditions (2), (3), and (4) are obviously satisfied. This phenomenon has been noticed by [18, 29, 30]. The following proposition states that under the conditions (2), (3), and (4), the sequence

$$\{\alpha(v_1/m_1, \cdots, v_k/m_k),\ v_j = 0, 1, \cdots, m_j \text{ for } j = 1, \cdots, k\}$$

can be illustrated by a copula function.

Proposition 2.6 *Under the conditions* (2), (3), *and* (4), *there exists a copula function* C *such that* (6) *holds for* $v_j = 0, \cdots, m_j,\ j = 1, \cdots, k$.

Proof Set

$$\begin{aligned} &C(u_1, \cdots, u_k) \\ &= \sum_{v_1=0}^{m_1-1} \cdots \sum_{v_k=0}^{m_k-1} \Delta_{1,\dots,k}\alpha(v_1/m_1, \cdots, v_k/m_k) \prod_{i=1}^{k} \min\{\max\{0, m_i(u_i - \frac{v_i}{m_i})\}, 1\}. \end{aligned} \tag{7}$$

We know that for $n_i = 0, \cdots, m_i$, where $i = 1, \cdots, k$,

$$\begin{aligned} C(\frac{n_1}{m_1}, \dots, \frac{n_k}{m_k}) &= \sum_{v_1=0}^{n_1-1} \cdots \sum_{v_k=0}^{n_k-1} \Delta_{1,\dots,k}\alpha(v_1/m_1, \cdots, v_k/m_k) \\ &= \sum_{v_1=0}^{n_1-1} \cdots \sum_{v_k=0}^{n_k-1} \sum_{\ell_1=0}^{1} \cdots \sum_{\ell_k=0}^{1} (-1)^{k+\ell_1+\cdots+\ell_k} \alpha(\frac{v_1+\ell_1}{m_1}, \cdots, \frac{v_k+\ell_k}{m_k}) \\ &= \sum_{\ell_k=0}^{1} \sum_{v_k=0}^{n_1-1} \cdots \sum_{\ell_1=0}^{1} \sum_{v_1=0}^{n_1-1} (-1)^{k+\ell_1+\cdots+\ell_k} \alpha(\frac{v_1+\ell_1}{m_1}, \cdots, \frac{v_k+\ell_k}{m_k}). \end{aligned}$$

Note that

$$\begin{aligned} &\sum_{\ell_1=0}^{1} \sum_{v_1=0}^{n_1-1} (-1)^{k+\ell_1+\cdots+\ell_k} \alpha(\frac{v_1+\ell_1}{m_1}, \cdots, \frac{v_k+\ell_k}{m_k}) \\ &= \sum_{\ell_1=0}^{1} (-1)^{k+\ell_1+\cdots+\ell_k} \left(\alpha(\frac{\ell_1}{m_1}, \frac{v_2+\ell_2}{m_2}, \cdots, \frac{v_k+\ell_k}{m_k}) \right. \\ &\qquad \left. + \cdots + \alpha(\frac{n_1-1+\ell_1}{m_1}, \frac{v_2+\ell_2}{m_2}, \cdots, \frac{v_k+\ell_k}{m_k}) \right) \\ &= (-1)^{k+\ell_2+\cdots+\ell_k} \left(\alpha(\frac{0}{m_1}, \frac{v_2+\ell_2}{m_2}, \cdots, \frac{v_k+\ell_k}{m_k}) + \cdots \right. \\ &\qquad \left. + \alpha(\frac{n_1-1}{m_1}, \frac{v_2+\ell_2}{m_2}, \cdots, \frac{v_k+\ell_k}{m_k}) \right) \end{aligned}$$

$$
\begin{aligned}
&-(-1)^{k+\ell_2+\cdots+\ell_k}\left(\alpha(\frac{1}{m_1},\frac{v_2+\ell_2}{m_2},\cdots,\frac{v_k+\ell_k}{m_k})+\cdots\right.\\
&\left.+\alpha(\frac{n_1}{m_1},\frac{v_2+\ell_2}{m_2},\cdots,\frac{v_k+\ell_k}{m_k})\right)\\
&=-(-1)^{k+\ell_2+\cdots+\ell_k}\alpha(\frac{n_1}{m_1},\frac{v_2+\ell_2}{m_2},\cdots,\frac{v_k+\ell_k}{m_k})\\
&+(-1)^{k+\ell_2+\cdots+\ell_k}\alpha(\frac{0}{m_1},\frac{v_2+\ell_2}{m_2},\cdots,\frac{v_k+\ell_k}{m_k})\\
&=(-1)^{k-1+\ell_2+\cdots+\ell_k}\alpha(\frac{n_1}{m_1},\frac{v_2+\ell_2}{m_2},\cdots,\frac{v_k+\ell_k}{m_k}).
\end{aligned}
$$

Then applying the above result we can get that

$$
\begin{aligned}
C(\frac{n_1}{m_1},\ldots,\frac{n_k}{m_k})&=\sum_{\ell_k=0}^{1}\sum_{v_k=0}^{n_k-1}\cdots\sum_{\ell_2=0}^{1}\sum_{v_2=0}^{n_2-1}(-1)^{k-1+\ell_2+\cdots+\ell_k}\alpha(\frac{n_1}{m_1},\cdots,\frac{v_k+\ell_k}{m_k})\\
&=\alpha(\frac{n_1}{m_1},\cdots,\frac{n_k}{m_k}).
\end{aligned}
\tag{8}
$$

Thus (6) holds. Next we will prove that $C(u_1,\cdots,u_k)$ is a copula function.

For fixed $0\le v_i\le m_i$, $i=1,\cdots,k$, $\min\{\max\{0,m_i(u_i-\frac{v_i}{m_i})\},1\}$ obviously is a univariate distribution function of $u_i\in[0,1]$. So $C(u_1,\cdots,u_k)$ is a linear combination of multivariate distribution functions according to the construction expression (7). From (8), we know that

$$C(1,\cdots,1)=\alpha(1,\cdots,1)=1.$$

Thus $C(u_1,\cdots,u_k)$ is a distribution function on $[0,1]^k$.

For $u_j\in[0,1]$, similarly as the proof process of Eq. (6) and from Eq. (4) we know that

$$
\begin{aligned}
&C(1,\cdots,1,u_j,1,\cdots,1)\\
&=\sum_{v_1=0}^{m_1-1}\cdots\sum_{v_k=0}^{m_k-1}\Delta_{1,\ldots,k}\alpha(\frac{v_1}{m_1},\cdots,\frac{v_k}{m_k})\min\{\max\{0,m_j(u_j-\frac{v_j}{m_j})\},1\}\\
&=\sum_{v_j=0}^{m_j-1}\min\{\max\{0,m_j(u_j-\frac{v_j}{m_j})\},1\}\sum_{\ell_j=0}^{1}(-1)^{\ell_j+k-(k-1)}\alpha(1,\cdots,1,\frac{v_j+\ell_j}{m_j},1,\cdots,1)\\
&=\sum_{v_j=0}^{m_j-1}\min\{\max\{0,m_j(u_j-\frac{v_j}{m_j})\},1\}\frac{1}{m_j}\\
&=u_j.
\end{aligned}
$$

Thus $C(u_1,\ldots,u_k)$ is a copula function. The proposition is proved.

2.3 Convergence of Bernstein Copulas

In the following, we consider the situation that the array α can be expressed as a copula function, that is, there exists a copula function C such that (6) holds. Then

$$C_B(u_1, \cdots, u_k) = \sum_{v_1=0}^{m_1} \cdots \sum_{v_k=0}^{m_k} C\left(\frac{v_1}{m_1}, \cdots, \frac{v_k}{m_k}\right) P_{v_1,m_1}(u_1) \cdots P_{v_k,m_k}(u_k). \quad (9)$$

Actually, given a copula C, we can use C_B in (9) to approximate the copula function C.

Reference [42] proved that in the case that C is a general continuous function, not limited to the family of copula functions, the first-order partial derivative of the function C is Lipschitz and when $m_1 = \cdots = m_k = m$, the following inequality

$$|C_B(u_1, \cdots, u_k) - C(u_1, \cdots, u_k)| \le M \sum_{j=1}^{k} \frac{u_j(1-u_j)}{2m}$$

holds, where M is a constant. It means that when C is a copula function, under some conditions the copula function $C_B(u_1, \cdots, u_k)$ tends to the copula function $C(u_1, \cdots, u_k)$ as m goes to infinity.

2.4 The Density Function of the BC

It is well known that

$$\frac{d}{du} P_{v,m}(u) = m[P_{v-1,m-1}(u) - P_{v-1,m-1}(u)], \text{ for } v = 0, \cdots, m,$$

with the convention $P_{-1,m-1}(u) = P_{m,m-1}(u) = 0$. By direct differentiation, [42] showed that the BC density can be expressed as

$$\begin{aligned} c_B(u_1, \cdots, u_k) &= \frac{\partial^k C_B}{\partial u_1 \cdots \partial u_k} \\ &= \sum_{v_1=0}^{m_1-1} \cdots \sum_{v_k=0}^{m_k-1} \Delta_{1,\ldots,k} C\left(\frac{v_1}{m_1}, \cdots, \frac{v_k}{m_k}\right) \times \left(\prod_{i=1}^{k} m_i P_{v_i,m_i-1}(u_i)\right). \end{aligned} \quad (10)$$

Suppose that $(V_1, \ldots, V_k)$ is a random vector with the cumulative distribution function C, here C is a copula satisfying (6). Let $K_i = \lfloor m_i V_i \rfloor$, the greatest integer that is less than $m_i V_i$, $i = 1, \cdots, k$. Then we have

$$\mathbb{P}(K_1 = v_1, \cdots, K_k = v_k) = \Delta_{1,\ldots,k} C\left(\frac{v_1}{m_1}, \cdots, \frac{v_k}{m_k}\right), \quad (11)$$

where $v_j = 0, \cdots, m_j - 1,\ j = 1, \cdots, k$, and the BC density function can also be expressed as

$$c_B(u_1, \cdots, u_k) = \sum_{v_1=0}^{m_1-1} \cdots \sum_{v_k=0}^{m_k-1} \mathbb{P}(K_1 = v_1, \cdots, K_k = v_k) \times \left(\prod_{i=1}^{k} m_i P_{v_i, m_i-1}(u_i) \right). \tag{12}$$

Note that $m P_{v,m-1}(u) = \dfrac{1}{B(v+1, m-v)} u^v (1-u)^{m-v-1}$ is the probability density function of Beta distribution, where $B(a, b)$ is the complete Beta function, i.e., $B(a, b) = \int_0^1 t^{a-1}(1-t)^{b-1} dt$. Thus the density function of the BC can be expressed as a mixture of the densities derived from Beta distributions.

Let $B(x; a, b) = \int_0^x t^{a-1}(1-t)^{b-1} dt$ denote the incomplete Beta function, $I_x(a, b) = \dfrac{B(x; a, b)}{B(a, b)}$ denote the regularized incomplete Beta function, and $F_{Bin(m,u)}$ be the binomial distribution function with parameter (m, u), $u \in [0, 1]$. It is known that the regularized incomplete Beta function $I_x(a, b)$ is the cumulative distribution function of the Beta distribution. For fixed $j = 0, 1, \ldots, m-1$, we know that

$$\begin{aligned} \overline{F}_{Bin(m,u)}(j) &:= 1 - F_{Bin(m,u)}(j) = \sum_{i=j+1}^{m} \binom{m}{i} u^i (1-u)^{m-i} \\ &= I_u(j+1, m-j) = m B_{j,m-1}(u),\ u \in [0, 1], \end{aligned} \tag{13}$$

where for $k = 0, 1, \ldots, m$,

$$b_{k,m}(t) = \binom{m}{k} t^k (1-t)^{m-k},\ t \in [0, 1]; \quad B_{k,m}(u) = \int_0^u b_{k,m}(t) dt.$$

Integrating Eq. (12), we can get an equivalent expression of the BC, i.e.,

$$\begin{aligned} C_B(u_1, \cdots, u_k) &= \int_0^{u_1} \cdots \int_0^{u_k} c_B(t_1, \cdots, t_k) dt_1 \cdots dt_k \\ &= \sum_{v_1=0}^{m_1-1} \cdots \sum_{v_k=0}^{m_k-1} \mathbb{P}(K_1 = v_1, \cdots, K_k = v_k) \times \left(\prod_{i=1}^{k} I_{u_i}(v_i + 1, m_i - v_i) \right) \\ &= \sum_{v_1=0}^{m_1-1} \cdots \sum_{v_k=0}^{m_k-1} \mathbb{P}(K_1 = v_1, \cdots, K_k = v_k) \times \left(\prod_{i=1}^{k} \overline{F}_{Bin(m_i, u_i)}(v_i) \right). \end{aligned}$$

Note that if we change the subscripts, the above equation has a form corresponding to the order-statistics representation presented in Sect. 2.6. That is,

$$C_B(u_1, \cdots, u_k) = \sum_{v_1=1}^{m_1} \cdots \sum_{v_k=1}^{m_k} \mathbb{P}(K_1 + 1 = v_1, \cdots, K_k + 1 = v_k) \times \left(\prod_{i=1}^{k} \overline{F}_{Bin(m_i, u_i)}(v_i - 1) \right). \tag{14}$$

The random vector $(K_1, \cdots, K_k)$ is called the discrete skeleton of the BC in [14]. From (10) (or (12)), we see that the Bernstein density function is a mixture of independent Beta densities. This representation has been used by [42] for the expression of the BC's Spearman's rho. References [14, 38] took this representation to model the dependence structure of observed data via checkerboard copulas (also called grid-type copulas in [37]) and empirical contingency tables. Reference [15] made use of this form to generate random samples from the BC.

2.5 *Dependence Measures*

As [19] pointed out, correlation index and other dependence measures are central in finance theory, and [19] also proposed to use rank correlations and the coefficients of tail dependence in the non-elliptical world. Sometimes, these measures cannot be written in closed form or can be available but in very complicated form. In these cases, we can replace the original copula by a BC due to the uniform convergence of Bernstein's approximation and the BC's computational feasibility.

The coefficient of the tail dependence is an important measure. Unfortunately, the BC fails to capture the extreme tail behavior ([1, 18, 42]). Actually, extreme tail behavior can be modeled with composite Bernstein copulas ([48]), which we will introduce in details in Sect. 3.

When $k = 2$, $m_1 = m_2 = m$, various representations of Spearman's rho and Kendall's tau of the BC can be founded in [1, 18, 40, 42]. We use the version from [1]. The Spearman's rank correlation ρ_s is given by

$$\rho_s = 12 \int_0^1 \int_0^1 C_B(u, v) du dv - 3 = \frac{12}{(m+1)^2} \sum_{i=1}^{m} \sum_{j=1}^{m} c\left(\frac{i}{m}, \frac{j}{m}\right) - 3.$$

The Kendall's tau is given by

$$\begin{aligned}
\tau &= 4 \int_0^1 \int_0^1 C_B(u, v) dC_B(u, v) - 1 \\
&= 1 - 4 \int_0^1 \int_0^1 \frac{\partial C_B(u, v)}{\partial u} \frac{\partial C_B(u, v)}{\partial v} du dv \\
&= 1 \\
&\quad - \left(\sum_{i=1}^{m} \sum_{j=1}^{m} \sum_{r=1}^{m} \sum_{s=1}^{m} c\left(\frac{i}{m}, \frac{j}{m}\right) c\left(\frac{r}{m}, \frac{s}{m}\right) \frac{\binom{m}{i}\binom{m}{j}\binom{m}{r}\binom{m}{s}(i-r)(s-j)}{(2m-i-r)(2m-j-s)\binom{2m-1}{i+r-1}\binom{2m-1}{j+s-1}} \right).
\end{aligned}$$

2.6 Baker's Distribution—One Family of Distributions with BCs

Reference [5] proposed a new method of constructing multivariate distributions with fixed marginal distributions via the random chosen pairs of the order statistics from marginal distributions. These distributions are called Baker's distributions whose copulas are just the Bernstein copulas.

Let $X_{(1)} \leq \cdots \leq X_{(m)}$ denote the order statistics of $X_1, \cdots, X_m$, a random sample from the population F. Define $F_{\ell,m}(x)$ as the distribution function of $X_{(\ell)}$, $1 \leq \ell \leq m$. For $j \geq 0$, it is known that $\overline{F}_{Bin(m,u)}(j) = I_u(j+1, m-j)$ is the distribution function of the $(j+1)$th smallest order statistic of m i.i.d. uniform [0,1] random variables ([39]). Thus

$$F_{\ell,m}(x) = \overline{F}_{Bin(m,F(x))}(\ell - 1) = I_{F(x)}(\ell, m - \ell + 1), \; 1 \leq \ell \leq m.$$

Suppose that $Y_1, \cdots, Y_k$ are k independent random variables. Let $F_j(x)$ denote, respectively, the distribution functions of Y_j, $j = 1, \cdots, k$, Y_{ij}, $i = 1, \cdots, m_j$, $j = 1, \cdots, k$ denote independent samples from these distributions, and $Y^j_{(i)}$, $i = 1, \cdots, m_j$, $j = 1, \cdots, k$ denote the corresponding order statistics.

The random vector $(K_1, \cdots, K_k)$ is defined as in Eq. (11) of Sect. 2.4, and independent of the random samples Y_{ij}, $i = 1, \cdots, m_j$, $j = 1, \cdots, k$. Thus the $k-$dimensional Baker's distribution can be defined as follows.

Definition 2.7 The joint distribution of $(Y^1_{(K_1+1)}, \cdots, Y^k_{(K_k+1)})$ is called *Baker's distribution.*

Based on the above mathematical preparation, we can derive the expression of the Baker's distribution as

$$\begin{aligned} H(x_1, \cdots, x_k) &= \mathbb{P}(Y^j_{(K_j+1)} \leq x_j, j = 1, \cdots, k) \qquad (15) \\ &= \sum_{v_1=1}^{m_1} \cdots \sum_{v_k=1}^{m_k} \mathbb{P}(K_1 + 1 = v_1, \cdots, K_k + 1 = v_k) \times \prod_{j=1}^{k} F_{v_j,m_j}(x_j) \\ &= \sum_{v_1=1}^{m_1} \cdots \sum_{v_k=1}^{m_k} \mathbb{P}(K_1 + 1 = v_1, \cdots, K_k + 1 = v_k) \times \prod_{j=1}^{k} \overline{F}_{Bin(m_j,F_j(x_j))}(v_j - 1). \end{aligned}$$

Substituting $F_j(x_j)$ with u_j on the above equation, we obtain Eq. (14), the Bernstein copula. Baker's distribution and its generalizations are investigated in [4–7, 16, 17, 31].

2.7 Random Numbers Generation

There are three methods to generate random samples from a BC in the literature. The first is order-statistics approach ([5]) focusing on the representation (15), the second is the multivariate acceptance–rejection method ([38]), and the third is applying Eq. (12) to generate random samples from the BC ([15]).

- Order-statistics approach

(1) Generate k sets of order statistics $\{U^j_{(i)}, i = 1, \cdots, m_j,\ j = 1, \cdots, k\}$ from independent uniformly distributed random numbers.
(2) Generate discrete skeleton $(K_1, \cdots, K_k) \in \prod_{j=1}^k \{0, \cdots, m_j - 1\}$ with probability (11).
(3) Use $\left(U^1_{(K_1+1)}, \cdots, U^k_{(K_k+1)}\right)$ as a sample from the Bernstein copula.

- Multivariate acceptance–rejection method

Suppose Bernstein copula densities are bounded by $M > 0$ over the unit cube $[0, 1]^k$.
(1) Generate $k + 1$ independent uniformly distributed random numbers $u_1, \cdots, u_{k+1}$.
(2) Check whether $c_B(u_1, \cdots, u_k) > M u_{k+1}$. If so, go to Step (3), otherwise go to Step(1).
(3) Use $(u_1, \cdots, u_k)$ as a sample from the Bernstein copula.

- Mixture of independent Beta distributions

(1) Sample $(K_1, \cdots, K_k) \in \prod_{j=1}^k \{0, \cdots, m_j - 1\}$ with probability (11).
(2) Sample $(Z_1, \cdots, Z_k)$, with independent $Z_i \sim Beta(K_i + 1, m_i - K_i)$ for $i = 1, \cdots, k$.

3 Composite Bernstein Copula

3.1 Definition of Composite Bernstein Copula

Let $F^{\leftarrow}_{Bin(m,u)}$ denote the left-continuous inverse function of $F_{Bin(m,u)}$. Note that $P_{v_j,m_j}(u_j) = \mathbb{P}(F^{\leftarrow}_{Bin(m_j,u_j)}(U) = v_j)$ for a random variable $U \sim \text{U}[0, 1]$. Actually, (9) can be expressed as

$$C_B(u_1, \ldots, u_k) = \mathbb{E}\left[C\left(\frac{F^{\leftarrow}_{Bin(m_1,u_1)}(U_1)}{m_1}, \cdots, \frac{F^{\leftarrow}_{Bin(m_k,u_k)}(U_k)}{m_k}\right)\right],$$

where $u_i \in [0, 1],\ i = 1, \cdots, k$, and $U_1, \cdots, U_k$ are independent Uniform [0,1] random variables.

Let $N_{m,u}$ be a binomial random variable with distribution $F_{Bin(m,u)}$. For any copula C, its survival copula is denoted as $\overline{C}$. Then

$$\overline{C}(u_1, \cdots, u_k) = \mathbb{P}(1 - V_i \le u_i,\ i = 1, \cdots, k),$$

where $(V_1, \cdots, V_k)$ is a random vector with distribution function C.

For a given C, by incorporating the information of another copula D, with the given positive integers m_i, $i = 1, \cdots, k$, [48] presented a new function $C_{m_1,\cdots,m_k}(u_1, \cdots, u_k|C, D)$, $u_i \in [0, 1]$, $i = 1, \cdots, k$ as follows:

$$C_{m_1,\cdots,m_k}(u_1, \cdots, u_k|C, D) \\ =\mathbb{E}\left[C\left(\frac{F^{\leftarrow}_{Bin(m_1,u_1)}(U_1^D)}{m_1}, \cdots, \frac{F^{\leftarrow}_{Bin(m_k,u_k)}(U_k^D)}{m_k} \right) \right], \tag{16}$$

where $(U_1^D, \cdots, U_k^D)$ is a random vector with distribution function $\overline{D}$.

From the above expression, when D is chosen as the independent copula, i.e.,

$$D(u_1, u_2, \cdots, u_k) = \overline{D}(u_1, u_2, \cdots, u_k) = \prod_{i=1}^{k} u_i, \quad u_i \in [0, 1], \ i = 1, \cdots, k,$$

Eq. (16) becomes the Bernstein copula (9) and it can be seen as a generalization of the BC, with possibly different features.

Theorem 3.1 ([48]) *Suppose C and D are two copulas. Then the followings hold:*

(1) $C_{m_1,\cdots,m_k}(u_1, \cdots, u_k|C, D)$, $u_1, \cdots, u_k \in [0, 1]$ is a copula function.
(2) $C_{1,\cdots,1}(u_1, \cdots, u_k|C, D) = D(u_1, \cdots, u_k)$ for any $u_1, \cdots, u_k \in [0, 1]$.
(3) As $\underline{m} := \min\{m_1, \cdots, m_k\} \to \infty$, for $u_1, \cdots, u_k \in [0, 1]$ we have that

$$C_{m_1,\cdots,m_k}(u_1, \cdots, u_k|C, D) \to C(u_1, \cdots, u_k)$$

uniformly and the convergence rate is bounded by

$$|C_{m_1,\cdots,m_k}(u_1, \cdots, u_k|C, D) - C(u_1, \cdots, u_k)| \leq \sum_{i=1}^{k} \sqrt{\frac{u_i(1-u_i)}{m_i}}.$$

(4) $C_{m_1,\cdots,m_k}(\cdot|C, D)$ admits a density on $[0, 1]^k$ if D admits a density on $[0, 1]^k$.

Definition 3.2 The function (16) is called the *composite Bernstein copula* (CBC). Here the copula function C is called the *target copula* and the copula D is called the *base copula*.

Some equivalent statements for the CBC are given in [20].

Theorem 3.3 ([20]) *For the given copula functions C and D, and the positive integers $m_i, i = 1, \ldots, k$, the following statements are equivalent:*

(1) For a random vector $(U_1^D, \ldots, U_k^D)$ with the distribution function $\overline{D}$,

$$C_{m_1,\ldots,m_k}(u_1, \ldots, u_k|C, D) = \mathbb{E}[C(\frac{F^{\leftarrow}_{Bin(m_1,u_1)}(U_1^D)}{m_1}, \ldots, \frac{F^{\leftarrow}_{Bin(m_k,u_k)}(U_k^D)}{m_k})]. \tag{17}$$

(2) *For a random vector* $(V_1, \dots, V_k)$ *with the distribution function* C,

$$C_{m_1,\dots,m_k}(u_1, \dots, u_k|C, D) = \mathbb{E}[D(\overline{F}_{Bin(m_1,u_1)}(m_1V_1), \dots, \overline{F}_{Bin(m_k,u_k)}(m_kV_k))]. \tag{18}$$

(3) *For a k-discrete random vector* $(L_1, \dots, L_k)$ *defined on the grids* $\prod_{i=1}^{k}\{1, 2, \dots, m_i\}$ *with the uniform marginal laws,*

$$\begin{aligned} &C_{m_1,\dots,m_k}(u_1, \dots, u_k|C, D)\\ &=\sum_{\ell_1=1}^{m_1}\cdots\sum_{\ell_k=1}^{m_k} D(m_1B_{\ell_1-1,m_1-1}(u_1), \dots, m_lB_{\ell_k-1,m_k-1}(u_k))\\ &\times \mathbb{P}(L_1 = \ell_1, \dots, L_k = \ell_k), \end{aligned} \tag{19}$$

where $L_i = \lfloor m_iV_i \rfloor + 1, i = 1, \dots, k$, *and* $(V_1, \dots, V_k)$ *is a random vector with the distribution* C.

Theorem 3.3 presents three equivalent statements for the CBC from the different viewpoints, and it can be applied for discussing the CBC and connecting the CBC with some known copula families in the literature.
(1) Expression (17) can be seen by adding noise on the target copula C. Note that we can express (17) as follows:

$$\begin{aligned} &C_{m_1,\dots,m_k}(u_1, \dots, u_k|C, D)\\ &= \mathbb{E}[C(u_1 + \frac{F^{\leftarrow}_{Bin(m_1,u_1)}(U_1^D)}{m_1} - u_1, \dots, u_k + \frac{F^{\leftarrow}_{Bin(m_k,u_k)}(U_k^D)}{m_k} - u_k)]. \end{aligned}$$

Thus the random noise vector

$$(\frac{F^{\leftarrow}_{Bin(m_1,u_1)}(U_1^D)}{m_1} - u_1, \dots, \frac{F^{\leftarrow}_{Bin(m_k,u_k)}(U_k^D)}{m_k} - u_k)$$

with mean zero and copula function $\overline{D}$ is added on the copula function $C(u_1, \dots, u_k)$.
(2) Expression (18) is obtained from the base copula function D by changing its components with the random vector

$$(\overline{F}_{Bin(m_1,u_1)}(m_1V_1), \dots, \overline{F}_{Bin(m_k,u_k)}(m_kV_k)).$$

When V_i is given, $\overline{F}_{Bin(m_i,u_i)}(m_iV_i)$ is the distribution function of the $(\lfloor m_iV_i \rfloor + 1)$th smallest order statistic from m_i i.i.d. Uniform [0,1] random variables.
(3) Expression (19) is motivated by [17]. Actually, for two-dimensional case, if we let $r_{n,\ell} = \mathbb{P}\{L_1 = n,\ L_2 = \ell\}, n = 1, \dots, m_1,\ \ell = 1, \dots, m_2$, then the CBC can be expressed as

$$C_{m_1,m_2}(u, v|C, D) = \sum_{n=1}^{m_1} \sum_{\ell=1}^{m_2} D(m_1 B_{n-1,m_1-1}(u), m_2 B_{\ell-1,m_2-1}(v)) r_{n,\ell}. \quad (20)$$

Note that Eq. (20) only involves the partial information of the copula function C, and it shows that the CBC can be expressed in a mixture form.

By Theorem 3.1, the copula function $C_{m_1,\cdots,m_k}(\cdot|C, D)$ is close to C as $\min\{m_i, i = 1, \cdots, k\} \to \infty$, hence the copula $C_{m_1,\cdots,m_k}(\cdot|C, D)$ can be used to approximate the copula C, as mentioned in [42]. On the other hand, when $m_1, \cdots, m_k$ are close to 1, the defined $C_{m_1,\cdots,m_k}(\cdot|C, D)$ is close to the copula D. For the above reasons, we call C a *target copula* and D a *base copula*.

3.2 Properties of the Composite Bernstein Copula

In this section, we introduce some properties of the CBC concerning continuity, linearity, and symmetry.

Proposition 3.4 ([48])

(1) If the sequence of copulas C_n converges to a copula C uniformly as n goes to infinity, then $C_{m_1,\cdots,m_k}(\cdot|C_n, D)$ converges to $C_{m_1,\cdots,m_k}(\cdot|C, D)$ uniformly.
(2) If the sequence of copulas D_n converges to a copula D uniformly as n goes to infinity, then $C_{m_1,\cdots,m_k}(\cdot|C, D_n)$ converges to $C_{m_1,\cdots,m_k}(\cdot|C, D)$ uniformly.
(3) Suppose that two target copulas C_1 and C_2 satisfy that $C_1 \le C_2$, then we have that $C_{m_1,\cdots,m_k}(\cdot|C_1, D) \le C_{m_1,\cdots,m_k}(\cdot|C_2, D)$.
(4) Suppose that two base copulas D_1 and D_2 satisfy that $D_1 \le D_2$, then we have that $C_{m_1,\cdots,m_k}(\cdot|C, D_1) \le C_{m_1,\cdots,m_k}(\cdot|C, D_2)$.

From the above proposition, we can see that the CBC is quite robust with respect to the target and base copulas. For a given target (base) copula, different base (target) copulas can be chosen to adjust the value of CBC. Moreover, a linear combination of base copulas can be chosen to further adjust the value of CBC conveniently, as shown in the following proposition.

Proposition 3.5 ([48]) *Suppose $\lambda \in [0, 1]$ is a constant.*

(1) Suppose C_1, C_2 are two copulas and $C = \lambda C_1 + (1 - \lambda)C_2$, then for any base copula D,

$$C_{m_1,\cdots,m_k}(\cdot|C, D) = \lambda C_{m_1,\cdots,m_k}(\cdot|C_1, D) + (1 - \lambda)C_{m_1,\cdots,m_k}(\cdot|C_2, D).$$

(2) Suppose D_1, D_2 are two copulas and $D = \lambda D_1 + (1 - \lambda)D_2$, then for any target copula C,

$$C_{m_1,\cdots,m_k}(\cdot|C, D) = \lambda C_{m_1,\cdots,m_k}(\cdot|C, D_1) + (1 - \lambda)C_{m_1,\cdots,m_k}(\cdot|C, D_2).$$

Remark 3.6 We can see that $C_{m_1,\cdots,m_k}$ is a mapping from $\mathscr{C}_k \times \mathscr{C}_k$ to $\mathscr{C}_k$, where $\mathscr{C}_k$ is the space of k-copulas. The above proposition shows that the CBC admits linearity in terms of base copulas and target copulas. In summary, $C_{m_1,\cdots,m_k} : \mathscr{C}_k \times \mathscr{C}_k \to \mathscr{C}_k$ is a monotone, bilinear, and continuous functional.

3.3 Reproduction Property

For a given target copula C, it is interesting to see whether there exists base copula D such that the corresponding CBC can reproduce the target copula C, i.e.,

$$C_{m_1,\cdots,m_k}(\cdot|C, D) = C$$

holds for some positive integers $m_1, \cdots, m_k$. The importance of the reproduction property is that the chosen base copula D makes no influence on the target copula C.

For the simplest case $m_1 = \cdots = m_k = 1$, $D = C$ is equivalent to $C_{1,\cdots,1}(\cdot|C, D) = C$ by Theorem 3.1 (2). However, for the other values of $m_1, \cdots, m_k$, $D = C$ is not sufficient for $C_{m_1,\cdots,m_k}(\cdot|C, D) = C$ in general. We find that in the case $m_1 = \cdots = m_k = m$, for the three fundamental copula functions: Fréchet upper bound

$$M(u_1, \cdots, u_k) = \min\{u_i, i = 1, \cdots, k\},\ u_1, \cdots, u_k \in [0, 1],$$

independent copula

$$\Pi(u_1, \cdots, u_k) = \prod_{i=1}^{k} u_i,\ u_1, \cdots, u_k \in [0, 1]$$

and Fréchet lower bound $W(u, v) = \max\{u + v - 1, 0\}$, $u, v \in [0, 1]$ (Fréchet lower bound is a copula only in the bivariate case), the condition $D = C$ is sufficient. In the two-dimensional case, it is known that

$$W(u_1, u_2) \le C(u_1, u_2) \le M(u_1, u_2).$$

Proposition 3.7 ([48]) *In (1) and (2), all copulas are k-copulas. In (3), all copulas are 2-copulas.*

(1) $C_{m,\cdots,m}(\cdot|M, D) = M$ *if* $D = M$;
(2) $C_{m_1,\cdots,m_k}(\cdot|\Pi, D) = \Pi$ *if* $D = \Pi$;
(3) $C_{m,m}(\cdot|W, D) = W$ *if* $D = W$.

Proposition 3.7 states that the CBC has the reproduction property for the Fréchet upper bound, the independent copula, and the bivariate Fréchet lower bound, which correspond to the three important dependence structures in insurance and finance:

comonotonicity, independence, and countermonotonicity. Thus, the CBC shows its advantage for modeling these special dependence structures. One open question is that under what circumstance the equality $C_{m_1,\cdots,m_k}(\cdot|C, D) = C$ holds.

3.4 Bivariate Tail Dependence

Tail dependence (see, e.g., [26]) describes the significance of dependence in the tail of a bivariate distribution; see also [44]. The lower tail dependence coefficient of a copula C is defined as $\lambda_L^C := \lim_{u\downarrow 0} \frac{C(u,u)}{u}$, and the upper tail dependence coefficient of a copula C is defined as $\lambda_U^C := \lim_{u\downarrow 0} \frac{\overline{C}(u,u)}{u}$. As we mentioned before, the BC is also unable to capture tail dependence. Note that $|C_m - C| \to 0$ uniformly as $m \to \infty$ does not imply

$$\lambda_L^{C_m} \to \lambda_L^C \quad \text{or} \quad \lambda_U^{C_m} \to \lambda_U^C.$$

The CBC is able to capture tail dependence by choosing appropriate base copulas. We have the following theorem.

Theorem 3.8 ([48]) *Assume that the tail dependence coefficients λ_L^D and λ_U^D of the base copula D exist.*

(1) The lower tail dependence coefficient

$$\lambda_L^{C_{m,m}(\cdot|C,D)} = m \times C(\frac{1}{m}, \frac{1}{m}) \times \lambda_L^D.$$

(2) The upper tail dependence coefficient

$$\lambda_U^{C_{m,m}(\cdot|C,D)} = m \times \overline{C}(\frac{1}{m}, \frac{1}{m}) \times \lambda_U^D.$$

(3) Assume that the tail dependence coefficients λ_L^C and λ_U^C of the target copula C also exist. Then, as $m \to \infty$,

$$\lambda_L^{C_{m,m}(\cdot|C,D)} \to \lambda_L^C \lambda_L^D, \quad \lambda_U^{C_{m,m}(\cdot|C,D)} \to \lambda_U^C \lambda_U^D.$$

Remark 3.9 Theorem 3.8 implies the fact that the BC always has zero tail dependence coefficients since the base copula D of a BC is the independent copula, with $\lambda_U^D = \lambda_L^D = 0$. This phenomenon has been pointed out by [42] without proof. In the CBC family, we can choose base copulas D with $\lambda_U^D = \lambda_L^D = 1$ (such as the Fréchet upper bound M) to preserve the tail dependence coefficients asymptotically.

3.5 Probabilistic Expressions for the CBC

For fixed $j = 0, 1, \ldots, m-1$, we know that $\overline{F}_{Bin(m,u)}(j)$ in Eq. (13) is the distribution function of the $(j+1)$th smallest order statistic of m i.i.d. Uniform [0,1] random variables. For convenience, we write $\overline{F}_{Bin(m,u)}(j)$ as $\Lambda_{m,j+1}(u)$ in the following. Obviously for fixed j, $\Lambda_{m,j+1}(u)$ is a strictly increasing and continuous function in $u \in [0,1]$, whose inverse function is denoted by $\Lambda^{\leftarrow}_{m,j+1}(x)$, $x \in [0,1]$. Thus for each fixed m, we have a sequence of distributions

$$\Lambda_{m,1}(u),\ \Lambda_{m,2}(u),\ \ldots,\ \Lambda_{m,m}(u),\ u \in [0,1],$$

and the corresponding inverse functions

$$\Lambda^{\leftarrow}_{m,1}(x), \Lambda^{\leftarrow}_{m,2}(x), \ldots, \Lambda^{\leftarrow}_{m,m}(x), \quad x \in [0,1].$$

In order to give probabilistic expressions for the CBC, we first introduce two independent random vectors $(V_1, V_2, \ldots, V_k)$ and $(U_1, U_2, \ldots, U_k)$, where the random vector $(V_1, V_2, \ldots, V_k)$ is from the target copula C and the random vector $(U_1, U_2, \ldots, U_k)$ is from the base copula D. The random vector $\mathbf{X}_{m_1,\ldots,m_k} = (X^{(1)}_{m_1}, \ldots, X^{(k)}_{m_k})$ is defined as follows:

$$\mathbf{X}_{m_1,\ldots,m_k} = (\Lambda^{\leftarrow}_{m_1,\lfloor m_1 V_1\rfloor+1}(U_1), \ldots, \Lambda^{\leftarrow}_{m_k,\lfloor m_k V_k\rfloor+1}(U_k)).$$

Actually, we can express $\mathbf{X}_{m_1,\ldots,m_k}$ as

$$\mathbf{X}_{m_1,\ldots,m_k} = \sum_{j_1=0}^{m_1-1} \cdots \sum_{j_k=0}^{m_k-1} (\Lambda^{\leftarrow}_{m_1,j_1+1}(U_1), \ldots, \Lambda^{\leftarrow}_{m_k,j_k+1}(U_k)) I_{\{j_i \le m_i V_i < j_i+1, i=1,2,\ldots,k\}}.$$

Theorem 3.10 ([20]) *(1) The random vector $\mathbf{X}_{m_1,\ldots,m_k}$ has distribution*

$$C_{m_1,\ldots,m_k}(u_1, \ldots, u_k|C, D).$$

Moreover, $\mathbf{X}_{1,\ldots,1} = (U_1, \ldots, U_k)$ *and*

$$\mathbf{X}_{\infty,\ldots,\infty} := \lim_{m_i\to\infty, i=1,\ldots,k} \mathbf{X}_{m_1,\ldots,m_k} = (V_1, \ldots, V_k), \text{ a.s..}$$

That is, $\mathbf{X}_{1,\ldots,1}$ *has the copula* D *and* $\mathbf{X}_{\infty,\ldots,\infty}$ *has the copula* C;
(2) $\mathbf{X}_{m_1,\ldots,m_k}$ *is a function of* $\mathbf{X}_{1,\ldots,1}$ *and* $\mathbf{X}_{\infty,\ldots,\infty}$. *Moreover, for each* $1 \le i \le k$,

$$X^{(i)}_{m_i} = \Lambda^{\leftarrow}_{m_i,\lfloor m_i X^{(i)}_{\infty}\rfloor+1}(X^{(i)}_1).$$

The roles of the base copula D and the target copula C can be explained in Theorem 3.10. The sequence $\mathbf{X}_{m_1,\ldots,m_k}$ starts from $(U_1, \ldots, U_k)$ having the base copula D

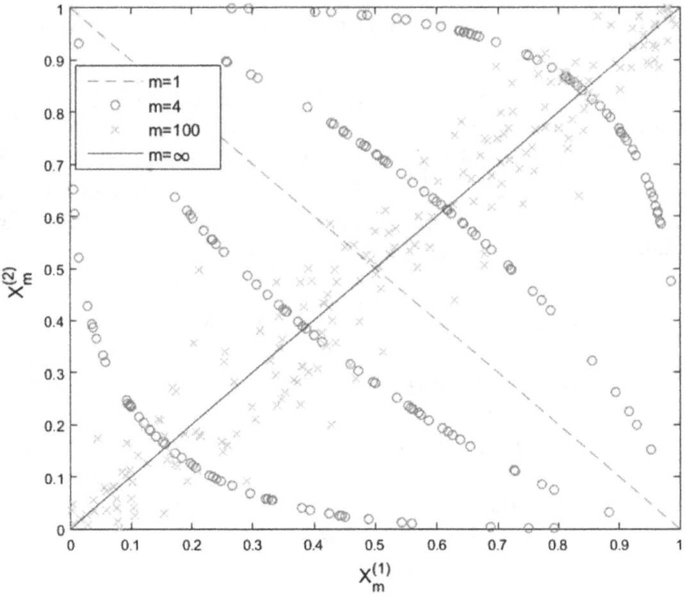

Fig. 1 Two random samples (circles and crosses) from $C_{m,m}(\cdot|M, W)$ with $m = 4$ and 100 and two supports (dashed line and solid line) with $m = 1$ and ∞

and tends to the limit $(V_1, \ldots, V_k)$ having the target copula C. Each random vector in the sequence can be expressed as a function of the starting vector and the limiting vector.

3.6 *The Simulation Method for CBC*

The probabilistic expression for the CBC presented in Theorem 3.10 can be applied to generate random vector from the CBC:

(i) Generate $(U_1, \ldots, U_k)$ from the copula D, and $(V_1, \ldots, V_k)$ from the copula C, where $(U_1, \ldots, U_k)$ and $(V_1, \ldots, V_k)$ are independent.
(ii) Calculate $(\Lambda^{\leftarrow}_{m_1,\lfloor m_1 V_1 \rfloor+1}(U_1), \ldots, \Lambda^{\leftarrow}_{m_k,\lfloor m_k V_k \rfloor+1}(U_k))$ to get a random number from the CBC.

The remaining computational issue is on how to calculate $\Lambda^{\leftarrow}_{m,\lfloor mv \rfloor+1}(u)$ efficiently. Since $\lfloor mv \rfloor$ is an integer range from 0 to $m - 1$, one can first discretize u and build two-dimensional grids to pre-calculate values for function $\Lambda^{\leftarrow}_{m,j+1}(u)$ for $j = \lfloor mv \rfloor$. There is no extra computational burden for the CBC after giving the table of $\Lambda^{\leftarrow}_{m,j+1}(u)$. Figure 1 ([20]) shows the scatter plots of two random samples of size 200 from $C_{m,m}(\cdot|M, W)$ when $m = 4$ and 100.

3.7 Some Families of the CBC

Consider the two-dimensional case. Suppose that the random vector (X, Y) has the joint distribution function $H(x, y)$ with marginal distributions $F(x)$ and $G(y)$, respectively. And we assume that the random variables X and Y have the copula $C_{m_1,m_2}(u, v|C, D)$ in (20).

Let $F_{n,m}(x)$ be the distribution function of the nth smallest order statistic from m i.i.d. random variables with the marginal distribution F, similarly the corresponding distribution of the nth smallest order statistic from the marginal distribution G is denoted by $G_{n,m}(y)$. Then the joint distribution $H(x, y)$ of (X, Y) can be expressed as

$$\begin{aligned} H(x, y) &= C_{m_1,m_2}(F(x), G(y)|C, D) \\ &= \sum_{n=1}^{m_1}\sum_{\ell=1}^{m_2} D(m_1 B_{n-1,m_1-1}(F(x)), m_2 B_{\ell-1,m_2-1}(G(y)))r_{n,\ell}. \end{aligned}$$

Using $m B_{n-1,m-1}(F(x)) = F_{n,m}(x)$, the above joint distribution function can also be written as

$$H(x, y) = \sum_{n=1}^{m_1}\sum_{\ell=1}^{m_2} D(F_{n,m_1}(x), G_{\ell,m_2}(y))r_{n,\ell}. \tag{21}$$

For some special target copulas and base copulas in the CBC $C_{m_1,m_2}(u, v|C, D)$, we can get some copula families presented in the literature.

(1) When the base copula $D = \Pi$, from Eq. (21) we have

$$C_{m_1,m_2}(F(x), G(y)|C, \Pi) = \sum_{n=1}^{m_1}\sum_{\ell=1}^{m_2} F_{n,m_1}(x)G_{\ell,m_2}(y)r_{n,\ell}.$$

This type of distribution is called Baker's bivariate distribution in [17], and it had been discussed thoroughly in [5, 16, 17, 31]. Especially, when $C = M$ and $m_1 = m_2 = m$,

$$C_{m,m}(F(x), G(y)|M, \Pi) = m^{-1}\sum_{n=1}^{m} F_{n,m}(x)G_{n,m}(y) =: H_+^{(m)}(x, y), \tag{22}$$

and when $C = W$,

$$C_{m,m}(F(x), G(y)|W, \Pi) = m^{-1}\sum_{n=1}^{m} F_{n,m}(x)G_{m+1-n,m}(y) =: H_-^{(m)}(x, y). \tag{23}$$

Here $H_+^{(m)}(x, y)$ and $H_-^{(m)}(x, y)$ are also called Baker's bivariate distribution in [4, 16, 31].

(2) When the target copula $C = M$ or W, $m_1 = m_2 = m$,

$$C_{m,m}(F(x), G(y)|M, D) = m^{-1}\sum_{n=1}^{m} D(F_{n,m}(x), G_{n,m}(y)) =: G_{+}^{(m)}(x, y),$$

and

$$C_{m,m}(F(x), G(y)|W, D) = m^{-1}\sum_{n=1}^{m} D(F_{n,m}(x), G_{m+1-n,m}(y)) =: G_{-}^{(m)}(x, y),$$

where $G_{+}^{(m)}(x, y)$ and $G_{-}^{(m)}(x, y)$ are called Baker's Type II BB distribution in [7]. (3) Reference [5] presented a construction of Farlie–Gumbel–Morgenstern (FGM)-type distribution with higher order grade correlation, which is defined as

$$H_{\pm,q}^{(m)}(x, y) = (1-q)F(x)G(y) + qH_{\pm}^{(m)}(x, y). \tag{24}$$

Applying (22) and (23), the above distributions can be rewritten as

$$H_{+,q}^{(m)}(x, y) = (1-q)F(x)G(y) + qC_{m,m}(F(x), G(y)|M, \Pi),$$

and

$$H_{-,q}^{(m)}(x, y) = (1-q)F(x)G(y) + qC_{m,m}(F(x), G(y)|W, \Pi).$$

Thus the copula function of the joint distribution $H_{+,q}^{(m)}(x, y)$ in (24) can be expressed as

$$C_{+,q}^{(m)}(u, v) = (1-q)uv + qC_{m,m}(u, v|M, \Pi),$$

and similarly the copula function of $H_{-,q}^{(m)}(x, y)$ can be expressed as

$$C_{-,q}^{(m)}(u, v) = (1-q)uv + qC_{m,m}(u, v|W, \Pi).$$

By applying the reproduction and linearity properties of the CBC, $C_{+,q}^{(m)}(u, v)$ and $C_{-,q}^{(m)}(u, v)$ can be written as

$$\begin{aligned} C_{+,q}^{(m)}(u, v) &= (1-q)C_{m,m}(u, v|\Pi, \Pi) + qC_{m,m}(u, v|M, \Pi) \\ &= C_{m,m}(u, v|(1-q)\Pi + qM, \Pi), \end{aligned}$$

and

$$C_{-,q}^{(m)}(u, v) = C_{m,m}(u, v|(1-q)\Pi + qW, \Pi).$$

Thus, the mixture copulas in [6] are special cases of the CBC family with target copulas $(1-q)\Pi + qM$ and $(1-q)\Pi + qW$, respectively.

4 Simulation Study and Stock Data Analysis

The *empirical composite Bernstein copula* (ECBC) has been introduced in [48] to estimate unknown copula based on CBCs. In this section, we examine the finite sample performance of ECBCs, especially study its behavior on modeling the data with tail dependence. The research results in [48] have also been incorporated into this section to show the advantage of ECBCs.

We will do our numerical analysis on an illustration in bivariate dimension, and we will also emphasize the role played by the quantile dependence functions ([36]). The main points are stated in the following:

1. In the simulation study, we use a true copula with negative correlation to generate random samples. The true copulas in [48] all have positive correlation. Thus the formulas of the $\mathscr{L}^1$-error in this paper are a little different from those in [48].
2. In the data analysis, the data are daily returns of NASDAQ and SPY500 from January 2, 1997 to December 29, 2017. The date in [48] is from January 29, 1993 to January 9, 2013.
3. We use two criteria to evaluate the performance of estimated copulas of the real data. One is $\mathscr{L}^1$-error in [48] and the other is quantile dependence function. They are used to compare copula estimators from different perspectives.
4. We carry out the calculation of $\mathscr{L}^1$-error with α ranging from 0.01 to 0.1 by step 0.01 and a special value 1, while [48] concentrated on the values $\alpha = 1, 0.05, 0.01$.

4.1 A Simulation Study

Let $C_N(\mathbf{u}), \mathbf{u} \in [0,1]^k$ be the empirical copula of sample $\mathbf{V}_1, \cdots, \mathbf{V}_N \in [0,1]^k$ from a copula C, i.e.,

$$C_N(\mathbf{u}) = \frac{1}{N}\sum_{j=1}^{N} \mathbf{1}_{\{\mathbf{V}_j \leq \mathbf{u}\}},$$

where the "$\leq$" is a component-wise inequality.

The ECBC [48] is defined as

$$\tilde{C}_{m_1,\cdots,m_k}(\mathbf{u}|N, D) := C_{m_1,\cdots,m_k}(\mathbf{u}|C_N, D), \quad \mathbf{u} \in [0,1]^k,$$

where the target copula C in the CBC is replaced by the empirical copula C_N. When the independent copula Π is chosen as the base copula, the ECBC turns to be an EBC, which is introduced by [42]. In the following, $\tilde{C}_{m,m}(\cdot|D)$ is used as a simplified notation of $\tilde{C}_{m,m}(\cdot|, D)$.

In this subsection, we carry out a simulation study in the bivariate case to compare the effects of different base copulas on the tail property of the ECBC. We generate $r = 1080$ random samples of size $N = 100$ from the Student's t-copula with the

Table 1 $\mathscr{L}^1$-error distance with α=1

$\alpha = 1$	$R^1(C_N, C) = 0.0216$		$R^1(\tilde{C}, C) = 0.0038$	
m	$\tilde{C}_{m,m}(\cdot\mid M)$	$\tilde{C}_{m,m}(\cdot\mid \Pi)$	$\tilde{C}_{m,m}(\cdot\mid W)$	$\tilde{C}_{m,m}(\cdot\mid C_N)$
3	0.0735	0.0354	0.0140	0.0192
5	0.0502	0.0263	0.0151	0.0195
10	0.0304	0.0190	0.0162	0.0196
20	0.0216	0.0174	0.0171	0.0195
50	0.0191	0.0183	0.0184	0.0200

correlation parameter $\rho = -0.75$ and the degree of freedom parameter $\nu = 3$. In each repetition, we use the sample to construct estimated copula, denoted as C^E. Since the true copula has negative correlation, which means the sample points tend to gather in the left-upper and right-lower corners of the unit square $[0, 1]^2$, we use the $\mathscr{L}^1$-error distance below to evaluate the estimation error on the square $[0, \alpha]^2$,

$$R^{\alpha}_{left-upper}(C^E, C) = \mathrm{E}\left\{\frac{1}{(K-1)^2}\sum_{i=1}^{K-1}\sum_{j=1}^{K-1}\left|C^E\left(\frac{i}{K}\alpha, 1-\alpha(1-\frac{j}{K})\right) - C\left(\frac{i}{K}\alpha, 1-\alpha(1-\frac{j}{K})\right)\right|\right\},$$

$$R^{\alpha}_{right-lower}(C^E, C) = \mathrm{E}\left\{\frac{1}{(K-1)^2}\sum_{i=1}^{K-1}\sum_{j=1}^{K-1}\left|C^E\left(1-\alpha(1-\frac{i}{K}), \frac{j}{K}\alpha\right) - C\left(1-\alpha(1-\frac{i}{K}), \frac{j}{K}\alpha\right)\right|\right\},$$

where $K = 100$ and C is the true copula. It is worth noting that when $\alpha = 1$, the two error distances are the same, which measure the overall approximation error. When α is close to 0, the two error distances measure tail errors in left-upper corner and right-lower corner, respectively.

We select the following several estimators to conduct our comparisons:

(a) The MLE of Student's t-copula: $\tilde{C}$.
(b) The empirical copula: C_N.
(c) The ECBCs with three base copulas M, Π and W: $\tilde{C}_{m,m}(\cdot\mid M)$, $\tilde{C}_{m,m}(\cdot\mid \Pi)$, $\tilde{C}_{m,m}(\cdot\mid W)$.
(d) The ECBC based on the empirical copula C_N: $\tilde{C}_{m,m}(\cdot\mid C_N)$.

All these calculations are carried out with $m \in \{3, 5, 10, 20, 50\}$ and α ranging from 0.01 to 0.1 by step 0.01 and a special value 1 in addition. Some simulation results are reported in the subsequent table and figures.

From Table 1 and Figs. 2, 3, 4, 5, 6, and 7, we can draw some conclusions below:

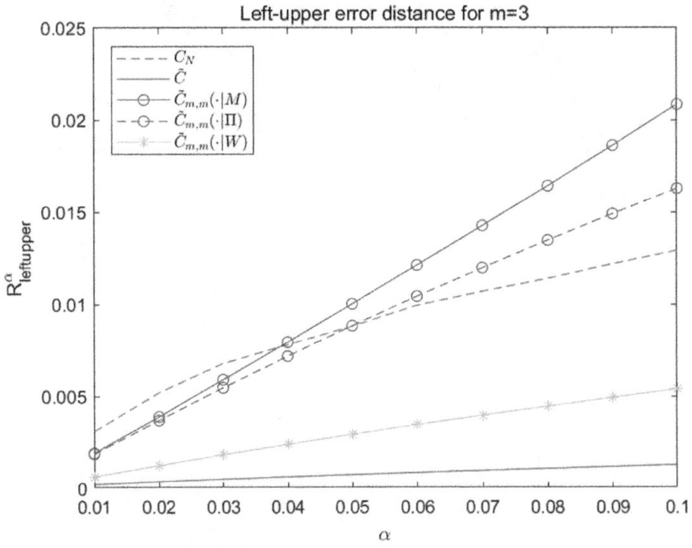

Fig. 2 Left-upper error for m = 3

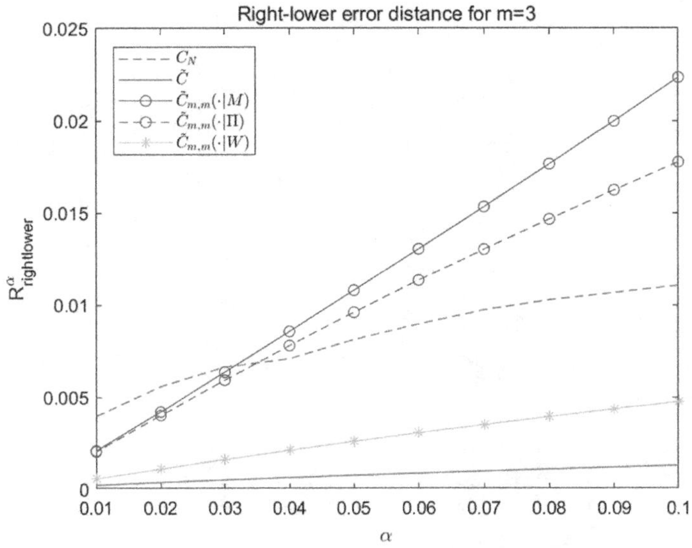

Fig. 3 Right-lower error for m = 3

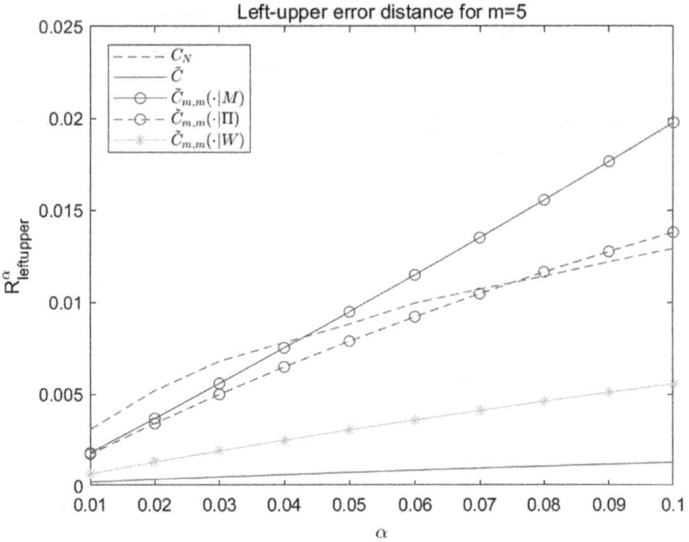

Fig. 4 Left-upper error for m = 5

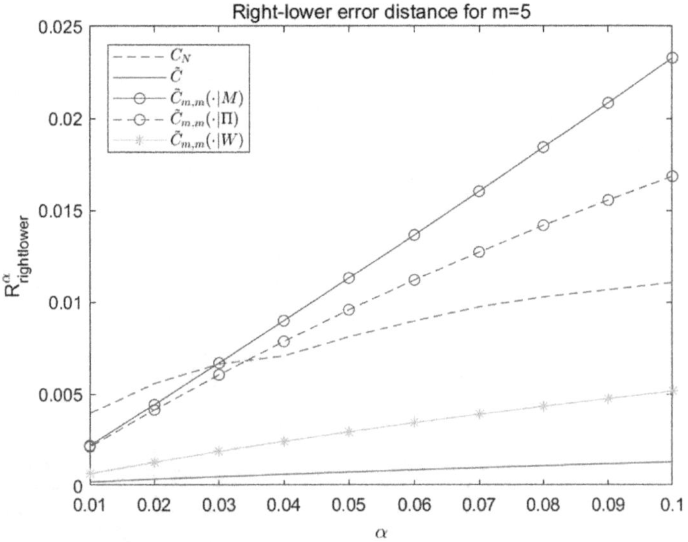

Fig. 5 Right-lower error for m = 5

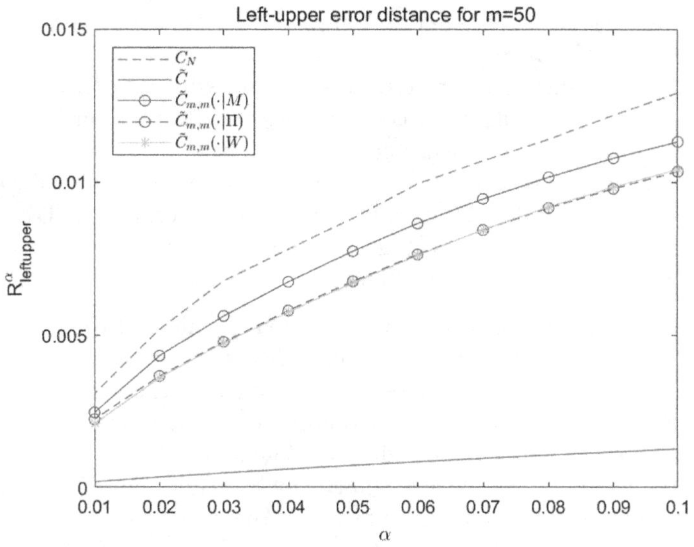

Fig. 6 Left-upper error for m = 50

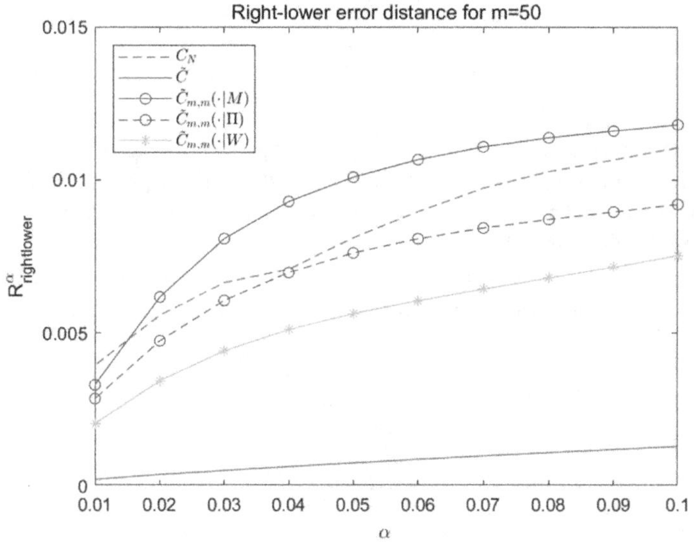

Fig. 7 Right-lower error for m = 50

i. The parametric MLE $\tilde{C}$ performs best under both overall error distance and tail error distance. It is understandable since the sample is drawn from a t-copula.
ii. Among all the ECBC methods, the ECBC with the base copula W performs superiorly to the others as expected, since W has negative correlation. Reversely,

the ECBC with the base copula M performs worst. The EBC's performance falls in between of the two cases.

iii. With the increasing of m, the error distances of all three ECBCs get smaller, even for the base copula M, which coincides with the phenomenon that CBCs converge to target copulas when m tends to infinity.

iv. When the target copula and the base copula both are the empirical copula C_N, the ECBC shows little difference from the behavior of the copula C_N and the smoothing parameter m almost takes no effect. We will delve it more deeply from the theoretical view in the future.

Reference [48] have used several bivariate copulas with positive correlation coefficients to conduct the experiments, including Gaussian copula, Clayton copula, t-copula, and Gumbel copula. They also observed similar phenomena as in this paper. In the case that the empirical copula C_N is chosen as the target copula, the above simulations confirm that, with the help of the information from the dependence summary statistics, it is possible to choose a comparatively proper base copula and smoothing parameter to improve the estimation precision. This is attracting, especially when there is no prior information about the true copula. In summary, the ECBC provides a competitively nonparametric estimation for the unknown copula.

4.2 Stock Data Analysis

In this subsection, we carry out a data analysis using the daily returns of NASDAQ and SPY500 from January 2, 1997 to December 29, 2017. Their prices and returns plots are shown in Fig. 8. The figures show a significant positive dependence relationship between NASDAQ and SPY500.

To analyze the dependence of the two index series, we use an AR(2) model to filter the daily log returns to eliminate autocorrelation. The sample plots of the

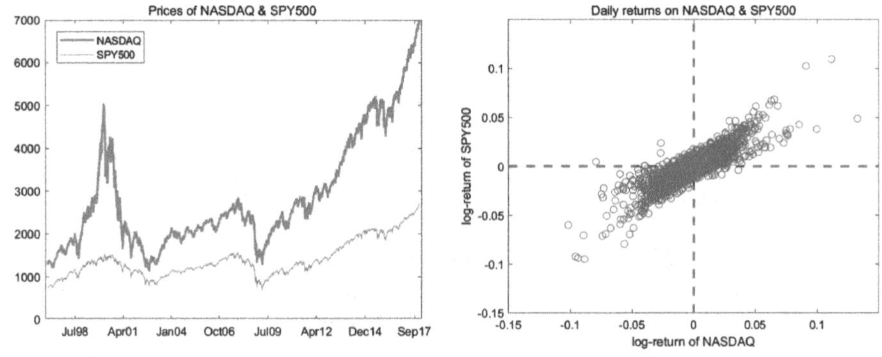

Fig. 8 The left panel is the stock prices from January 2, 1997 to December 29, 2017. The right panel is the scatter plot of daily returns. They both provide evidence for the existing of positive dependence between the two equity indices

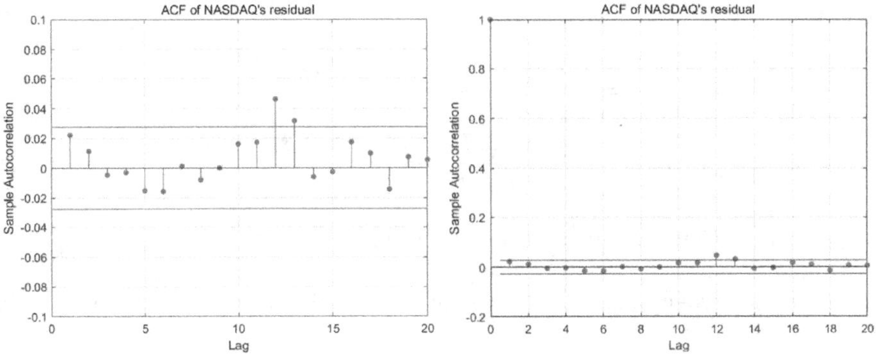

Fig. 9 ACF of NASDAQ's residuals

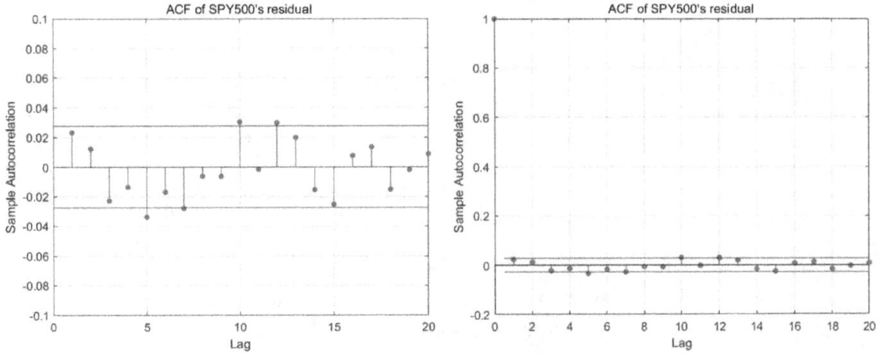

Fig. 10 ACF of SPY500's residuals

autocorrelation function (ACF) of the two filtered residuals are shown in Figs. 9 and 10, which reveal almost no autocorrelation. Besides, the related ARCH tests also show no ARCH effects existing, where the p-values are 0.6060 and 0.3107, respectively.

In the following, we analyze the dependence of the two residuals series by ECBCs (including EBC), in addition with Gaussian copula, Student's t-copula, Clayton copula, and Gumbel copula. Based on the filtered residuals series, we randomly choose 60% of the samples as the training dataset, the other (40%) as the test dataset. The repetition $r = 250$. Each time we use the training data to construct the empirical copula C_N, while the test data is used for estimating C^E.

We use two criteria to evaluate the performance of these copulas: the quantile dependence plot and the $\mathscr{L}^1$-error distance. For a copula C, its quantile lower and upper dependence functions ([36]) are defined as follows:

$$\lambda^q_{lower}(C) = P(U_1 \leq q | U_2 \leq q), \quad q \in (0, 0.5],$$
$$\lambda^q_{upper}(C) = P(U_1 > q | U_2 > q), \quad q \in [0.5, 1).$$

Note that

$$\lambda^q_{lower}(C) = \frac{P(U_1 \le q, U_2 \le q)}{q} = \frac{C(q,q)}{q},$$
$$\lambda^q_{upper}(C) = \frac{P(U_1 > q, U_2 > q)}{1-q} = \frac{1-2q+C(q,q)}{1-q}.$$

The quantile dependence functions are plotted for $q \in [0.025, 0.975]$ in our empirical illustration.

The second criterion is the $\mathscr{L}^1$-error distance, similar to what we used in the last subsection,

$$R^\alpha_{lower}(C^E, C_N) = \mathrm{E}\left\{\frac{1}{(K-1)^2}\sum_{i=1}^{K-1}\sum_{j=1}^{K-1}\left|C^E\left(\frac{i}{K}\alpha, \frac{j}{K}\alpha\right) - C_N\left(\frac{i}{K}\alpha, \frac{j}{K}\alpha\right)\right|\right\},$$
$$R^\alpha_{upper}(C^E, C_N) = \mathrm{E}\left\{\frac{1}{(K-1)^2}\sum_{i=1}^{K-1}\sum_{j=1}^{K-1}\left|C^E\left(1-\alpha(1-\frac{i}{K}), 1-\alpha(1-\frac{j}{K})\right)\right.\right.$$
$$\left.\left. -C_N\left(1-\alpha(1-\frac{i}{K}), 1-\alpha(1-\frac{j}{K})\right)\right|\right\},$$

where $K = 100$. The "right-lower" and "left-upper" are replaced by "lower" and "upper", respectively, due to the positive dependence of the data. And the benchmark copula C is replaced by the empirical copula C_N.

Our estimated copulas include four parametric copulas: Gaussian copula, Gumbel copula, Clayton copula, and Student's t-copula, in addition to the EBC and the ECBC with the base copula M. All the error calculations are carried out with $m \in \{3, 5, 10, 20, 50, 200\}$ and the same choices of α as in the last subsection.

For clarity, the quantile dependence plot has been split into two parts, Figs. 11 and 12. The upper $\mathscr{L}^1$-error distance and lower $\mathscr{L}^1$-error distance for the ECBC and the EBC with $m = 3, 5, 10, 200$ are presented in Figs. 13, 14, 15, 16, 17, and 18.

From Figs. 11–18, several observations can be obtained:

i. Figure 11 presents the quantile dependence results for several parametric copulas. The Gaussian copula and the Student's t-copula overwhelm the Clayton copula and the Gumbel copula, especially on modeling the tail parts. Considering the fact that the Clayton copula has zero upper tail dependence, while the Gumbel copula has zero lower tail dependence, this phenomenon is not strange at all. In Fig. 12, the comparison between the Gaussian copula and the Student's t-copula with the ECBC is given. We can easily get the judgement that the ECBC and the EBC show advantages in fitting this real data. For precision, we calculate the cumulative error distance

$$\int_{0.025}^{0.5} |\lambda^q_{lower}(C_N) - \lambda^q_{lower}(C^E)|dq + \int_{0.5}^{0.975} |\lambda^q_{upper}(C_N) - \lambda^q_{upper}(C^E)|dq$$

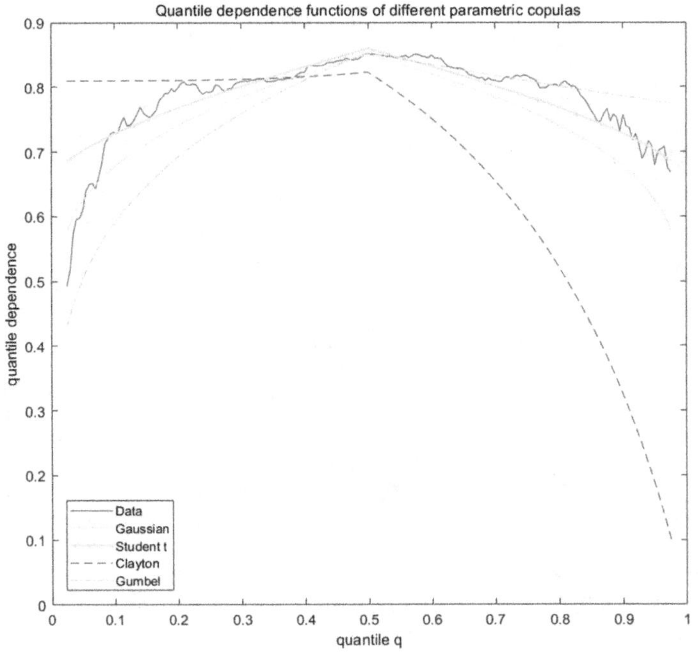

Fig. 11 Quantile dependence function of several parametric copulas with maximum likelihood estimators. Obviously, the Gaussian copula and the Student's t-copula fit the real data better, comparing the Gumbel copula and Clayton copula

between each curve and the data curve. The distances of the Gaussian copula and the Student's t-copula are 0.0341 and 0.0183, respectively, while those of ECBC and EBC are 0.0058 and 0.0163.

ii. ECBCs and EBCs with $m = 3, 5, 10$ have been together presented in Figs. 13 and 14. These two figures show that the ECBC with the base copula M performs better than the EBC in modeling the tail part of the data. The case $m = 10$ is a little superior to the cases $m = 3$ and $m = 5$. So we amplify to form Figs. 15 and 16, from which we can see that the ECBC with $m = 10$ is comparable to the Gumbel copula in describing the lower tail while inferior to the Gaussian copula in both tails. When $m = 200$, similar results can be drawn.

iii. We notice that in the $\mathscr{L}^1$-error distance plots the Gaussian copula is better. In general, the tail $\mathscr{L}^1$-error and the quantile dependence function measure the goodness of fitting of the copula C^E from different perspectives. For instance, the $\mathscr{L}^1$-error R^{α}_{lower} measures the distance of the two copulas C^E and C_N on the square $[0, \alpha]^2$, while the quantile dependence function λ^{α}_{lower} measures the ratio $\frac{C^E(\alpha, \alpha)}{\alpha}$. Therefore, even the Gaussian copula fails to capture the tail dependence of the data, it can still have a very good performance in approximating the empirical copula C_N in the corner region. Acceptably, the $\mathscr{L}^1$-errors of

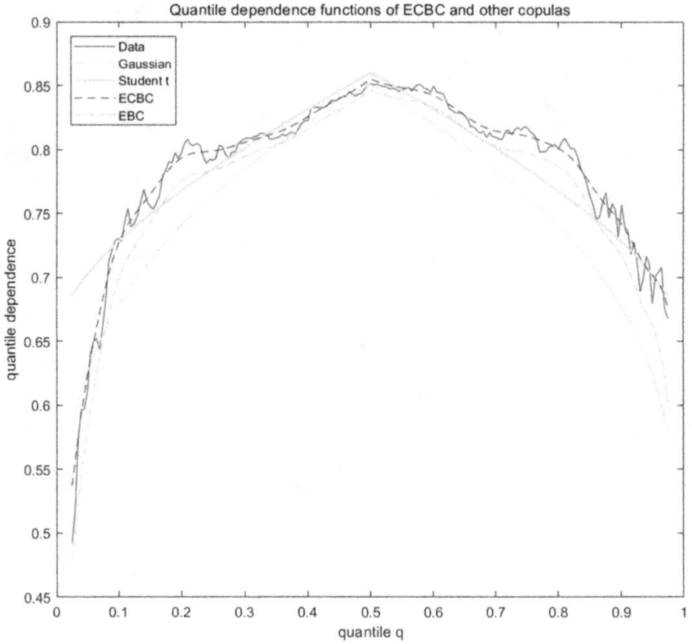

Fig. 12 Comparing to the Gaussian copula and the Student's t-copula, the ECBC and EBC perform more conspicuously

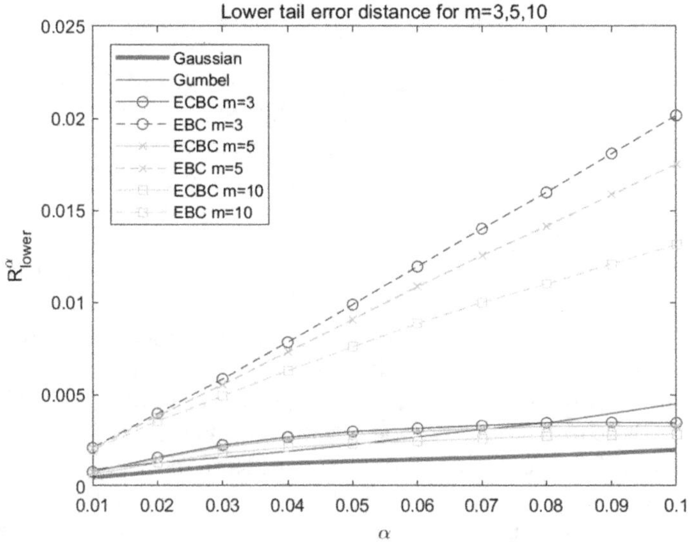

Fig. 13 Lower tail error for m = 3, 5, 10

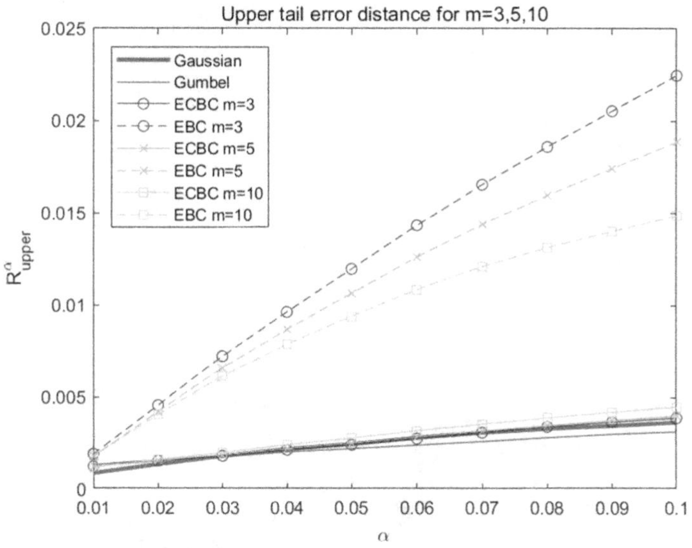

Fig. 14 Upper tail error for m = 3, 5, 10

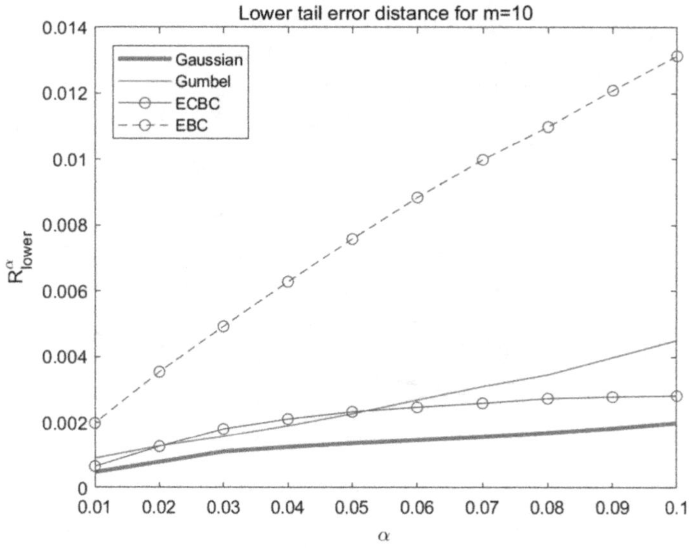

Fig. 15 Lower tail error for m = 10

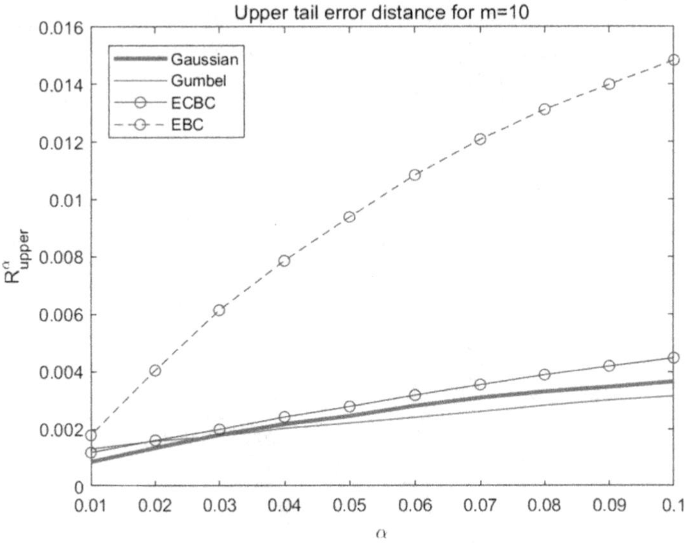

Fig. 16 Upper tail error for m = 10

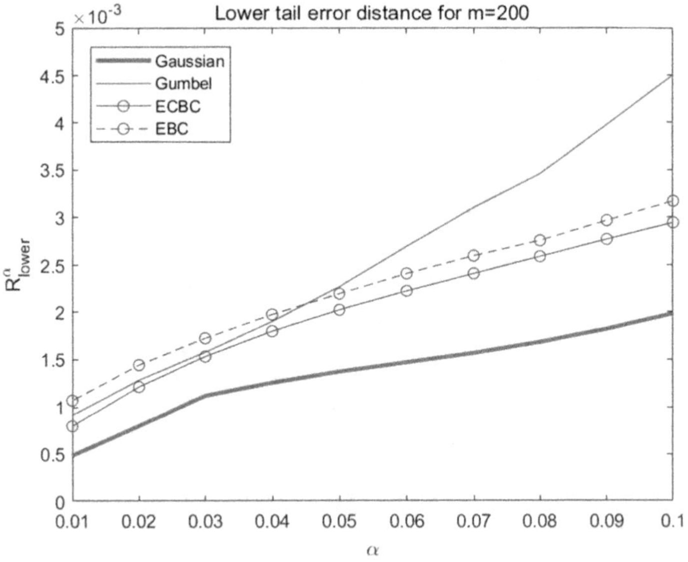

Fig. 17 Lower tail error for m = 200

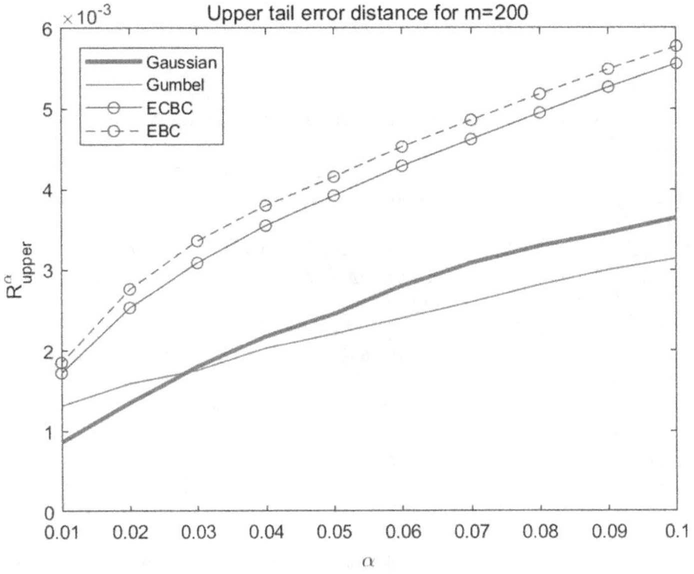

Fig. 18 Upper tail error for m = 200

ECBCs are in the same order as those of the Gaussian copula, which means the approximation of ECBCs is still effective.

5 Conclusions

In this paper, we introduced main results on the Bernstein copula [42] and the composite Bernstein copula [20, 48]. The Bernstein copula has the advantage of approximating each copula function. As a generalization of Bernstein copulas, the composite Bernstein copula shows its advantage on capturing tail dependence, and the base copula and the target copula play different roles. Composite Bernstein copulas include some copula families presented in the literature. The numerical results show the advantage of the ECBC on modeling tail dependence in small dimension.

Acknowledgements The authors would like to thank the anonymous reviewer for his constructive comments and suggestions that have led to improvements in the paper. The research work described in the paper was partly supported by the National Natural Science Foundation of China (Grants No.11671021, No.11471222), Capacity Building for Sci-Tech Innovation - Fundamental Scientific Research Funds (No. 025185305000/204) and Youth Innovative Research Team of Capital Normal University.

References

1. Abegaz, F., Mandrekar, V., Naik-Nimbalkar, U.: On computing approximation of correlations using Bernstein copula and probabilistic tools. In: Proceedings of the 58th World Statistical Congress, Dublin (2011)
2. Ahmadabadi, A., Ucer, B.H.: Bivariate nonparametric estimation of the Pickands dependence function using Bernstein copula with kernel regression approach. Comput. Stat. **32**(4), 1515–1532 (2017)
3. Avanzi, B., Cassar, L.C., Wong, B.: Modelling dependence in insurance claims processes with Lévy copulas. ASTIN Bull. **41**(02), 575–609 (2011)
4. Bairamov, I., Bayramoglu, K.: From the Huang-Kotz FGM distribution to Baker's bivariate distribution. J. Multivar. Anal. **113**, 106–115 (2013)
5. Baker, R.: An order-statistics-based method for constructing multivariate distributions with fixed marginals. J. Multivar. Anal. **99**(10), 2312–2327 (2008)
6. Baker, R.: Copulas from order statistics. 2014. Working paper, University of Salford. arXiv:1412.0948
7. Bayramoglu, K., Bayramoglu (Bairamov), I.: Baker-Lin-Huang type bivariate distributions based on order statistics. Commun. Stat.-Theory Methods **43**(10-12), 1992–2006 (2014)
8. Bouezmarni, T., Funke, B., Camirand Lemyre, F.: Regression estimation based on Bernstein density copulas (2014)
9. Bouezmarni, T., Ghouch, E., Taamouti, A.: Bernstein estimator for unbounded copula densities. Stat. Risk Model. **30**(4), 343–360 (2013)
10. Bouezmarni, T., Rombouts, J.V.K., Taamouti, A.: Asymptotic properties of the Bernstein density copula estimator for α-mixing data. J. Multivar. Anal. **101**(1), 1–10 (2010)
11. Bouezmarni, T., Rombouts, J.V.K., Taamouti, A.: Nonparametric copula-based test for conditional independence with applications to Granger causality. J. Bus. Econ. Stat. **30**(2), 275–287 (2012)
12. Carnicero, J.A., Wiper, M.P., Ausin, C.: Non-parametric methods for circular-circular and circular-linear data based on Bernstein copulas. Working paper (2011)
13. Cherubini, U., Luciano, E., Vecchiato, W.: Copula Methods in Finance. Wiley-Blackwell (2004)
14. Cottin, C., Pfeifer, D.: From Bernstein polynomials to Bernstein copulas. J. Appl. Funct. Anal **9**(3–4), 277–288 (2014)
15. Diers, D., Eling, M., Marek, S.D.: Dependence modeling in non-life insurance using the Bernstein copula. Insur.: Math. Econ. **50**(3), 430–436 (2012)
16. Dou, X., Kuriki, S., Lin, G.D.: Dependence structures and asymptotic properties of Baker's distributions with fixed marginals. J. Stat. Plan. Inference **143**(8), 1343–1354 (2013)
17. Dou, X., Kuriki, S., Lin, G.D., Richards, D.: EM algorithms for estimating the Bernstein copula. Comput. Stat. Data Anal. **93**, 228–245 (2016)
18. Durrleman, V., Nikeghbali, A., Roncalli, T.: Copulas approximation and new families. Working paper (2000)
19. Embrechts, P., McNeil, A., Straumann, D.: Correlation and dependence in risk management: properties and pitfalls. In: Risk Management: Value at Risk and Beyond, pp. 176–223. Cambridge University Press (2002)
20. Guo, N., Wang, F., Yang, J.: Remarks on composite Bernstein copula and its application to credit risk analysis. Insur.: Math. Econ. **77**, 38–48 (2017)
21. Hernández-Maldonado, V., Díaz-Viera, M., Erdely, A.: A joint stochastic simulation method using the Bernstein copula as a flexible tool for modeling nonlinear dependence structures between petrophysical properties. J. Pet. Sci. Eng. **90**, 112–123 (2012)
22. Janssen, P., Swanepoel, J., Veraverbeke, N.: Large sample behavior of the Bernstein copula estimator. J. Stat. Plan. Inference **142**(5), 1189–1197 (2012)
23. Janssen, P., Swanepoel, J., Veraverbeke, N.: A note on the asymptotic behavior of the Bernstein estimator of the copula density. J. Multivar. Anal. **124**, 480–487 (2014)
24. Janssen, P., Swanepoel, J., Veraverbeke, N.: Bernstein estimation for a copula derivative with application to conditional distribution and regression functionals. Test **25**(2), 351–374 (2016)

25. Janssen, P., Swanepoel, J., Veraverbeke, N.: Smooth copula-based estimation of the conditional density function with a single covariate. J. Multivar. Anal. **159**, 39–48 (2017)
26. Harry, J.: Multivariate Models and Dependence Concepts. Chapman and Hall/CRC (1997)
27. Kiriliouk, A., Segers, J., Tafakori, L.: An estimator of the stable tail dependence function based on the empirical beta copula. Extremes **21**(4), 581–600 (2018)
28. Kulpa, T.: On approximation of copulas. Int. J. Math. Math. Sci. **22**(2), 259–269 (1999)
29. Li, X., Mikusiński, P., Sherwood, H., Taylor, M.D.: On approximation of copulas. In: Distributions with Given Marginals and Moment Problems, pp. 107–116. Springer (1997)
30. Li, X., Mikusiński, P., Taylor, M.D.: Strong approximation of copulas. J. Math. Anal. Appl. **225**(2), 608–623 (1998)
31. Lin, G.D., Huang, J.S.: A note on the maximum correlation for Baker's bivariate distributions with fixed marginals. J. Multivar. Anal. **101**(9), 2227–2233 (2010)
32. Lorentz, G.G.: Bernstein Polynomials. University of Toronto Press (1953)
33. McNeil, A.J., Frey, R., Embrechts, P.: Quantitative Risk Management: Concepts, Techniques and Tools. Princeton University Press (2015)
34. Mendoza-Torres, F., Díaz-Viera, M.A., Erdely, A.: Bernstein copula modeling for 2d discrete fracture network simulations. J. Pet. Sci. Eng. **156**, 710–720 (2017)
35. Nelsen, R.B.: An Introduction to Copulas. Springer Science & Business Media (2006)
36. Patton, A.: Copula methods for forecasting multivariate time series. In: Handbook of Economic Forecasting, pp. 899–960. Elsevier B.V. (2013)
37. Pfeifer, D., Straßburger, D.: Dependence matters. In: 36th International ASTIN Colloquium, pp. 4–7. ETH Zürich (2005)
38. Pfeifer, D., Strassburger, D., Philipps, J.: Modelling and simulation of dependence structures in nonlife insurance with Bernstein copulas. Preprint (2009)
39. Reiss, R.D.: Approximate Distributions of Order Statistics: with Applications to Nonparametric Statistics. Springer Science & Business Media (2012)
40. Rose, D.: Modeling and estimating multivariate dependence structures with the Bernstein copula. Ph.D. Thesis, lmu (2015)
41. Sancetta, A.: Online forecast combinations of distributions: Worst case bounds. J. Econ. **141**(2), 621–651 (2007)
42. Sancetta, A., Satchell, S.: The Bernstein copula and its applications to modeling and approximations of multivariate distributions. Econ. Theory **20**(03), 535–562 (2004)
43. Scheffer, M., Weiß, G.N.F.: Smooth nonparametric Bernstein vine copulas. Quant. Financ. **17**(1), 139–156 (2017)
44. Schmidt, R.: Tail dependence. In: Statistical Tools for Finance and Insurance, pp. 65–91. Springer (2005)
45. Segers, J., Sibuya, M., Tsukahara, H.: The empirical beta copula. J. Multivar. Anal. **155**, 35–51 (2017)
46. Shi, P., Frees, E.W.: Dependent loss reserving using copulas. ASTIN Bull. **41**(02), 449–486 (2011)
47. Tavin, B.: Detection of arbitrage in a market with multi-asset derivatives and known risk-neutral marginals. J. Bank. Financ. **53**, 158–178 (2015)
48. Yang, J., Chen, Z., Wang, F., Wang, R.: Composite Bernstein copulas. ASTIN Bull. **45**(02), 445–475 (2015)

Wealth Transfers, Indifference Pricing, and XVA Compression Schemes

Claudio Albanese, Marc Chataigner and Stéphane Crépey

Abstract Since the 2008–2009 financial crisis, banks have introduced a family of XVA metrics to quantify the cost of counterparty risk and of its capital and funding implications: the credit/debt valuation adjustment (CVA and DVA), the costs of funding variation margin (FVA) and initial margin (MVA), and the capital valuation adjustment (KVA). We revisit from a wealth conservation and wealth transfer perspective at the incremental trade level the cost-of-capital XVA approach developed at the level of the balance sheet of the bank in [10]. Trade incremental XVAs reflect the wealth transfers triggered by the deals due to the incompleteness of counterparty risk. XVA-inclusive trading strategies achieve a given hurdle rate to shareholders in the conservative limit case that no new trades occur. XVAs represent a switch of paradigm in derivative management, from hedging to balance sheet optimization. This is illustrated by a review of possible applications of the XVA metrics.

Keywords Counterparty risk · Credit/Debt valuation adjustment (CVA and DVA) · Market incompleteness · Wealth transfer · Cost of capital · Funding valuation adjustment (FVA) · Margin valuation adjustment (MVA) · Cost of capital (KVA) · Funds transfer price (FTP)

This research has been conducted with the support of the Research Initiative "Modélisation des Marchés actions, obligations et dérivés" financed by HSBC France under the aegis of the Europlace Institute of Finance. The views and opinions expressed in this presentation are those of the author alone and do not necessarily reflect the views or policies of HSBC Investment Bank, its subsidiaries or affiliates. The research of Marc Chataigner is co-supported by a public grant as part of investissement d'avenir project, reference ANR-11-LABX-0056-LLH LabEx LMH.

C. Albanese
Global Valuation Ltd, London, UK
e-mail: claudio.albanese@global-valuation.com

M. Chataigner · S. Crépey (✉)
Université Paris-Saclay, CNRS, Univ Evry, Laboratoire de Mathématiques et Modélisation d'Évry, 91037 Évry, France
e-mail: stephane.crepey@univ-evry.fr

M. Chataigner
e-mail: marc.chataigner@univ-evry.fr

© Springer Nature Singapore Pte Ltd. 2020
Y. Jiao (ed.), *From Probability to Finance*, Mathematical Lectures
from Peking University, https://doi.org/10.1007/978-981-15-1576-7_5

Mathematics Subject Classification 91B25 · 91B26 · 91B30 · 91G20 · 91G40 · 91G60

JEL Classification D52 · D53 · G01 · G13 · G21 · G24 · G28 · G33 · M41 · P12

1 Introduction

In the aftermath of the financial crisis of 2008–2009, regulators launched in a major banking reform aimed at reducing counterparty risk by raising collateralisation and capital requirements and by incentivising central clearing (see [15]). The Basel III banking regulatory framework also set out guidelines for CVA and DVA (credit and debt valuation adjustments), which are valuation metrics for counterparty and own default risk in bilateral markets.

In bilateral as in centrally cleared transactions, collateral nowadays comes in two forms. The variation margin (VM), which is typically re-hypothecable, tracks the (counterparty-risk-free) value of a portfolio. The initial margin (IM) is an additional layer of margin, typically segregated, which is meant as a guarantee against the risk of slippage of a portfolio between default and liquidation. To quantify the respective costs of VM and IM, banks started to price into contingent claims funding and now margin valuation adjustments (FVA and MVA).

On a parallel track, the regulatory framework for the insurance industry has also been reformed, but on the basis of a different set of principles. Insurance claims are largely unhedged and markets are intrinsically incomplete. The cost of capital is reflected into prices. Solvency II focuses on regulating dividend distribution policies in such a way to ensure a sustainable risk remuneration to the shareholders over the lifetime of the portfolio. This is based on the notion of risk margin, which in banking parlance corresponds to a capital valuation adjustment (KVA).

In a related stream of papers, we develop a cost-of-capital XVA approach in incomplete counterparty risk markets. Albanese et al. [10] state the fundamental principles, rooted in the specificities of the balance sheet of a dealer bank. The application of these principles in continuous time leads in [9] to a progressive enlargement of filtration setup and nonstandard XVA backward stochastic differential equations (BSDEs) stopped before the bank default time. Crépey et al. [24] consider further the BSDEs "of the Mc Kean type" that arise when one includes the possibility for the bank to use its capital as a funding source for variation margin. Abbas-Turki et al. [1] deal with the numerical solution of our XVA equations by nested Monte Carlo strategies optimally implemented on graphics processing units (GPUs). Albanese et al. [8] apply what precedes to the concrete situation of a bank engaged into bilateral trade portfolios, demonstrating the feasibility of this approach on real banking portfolio involving thousands of counterparties and hundreds of thousands of trades. They also illustrate numerically the importance of the FVA mitigation provided, in the case of uncollateralized portfolios, by the use of capital as a funding source. Armenti and Crépey [12] and Albanese et al. [5] apply the generic principles of

[10] to the XVA analysis of centrally cleared portfolios: in [12], this is done under standard regulatory assumptions on the default fund and the funding strategies for initial margins, in order to compare in XVA terms bilateral versus centrally cleared trading networks. Albanese et al. [5], on the other hand, challenge these assumptions in the direction of achieving a greater efficiency (in XVA terms) of centrally cleared trading networks.

The aim of the present paper is to show how XVAs represent a switch of paradigm in derivative management, from hedging to balance sheet optimization. This is done by extensively relying on the notion of wealth transfer. This notion is already at the core of [10], but kind of implicitly and at the level of the whole portfolio of the bank, mainly used a posteriori and for interpretation purposes. By contrast, in this paper, we make it our main tool, exploited directly at the trade incremental level. The rewiring of the theory around the notion of wealth transfer allows re-deriving explicitly and "linearly", with virtually no mathematics, a number of the conceptual relations obtained in [9].

By the variety of situations that we consider, we try and demonstrate that the wealth transfer view yields a very practical and versatile angle on XVA analysis. The benefit of this angle is to bring in intuition and flexibility. The price for it lies in detail and accuracy, as, beyond the elementary static setup considered as a toy example in Sect. 4, going deeper into the arguments in order to obtain precise XVA equations brings back to the peculiarities in [9] and the follow up papers.

Under our cost-of-capital pricing approach, beyond CVA and DVA, the whole XVA suite is rooted on counterparty risk incompleteness. For alternative, replication-based, XVA approaches, see, for instance, [17–21] (without KVA), or, with a KVA meant as an additional liability like the CVA and the FVA (as opposed to a risk premium in our case), [30]. A detailed comparative discussion is provided in [9, Sect. 6].

The ***outline*** of the paper goes as follows. Section 2 sets our XVA pricing stage and delivers our main result Theorem 2.12 (which is just another perspective on the FTP formulas (32) or (40) in [10]). Starting from the limiting case of complete markets, the successive wealth transfers triggered by different forms of counterparty risk incompleteness or trading restrictions are reviewed in Sect. 3, along with the corresponding XVA implications. Section 4, which is a rewiring of Sect. 3 in [10], yields explicit XVA formulas and illustrates the XVA wealth transfer issue in a one-period static setup. Section 5 illustrates the switch of paradigm that XVAs represent in derivative management, from hedging to balance sheet optimization. Some connections with the Modigliani–Miller theory are discussed in Appendix "Connections with the Modigliani–Miller Theory", which develops the concluding paragraph of Sect. 3.5 in [10].

1.1 Abbreviations

Here is a recapitulative list of the abbreviations introduced in the course of the paper.

BCVA Bilateral CVA
BDVA Bilateral DVA

CA	Contra-assets valuation
CCP	Central counterparty
CDS	Credit default swap
CL	Contra-liabilities valuation
CSA	Credit support annex
CVA	Credit valuation adjustment (can be unilateral or bilateral)
CVA^{CL}	Contra-liability component of a unilateral CVA
DFC	Default fund contribution
DVA	Debt valuation adjustment (can be unilateral or bilateral)
FDA	Funding debt adjustment
FTP	Funds transfer price
FVA	Funding valuation adjustment
IAS	International accounting standard
IFRS	International financial reporting standards
KVA	Capital valuation adjustment
MDA	Margin debt adjustment
MtM	Mark-to-market
MVA	Margin valuation adjustment
OIS	Overnight index swap
RC	Reserve capital
REPO	Repurchase agreement
RM	Risk margin
SCR	Shareholder capital at risk
UCVA	Unilateral CVA
XVA	Generic "X" valuation adjustment

Also:

BA Value of the derivative portfolio of the bank to the bank as a whole (shareholders and bondholders)
SH Value of the derivative portfolio of the bank to the bank shareholders
BH Value of the derivative portfolio of the bank to the bank bondholders
CO Value of the derivative portfolio of the bank to the bank clients (corporate counterparties)

2 XVA Framework

2.1 Agents

Banks play a unique role in the industry as they accept deposits, make loans, and enter into risk transformation contracts with clients. Banks compete with each other to provide their services by offering the best prices to clients. A dealer bank is a price maker which cannot decide on asset selection: trades are proposed by clients and the market maker needs to stand ready to bid for a trade at a suitable price, no matter what the trade is and when it arrives.

When modeling a bank with defaultable debt, we need to consider its shareholders and creditors as separate entities. Specifically, we model a bank as a composite entity split into shareholders and bondholders. Shareholders make investment decisions up until the default of the bank, at which point they are wiped out. Bondholders instead represent the junior creditors of the bank, which have no decision power until the time of default, but are protected by laws such as pari-passu forbidding certain trades that would trigger wealth away from them to shareholders (or to more senior creditors) during the default resolution process of the bank.

Derivative clients (corporate counterparties of the bank) are also individual economic entities. Non-financial firms are characterized by a portfolio of real investments which is separate and in addition to their portfolio of financial contracts. In a reduced form model of the economy where we ignore the real investments portfolios of clients and model only their financial contracts, we lack the information required to decide whether a trade would be optimal to execute or not. Non-financial firms are just viewed as price takers that do not optimize and possibly accept to sustain a loss as a consequence of trading with banks.

The bank also needs an external "funder" to borrow unsecured, as required in last resort by its trading strategy (once all the internal sources of funding, such as received and rehypothecable variation margin, have been exhausted). This funder can be seen as a senior creditor of the bank, which in our setup enjoys an exogenous recovery rate upon bank default.

Last, the bank needs an access to the financial markets, e.g. repo markets, other banks (possibly via CCPs), etc., for setting up a hedge of its portfolio or, more precisely, of its mark-to-market component (as we assume jump-to-default risk hardly hedgeable in practice). We call abstractly the "hedger" of the bank the corresponding agent.

Hence, we consider an economy, with agents, labeled by an index a, coming in five different types:

- Bank shareholders (*sh*), who will only agree upon the bank entering a new trade if appropriately compensated by the client through the entry price of the deal, accounting for the costs of funding and cost of capital in particular;
- Bank *bondholders* (*bh*), who have no saying on trades but are protected by pari-passu type laws;
- Bank clients (or counterparties, *co*), who are price takers and willing to accept a loss in a trade for the sake of receiving (e.g. hedging) benefits that become apparent only once one includes their real investment portfolio, which is not explicitly modeled;
- Bank *funders* (*fu*), who agree to lend cash to the bank unsecured at some risky spread, which can be proxied by the CDS spread of the bank;
- Bank hedgers (*he*), who agree to provide a mark-to-market hedge (fully collateralised back-to-back hedge) of the derivative portfolio of the bank.

As these entities enter into contracts, wealth is transferred among them, as defined and explained in what follows.

2.2 Cash Flows

We assume that, at time 0, agents are already bound by contractual agreements between each other, which obligate them to exchange the related trading cash flows in the future. The cumulative cash flows up to time t received by entity a from all other agents in the economy assuming no new trade is entered (other than the ones initially planned at time 0, even though the latter may include forward starting contracts or dynamic hedging positions) is denoted by a (i.e. we identify agents with the corresponding cash flow processes), premium payments included.

We then assume that at time $t = 0$ a new trade is concluded. We prefix by Δ any trade incremental quantity of interest, e.g. Δa denotes the difference between the cumulative cash flow streams affecting the corresponding agent with and without the new deal.

There are also cash flows affecting our different economic agents, unrelated to the derivative portfolio of the bank. By definition, such cash flows are unchanged upon inclusion of the new deal in the bank portfolio. We assume that none of our economic agents is a monetary authority. Hence, money can neither be created nor destroyed and all relevant entities are included in the model. Under these conditions:

Assumption 2.1 We have that

$$\sum_{agents} \Delta a = 0. \tag{1}$$

Our objective is to assess the incremental impact of the new trade on counterparty risk and on the cost of debt financing and the cost of capital, in such a way that this information can be reflected into entry prices at a level making the shareholders indifferent (at least) to the deal. The analysis can then be repeated at each new trade as frequently as one wishes.

2.3 Valuation Operator

Under a cost-of-capital XVA approach, shareholders decide whether to invest depending on two inputs: a value function and incremental cost of capital. In this part we define the former.

We consider throughout the paper a pricing stochastic basis $(\Omega, \mathbb{G}, \mathbb{Q})$, with model filtration $\mathbb{G} = (\mathcal{G}_t)_{t \in \mathbb{R}_+}$ and pricing measure $\mathbb{Q}$ (the $\mathbb{Q}$ expectation is denoted by $\mathbb{E}$), such that all the processes of interest are $\mathbb{G}$ adapted.

Remark 2.2 In case markets are assumed to be complete, then there is one unique pricing measure $\mathbb{Q}$. In case markets are incomplete, several calibratable risk neutral probabilities can coexist and we need to recognize the related model risk. For this purpose, one can deal with a Bayesian-like prior distribution $\mu(d\mathbb{Q})$ in the space of risk neutral measures. Subjective views of price makers are embedded in the choice of the prior measure μ. In this paper, to keep things simple, we assume that μ is an atomic delta measure, i.e. we simply pick one possible risk neutral measure $\mathbb{Q}$ as it emerges from a calibration exercise and stick to it without including model risk.

However, we will introduce cost of capital, as a KVA risk-premium entering prices on top of risk-neutral $\mathbb{Q}$ valuation of the cash flows (shareholder cash flows, i.e. pre-bank default cash flows, for alignment of deals to shareholder interest, which drives the trading decisions of the bank as long as it is nondefault).

We denote by r a $\mathbb{G}$ progressive OIS rate process, i.e. overnight indexed swap rate, the best market proxy for a risk-free rate as well as the reference rate for the remuneration of cash collateral). Let $\beta = e^{-\int_0^{\cdot} r_t dt}$ be the corresponding risk-neutral discount factor. The representation of valuation by the traders of the bank is encoded into the following:

Definition 2.3 The (time 0) value of a cumulative cash flow stream $\mathscr{P}$ is given by

$$\mathbb{E}\int_{[0,\infty)} \beta_t d\mathscr{P}_t \tag{2}$$

(integral from time 0 included onward, under the convention that all processes are nil before time 0).

In particular, we call mark-to-market MtM of the new deal the value of its contractually promised cash flow stream ρ (i.e. MtM is the value of the deal ignoring the impact of counterparty risk and of its funding and capital implications). For each agent, we denote by the corresponding capitalized acronym $A = \text{SH, BH, CO, FU, HE}$, the value of a.

Definition 2.4 The wealth transfer triggered by the deal to a given agent is the difference between the values of the corresponding cash flow streams a accounting or not for the new deal, i.e. the value ΔA of Δa (by linearity of our valuation operator).

Recalling that the corresponding cash flow streams are premium inclusive (and that the time integration domain includes 0 in (2)):

Assumption 2.5 The risky funding assets and the hedging assets are "fairly" valued, in the sense that FU = HE = 0.

Hence we no longer report about FU and HE in the sequel.

Lemma 2.6 *We have*

$$\sum_{agents} \Delta \text{A} = 0 \tag{3}$$

and, more specifically,

$$\Delta \text{SH} + \Delta \text{BH} + \Delta \text{CO} = 0. \tag{4}$$

Proof Definition 2.4 and (1) immediately imply (3), which, under Assumption 2.5, specializes to (4). □

Hence, our setup is in line with the [43] law of conservation of investment value according to which, as a consequence of a financial trade among a number of entities who enter into a contract at time 0 to exchange future cash flows towards each other, the algebraic sum of all wealth transfers at time 0 among all entities involved is zero. But the wealth transfer amount ΔA is possibly non-zero, in general, for some entities $a(= sh, bh,$ and/or $co)$.

We also introduce $ba = sh + bh$ and $\text{BA} = \text{SH} + \text{BH}$ for the cash flows and value of the derivative portfolio to the bank as a whole. From a balance sheet interpretation point of view that is detailed in [10, Sect. 2], BA and SH correspond to the accounting equity and to the core equity Tier 1 capital of the bank (at least, at the trade incremental level).

2.4 *Contra-Assets and Contra-Liabilities*

In the case of an investment bank, counterparty risk entails several sources of market incompleteness, or trading restrictions (cf. Appendix "Connections with the Modigliani–Miller Theory"):

- Pari-passu rules, meant to guarantee to the bank bondholders the benefit of any residual value within the bank in case of default;
- Bank debt cannot possibly be fully redeemed by the bank shareholders;
- Client default losses cannot be perfectly replicated.

In order to focus on counterparty risk and XVAs, we assume throughout the paper that the market risk of the bank is perfectly hedged by means of perfectly collateralized back-to-back trades. That is, each client deal is replicated, in terms of market risk, by a perfectly collateralized back-to-back trade with another bank. Hence, all remains to be priced is counterparty risk and its capital and funding implications.

More precisely, netting the cash flows of the client portfolio and its hedge results in a set of counterparty risk related cash flows that can be subdivided into counterparty default exposures and funding expenditures, incurred by the bank as long as it is alive, and cash flows received by the bank from its default time onward, when shareholders have already been wiped out by the bondholders, which increase the realized recovery of the latter: See Sect. 4 in an illustrative static setup. Accordingly (cf. Eq. (14) and (17) in [9] for more technical detail):

Definition 2.7 We call contra-assets and contra-liabilities the synthetic liabilities and assets of the bank that emerge endogenously from its trading, through the impact of counterparty risk on its back-to-back hedged portfolio.

We denote by ca and cl the corresponding cash flow streams, with respective values CA and CL. As the counterparty risk related add-ons are not known yet, this is all "FTP excluded" (cf. Sect. 2.6), i.e. not accounting for the corresponding (to be determined) premium that will be paid by the client to the bank.

Note that, by Definition 2.4, contra-liabilities (such as the DVA) do not benefit to the wealth of shareholders but only to bondholders, because the corresponding cash flows come too late, when shareholders have already been wiped out by the bondholders.

The counterparty related cash flows, i.e. $(ca - cl)$, are composed of credit and funding cash flows. Under bilateral counterparty risk, there is a credit valuation adjustment (CVA) for each risky counterparty and a debt valuation adjustment (DVA) for default of the bank itself. Each CVA (respectively DVA) is equal to the cost of buying protection against the credit risk of that counterparty (respectively the symmetrical cost seen from the perspective of the counterparty toward the bank). Accordingly, credit cash flows are valued as $(\Delta\mathrm{BDVA} - \Delta\mathrm{BCVA})$. Note that we are dealing with bilateral BCVA and BDVA here, where the CVA and DVA related cash flows between the bank and each given counterparty are only considered until the first occurrence of a default of the two entities, consistent with the fact that later cash flows will not be paid.

Proposition 2.8 *We have*

$$\mathrm{CA} - \mathrm{CL} = \mathrm{BCVA} - \mathrm{BDVA}. \tag{5}$$

Proof In view of the above, this is an immediate consequence of $\Delta\mathrm{FU} = 0$ (by Assumption 2.5). □

2.5 *Cost of Capital*

Following the principles of Basel III and Solvency II, we handle (unhedgeable) counterparty risk by means of a combination of a reserve capital account, which is used by the bank to cover systematic losses, and capital at risk to cover exceptional losses.

Shareholders require a dividend premium as compensation for the risk incurred on their capital at risk. The level of compensation required on shareholder capital at risk (SCR) is driven by market considerations. Typically, investors in banks expect a hurdle rate h of about 10–12%.

When a bank charges cost of capital to clients, these revenues are accounted for as profits. Unfortunately, prevailing accounting standards for derivative securities are based on the theoretical assumption of market completeness. They do not envision a mechanism to retain these earnings for the purpose of remunerating capital across the entire life of transactions, which can be as long as decades. In complete markets, there is no justification for risk capital. Hence, profits are immediately distributable. A strategy of earning retention beyond the end of the ongoing accounting year (or quarter) is still possible as in all firms, but this would be regarded as purely a business decision, not subject to financial regulation under the Basel III accord.

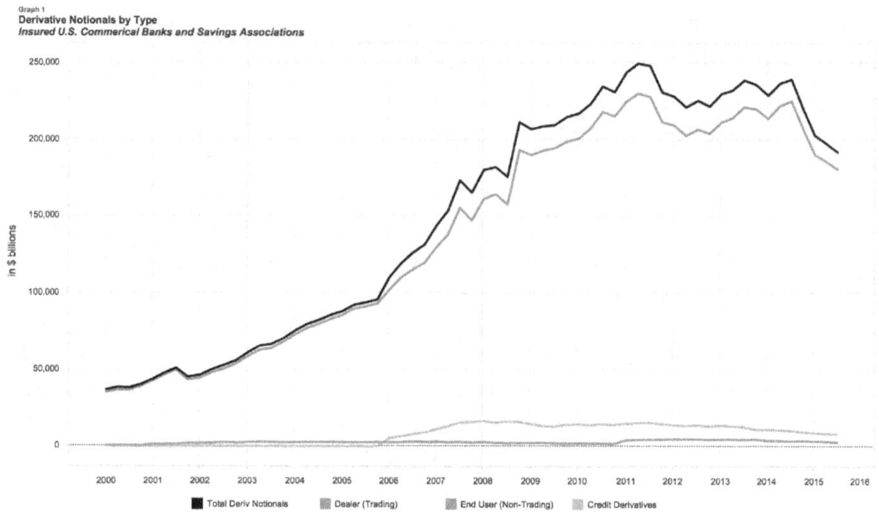

Fig. 1 2008 financial derivatives Ponzi scheme (*Source* Office of the comptroller of the currency, Q3 2015 quarterly bank trading revenue report)

This leads to an explosive instability characteristic of a Ponzi scheme. For instance, if a bank starts off today by entering a 30-year swap with a client, the bank books a profit. Assuming the trade is perfectly hedged, the profit is distributable at once. But, the following year, the bank still needs capital to absorb the risk of the 29-year swap in the portfolio. If the profits from the trade have already been distributed the previous year, the tentation for the bank to maintain shareholder remuneration levels is to lever up by selling and hedging another swap, booking a new profit and distributing the dividend to shareholders that are now posting capital for both swaps. As long as trading volumes grow exponentially, the scheme self-sustains. When exponential growth stops, the bank return on equity crashes. The global financial crisis of 2008–09 can be analyzed along these lines (cf. the 2008 financial derivative Ponzi scheme displayed in Fig. 1): In the aftermath of the crisis, the first casualty was the return on equity for the fixed income business, as profits had already been distributed and market-level hurdle rates could not be sustained by portfolio growth.

Interestingly enough however, in the insurance domain, the Swiss Solvency Test and Solvency II (see [23, 41]), unlike Basel III, do regulate the distribution of retained earnings through a mechanism tied to so called "risk margins" (see [26, 44], or [40] regarding the risk margin and cost of capital actuarial literature). The accounting standards set out in IFRS 4 Phase II (see [34, 35]) are also consistent with Solvency II and include a treatment for risk margins that has no analogue in the banking domain.

Under the KVA approach of [9, Sect. 3.6], which provides a continuous-time and banking analog of the above-mentioned insurance framework, earnings are retained into a risk margin (RM) account and distributed gradually and sustainably, at some "hurdle rate" h, to the shareholders. This yields a framework for assessing cost of capital for a bank, passing it on to the bank clients, and distributing it gradually to the bank shareholders through a dividend policy that would be sustainable even in the limit case of a portfolio held on a run-off basis, with no new trades ever entered in the future.

2.6 Funds Transfer Price

A strategy based on capital for unhedged risk and risk margin for dividend distribution allows one to setup an optimal portfolio management framework for a market maker.

Shareholders have decision power in a firm and make investment decisions with the purpose of optimising their own wealth. Accordingly, our broad XVA principle reads as follows:

Assumption 2.9 The bank will commit to an investment decision at time 0 and execute the trade in case ΔSH exceeds the incremental cost of capital or ΔKVA[1] for the trade.

By Definition 2.3, the mark-to-market MtM of the new deal corresponds to its value ignoring counterparty risk and its funding and capital consequences. In line with it:

Assumption 2.10 MtM is the amount required by the bank for setting up the fully collateralized back-to-back market hedge to the deal.

Hence, at inception of a new trade at time 0, the client is asked by the bank to pay the mark-to-market (MtM) of the deal (as the cost of the back-to-back hedge for the bank), plus an add-on, called funds transfer price (FTP), reflecting the incremental counterparty risk of the trade and its capital and funding implications for the bank. Accordingly, Assumption 2.9 is refined as follows.

Definition 2.11 The FTP is computed by the bank in order to achieve

$$\Delta \text{SH} = \Delta \text{KVA}. \tag{6}$$

We emphasize that, since trading cash flows include premium payments (see the beginning of Sect. 2.2), ΔSH contains the FTP (cf. Sect. 2.7 below). Hence, the rationale for Definition 2.11 is that, should a tentative price be less than MtM plus the FTP that is implicit in (6), then, in line with Assumption 2.9, the bank should refuse the deal at this price, as it would be detrimental to shareholders. We set an equality rather than $\geq$ in (6) in view of the competition between banks, which pushes them to accept from the client "the lowest price admissible to their shareholders": see the concluding paragraph of Sect. 3.3 in [10].

[1]cf. Sect. 3.6.

The FTP of a deal can be negative, meaning that the bank should effectively be ready to give some money back to the client (it may do it or not in practice) in order to account for a counteparty(/funding/capital) risk reducing feature of the deal.

2.7 A General Result

As we will see in all the concrete cases reviewed in Sect. 3, a dealer bank only incurs actual (nonnegative) costs before its defaults and actual (nonnegative) benefits from its default time onward, i.e. $bh = cl$, hence $\mathrm{BH} = \mathrm{CL}$. A common denominator to all our later formulas (whatever the detailed nature of the wealth transfers involved) is the following (cf. the formulas (32) or (40) in [10]):

Theorem 2.12 *Assuming that* $\Delta\mathrm{BH} = \Delta\mathrm{CL}$*, we have*

$$\Delta\mathrm{SH} = \mathrm{FTP} - \Delta\mathrm{CA}. \tag{7}$$

Hence, the investment criterion (6) *is equivalent to*

$$\mathrm{FTP} = \Delta\mathrm{CA} + \Delta\mathrm{KVA} \tag{8}$$

and

$$\begin{aligned} \Delta\mathrm{SH} &= \Delta\mathrm{KVA}, \\ \Delta\mathrm{BH} &= \Delta\mathrm{CL}, \\ \Delta\mathrm{CO} &= -(\Delta\mathrm{CL} + \Delta\mathrm{KVA}). \end{aligned} \tag{9}$$

Proof Disregarding the FTP, the trade incremental cash flows that affect the bank are $\Delta(ca - cl)$ and he (which includes MtM as the cost of setting up the fully collateralized back-to-back market hedge of ρ, cf. Assumption 2.10). Hence, by Definition 2.7,

$$\Delta\mathrm{BA} = \mathrm{FTP} - (\Delta\mathrm{CA} - \Delta\mathrm{CL}). \tag{10}$$

As a consequence, the assumption that $\Delta\mathrm{BH} = \Delta\mathrm{CL}$ implies

$$\Delta\mathrm{SH} = \mathrm{FTP} - \Delta(\mathrm{CA} - \mathrm{CL}) - \Delta\mathrm{BH} = \mathrm{FTP} - \Delta\mathrm{CA},$$

which is (7). Therefore, the investment criterion (6) is rewritten as (8). The last identity in (9) readily follows from the two first ones and from (4). □

As a concluding remark to this section, let us now assume, for the sake of the argument (cf. [10, Sect. 3.5]), that the bank, all other things being equal, would be able to hedge (or monetize) its own jump-to-default exposure through a further deal corresponding to the delivery by the bank of a cash flow stream *cl* in exchange of a premium fee "fairly" priced as its valuation CL. Then, switching from the FTP given by (8) to a modified

$$\mathrm{FTP} = \Delta\mathrm{CA} - \Delta\mathrm{CL} + \Delta\mathrm{KVA} = \Delta\mathrm{BCVA} - \Delta\mathrm{BDVA} + \Delta\mathrm{KVA}, \tag{11}$$

but accounting for the extra income to shareholders provided by the premium fees of the hedge in each portfolio including and excluding the new deal, all shareholder cash flows are now exactly the same as before (in each portfolio including and excluding the new deal), but the bondholders are entirely wiped out by the hedge. Hence we have, instead of (9),

$$\begin{aligned} \Delta\mathrm{SH} &= \Delta\mathrm{KVA}, \\ \Delta\mathrm{BH} &= 0, \\ \Delta\mathrm{CO} &= -\Delta\mathrm{KVA}. \end{aligned} \tag{12}$$

This shows that, under the trading strategy including the hedge, our investment criterion (6) becomes equivalent to the modified pricing rule (11), which supersedes (8), and the corresponding wealth transfers are given by (12).

If, in addition, the trading loss of the bank shareholders was also hedged out, then there would be no more risk for the bank and therefore no capital at risk required, hence the KVA would vanish and we would be left with the "complete market formulas"

$$\mathrm{FTP} = \Delta\mathrm{BCVA} - \Delta\mathrm{BDVA} \text{ and } \Delta\mathrm{SH} = \Delta\mathrm{BH} = \Delta\mathrm{CO} = 0$$

(cf. Sect. 3.1).

However, in practice, a bank cannot hedge its own default nor replicate its counterparty default exposure, hence we remain with the outputs of Theorem 2.12. The exact specification of the different terms in (8)–(9) depends on the trading setup, as our next sections illustrate.

3 Wealth Transfers Triggered by Market Incompleteness

In this section we provide more detailed (but still conceptual) formulas for the FTP and the different wealth transfers involved, under more and more realistic assumptions regarding the trading restrictions and market incompleteness faced by a dealer bank. See the papers commented upon in Sect. 1 (or see Sect. 4 in a static static setup) for definite XVA formulas or equations in various concrete trading setups.

3.1 The Limiting Case of Complete Markets

In the special case of complete markets (without trading restrictions, in particular), wealth transfers between bank shareholders and bank creditors can occur but are irrelevant to investment decisions and they have no impact on prices, i.e. there is no point to distinguish shareholders from bondholders. The (complete market Modigliani–Miller form of) justification of this statement (cf. Sects. 2.7 and Appendix "Connections with the Modigliani–Miller Theory") is that, in case

$$\Delta \mathrm{BA} = \Delta \mathrm{SH} + \Delta \mathrm{BH} \geq 0$$

but "it seems that" $\Delta \mathrm{SH} < 0$, shareholders would still be able to increase their wealth (canceling out the wealth transfer from them to bondholders triggered by the deal) by buying the firm debt prior to executing the trade, making $\Delta \mathrm{SH}$ nonnegative (and $\Delta \mathrm{BH}$ diminished by the same amount, in line with the conservation law (3)). Moreover, in complete markets, there is no justification for capital at risk, so that there is no cost of capital either. Hence, the criterion (6) for an investment decision is rewritten as

$$\Delta \mathrm{BA} = 0. \tag{13}$$

In complete markets without trading restrictions, funding comes for free because unsecured derivatives can be REPOed, i.e. posted as a guarantee against the corresponding funding debt, which therefore is free of credit risk. That is, $fu = 0$.

Hence, consistent with (10) and (5) (but with even $fu = 0$ here, instead of only $\mathrm{FU} = 0$ in general), we obtain

$$\Delta \mathrm{BA} = \Delta \mathrm{BDVA} - \Delta \mathrm{BCVA} + \mathrm{FTP}. \tag{14}$$

An application of the investment criterion (13) to (14) yields

$$\mathrm{FTP} = \Delta \mathrm{BCVA} - \Delta \mathrm{BDVA} \tag{15}$$

(and we have $\Delta \mathrm{SH} = \Delta \mathrm{BH} = \Delta \mathrm{CO} = 0$), which flows into the reserve capital account maintained by the bank for dealing with counterparty risk in expectation (and no capital at risk, hence no risk margin, are required).

As discussed in Sects. 2.7 and Appendix "Connections with the Modigliani–Miller Theory", the (complete market form of Modigliani–Miller) argument whereby shareholders buy hedge the default of the bank is crucial in relation to the pricing rule (15). If shareholders cannot do so, then this pricing rule triggers wealth transfers from shareholders to bondholders by the amount

$$\Delta \mathrm{SH} = -\Delta \mathrm{BDVA}, \qquad \Delta \mathrm{BH} = \Delta \mathrm{BDVA}. \tag{16}$$

The reason for this wealth transfer is that gains conditional to the default of the bank represent cash flows which are received by the bank bondholders at time of default. They do not benefit the shareholders, which at that point in time are wiped out, unless precisely these gains can be monetized before the default through hedging by the bank (see the computations in Sect. 2.7).

Accounting standards for derivative securities such as IAS 39 followed by IFRS 9 have been designed around a concept of "fair valuation", which implicitly depends on the complete market hypothesis (cf. Sect. 2.5). Under this understanding the valuation of a bilateral contract is "fair" if both parties agree on the valuation independently: This is the case if there exists a replication strategy that precisely reproduces the cash flow stream of a given derivative contract.

In complete markets, there is no distinction between price maker and price taker. There is also no distinction between entry price and exit price of a derivative contract. The fair valuation of a derivative asset is the price at which the asset can be sold. All buyers would value the derivative at the exact same level, as any deviation from the cost of replication would lead to an arbitrage opportunity. In particular, fair valuations are independent of endowments and any other entity specific information.

The most glaring omission in a complete market model for a bank is a justification for equity capital as a loss absorbing buffer. Capital may still be justified as required to finance business operations. A bank is justified to charge fees for services rendered, but these fees should not depend on the risk profile of the trades. They should only be proportional to the operational workload (in the sense of a volume based fee) of the bank. Once the fees are received, a portion is allocated to cover operational costs and the remainder is released into the dividend stream.

3.2 DVA Wealth Transfer Triggered by Shareholders Not Being Able to Redeem Bank Debt

As explained at different levels in Sects. 2.7, 3.1 and Appendix "Connections with the Modigliani–Miller Theory", if shareholders were able to freely trade bank debt, the interest of shareholders would be aligned with the interests of the firm as a whole (shareholders and bondholders altogether). However, in reality, trading restrictions prevent shareholders from effectively offsetting wealth transfers to bondholders by buying bank debt. Hence these wealth transfers can only compensated by clients in the form of suitable valuation adjustments at deal inception. The debt valuation adjustment (DVA) illustrates well this phenomenon. As a bank enters a new trade, they pass on to clients the credit valuation adjustment (CVA) as a compensation against the counterparty default risk. The DVA is the CVA the counterparty assigns to the bank. If valuations were fair, a bank should symmetrically recognise also a DVA benefit to clients. However, since the DVA can be monetised by the bank only by defaulting, managers are reluctant to recognise a DVA benefit to clients. If they did, they would effectively trigger a wealth transfer as in (16) from shareholders to bondholders that they would not be able to hedge. In order to ensure $\Delta\text{SH} = 0$ (we are not considering KVA yet) in spite of this, a bank should charge, instead of (15),

$$\text{FTP} = \Delta\text{BCVA}, \tag{17}$$

so that the wealth transfer to bondholders becomes borne by the client of the deal instead of the bank shareholders. This yields, instead of (16),

$$\begin{aligned} \Delta\text{SH} &= 0, \\ \Delta\text{BH} &= \Delta\text{BDVA}, \\ \Delta\text{CO} &= -\Delta\text{BH}. \end{aligned} \tag{18}$$

Conceivable DVA hedges would involve the bank selling credit protection on itself, an impossible trade, or violations of the pari passu rule on debt seniority. Following up to these considerations, regulators have started to de-recognize the DVA as a contributor to core equity tier I capital, the metric that roughly represents the value of the bank to the shareholders (see [16]).

3.3 CVA^{CL} Wealth Transfer Triggered by Shareholders Bankruptcy Costs

The [16] went even further and decided that banks should compute a unilateral CVA, which we denote by UCVA. This can be interpreted as saying that shareholders face a bankruptcy cost, equal to UCVA at default time, which goes to benefit bank bondholders. This bankruptcy cost corresponds to the transfer to bondholders of the residual amount on the reserve capital account of the bank in case the latter defaults. Accounting for this feature, upon entering a new trade, if entry prices are struck at the indifference level for shareholders (still ignoring capital and its KVA implication, here and until Sect. 3.6), we obtain $\text{FTP} = \Delta\text{UCVA}$ and

$$\begin{aligned} \Delta\text{SH} &= 0, \\ \Delta\text{BH} &= \Delta\text{BDVA} + \Delta\text{CVA}^{\text{CL}}, \\ \Delta\text{CO} &= -\Delta\text{BH}. \end{aligned} \tag{19}$$

Here, by definition, CVA^{CL} is the difference between the unilateral UCVA and BCVA, i.e. the valuation of the counterparty default losses occurring beyond the default of the bank itself. The acronym CL in CVA^{CL} stands for contra-liability (see Definition 2.7), because CVA^{CL} indeed corresponds to a "contra-liability component" of the UCVA.

3.4 FVA Wealth Transfer Triggered by the Impossibility to REPO Derivatives

A related form of wealth transfer occurs in the case of costs of funding for variation margin. The acquisition of assets funded with unsecured debt triggers a wealth transfer from shareholders to bondholders. This happens because shareholders sustain a cost of carry for unsecured debt while bondholders benefit out of having a claim on the asset in case the bank defaults.

Remark 3.1 In this regard, REPO contracts are a more efficient method for funding asset acquisitions since shareholders sustain far lower funding rates (close to risk-free rates); on the flip side however, bondholders do not have a claim on an asset passed as collateral in a REPO transaction.

Applying [43]'s wealth conservation principle to unsecured debt valuation (cf. Sect. 2.3), the cost of carry of debt to shareholders equals the gain to bondholders induced by the non reimbursement by the bank of the totality of its funding debt to the funder if it defaults. In [2, 3] and [10], the wealth transferred from shareholders to bondholders by the cost of unsecured debt is called funding valuation adjustment (FVA), while the wealth received by bondholders through the accordingly increased recovery rate is denoted with FDA. Wealth conservation implies that FDA = FVA (akin to FU = 0 in Assumption 2.5). So, in accounts, the fair valuation of the derivative portfolio of the bank (i.e. of the counterparty risk related cash flows, under our back-to-back hedge assumption of market risk) should also contain a term FDA equal to FVA in absolute value, but contributing to the bank fair valuation with an opposite sign, i.e. appearing as an asset, of the contra-liability kind, in the bank balance sheet (see [10, Fig. 1 in Sect. 2] for the detailed balance sheet perspective on the XVA metrics). This way, the accounting equity of the bank, i.e. the wealth of the bank as a whole, does not depend on the funding spreads or funding policies of the bank, a result in the line of [43]'s law, which holds independently of whether markets are complete or not (cf. our general identity (5)).

However, the resulting notion of fair valuation of the portfolio (or wealth of the bank as a whole) is only relevant in a complete market. If markets are incomplete in the sense that shareholders are forbidden to buy bank bonds, then one needs to refocus on shareholder interest. As the FVA subtracts from shareholder value (because the corresponding cash flows are pre-bank default), investment decisions do depend on funding strategies. That is, in order for an investment to be acceptable, the bank needs to ensure that the change in FVA is passed on to the client. The same amount is then also transferred as a net benefit to bank bondholders.

As a result, FTP and wealth transfers for indifference entry prices in the case of unsecured funding of variation margin are given by $\mathrm{FTP} = \Delta\mathrm{UCVA} + \Delta\mathrm{FVA}$ and

$$\begin{aligned}\Delta\mathrm{SH} &= 0,\\ \Delta\mathrm{BH} &= \Delta\mathrm{BDVA} + \Delta\mathrm{CVA}^{\mathrm{CL}} + \Delta\mathrm{FVA},\\ \Delta\mathrm{CO} &= -\Delta\mathrm{BH}.\end{aligned} \tag{20}$$

3.5 MVA Wealth Transfer Triggered by Different Funding Policies for Initial Margins

Initial margin (IM) offers a fourth example of wealth transfer. When the IM posted by the bank is funded using debt unsecured to the funder, an additional wealth transfer ΔMVA from shareholders to bondholders is triggered (unless the bank could hedge its own default, see Sects. 2.7 and Appendix "Connections with the Modigliani–Miller Theory"). Hence, at indifference, we have

$$\mathrm{FTP} = \Delta\mathrm{UCVA} + \Delta\mathrm{FVA} + \Delta\mathrm{MVA} \tag{21}$$

and

$$\begin{aligned} &\Delta\mathrm{SH} = 0, \\ &\Delta\mathrm{BH} = \Delta\mathrm{BDVA} + \Delta\mathrm{CVA}^{\mathrm{CL}} + \Delta\mathrm{FVA} + \Delta\mathrm{MVA}, \\ &\Delta\mathrm{CO} = -\Delta\mathrm{BH}. \end{aligned} \tag{22}$$

It is debatable whether or not unsecured collateral funding strategies are unavoidable. As a rule, wealth transfers, entry prices for clients, and investment decisions by banks depend on the collateral strategies which are enacted. For instance, in case initial margin posting would be delegated to a non-banking specialist lender without funding costs and recovering the portion of IM unused to cover losses if the bank defaults, then, since the IM is sized as a large quantile of the return distribution over the margin period, the corresponding MVA is bound to be much smaller than the one resulting from unsecured borrowing at the bank CDS spread. See [8, Sect. 4.3] and [5, Sect. 5] for details.

Remark 3.2 Strategies to achieve secured funding for variation margin are discussed in [7] and [4]. However, such funding schemes are much more difficult to implement for VM because VM is far larger and more volatile than IM.

3.6 KVA Wealth Transfer Triggered by The Cost of Capital Which Is Required by the Impossibility of Hedging out Counterparty Default Losses

The formulas (21)–(22) do not account for the cost of the capital earmarked to absorb exceptional losses (beyond the expected losses already accounted for by reserve capital). In our framework we include this cost as a capital valuation adjustment (KVA), which is dealt with separately as a risk premium, flowing into a risk margin account distinct from the reserve capital account. Specifically, we define our KVA as the cost of remunerating shareholders at some constant hurdle rate $h > 0$ for their capital at risk. The hurdle rate is the instantaneous remuneration rate of one unit of shareholder capital at risk, which can be interpreted as a risk aversion parameter of the shareholders (cf. the concluding paragraph of Sect. 3.3 in [10]). Such a KVA is the investment banking analog of the Solvency II notion of risk margin, or market value margin.

Accounting further for cost of capital, indifference in the sense of (6) for shareholders corresponds to

$$\mathrm{FTP} = \Delta\mathrm{UCVA} + \Delta\mathrm{FVA} + \Delta\mathrm{MVA} + \Delta\mathrm{KVA} \tag{23}$$

and

$$\Delta\mathrm{SH} = \Delta\mathrm{KVA}, \tag{24}$$

$$\Delta\mathrm{BH} = \Delta\mathrm{BDVA} + \Delta\mathrm{CVA}^{\mathrm{CL}} + \Delta\mathrm{FVA} + \Delta\mathrm{MVA}, \tag{25}$$

$$\Delta\mathrm{CO} = -\Delta\mathrm{BH} - \Delta\mathrm{KVA}. \tag{26}$$

If the deal occurs, the sum of the first three incremental amounts in (23), i.e. $\Delta\mathrm{UCVA} + \Delta\mathrm{FVA} + \Delta\mathrm{MVA}$, accrues to reserve capital, while the last term $\Delta\mathrm{KVA}$ accrues to the risk margin account.

We emphasize that the FTP formula (23) makes the price of the deal both entity-dependent (via, for instance, the CDS funding spread of the bank, which is a major input to the FVA) and portfolio-dependent (via the trade incremental feature of the FTP), far away from the law of one price, the complete market notion of fair valuation, and the (complete market form of) the Modigliani–Miller invariance principle.

4 XVA Formulas and Wealth Transfers in a Static Setup

In this section, which is a rewiring, around the notion of wealth transfer, of Sect. 3 in [10], we illustrate the XVA wealth transfer issues in an elementary static one-year setup, with r set equal to 0.

Assume that at time 0 a bank, with no prior endowment and equity E corresponding to its initial wealth, enters a derivative position with a client. We drop the $\Delta\cdot$ notation in this section, where every quantity of interest prior to the deal is simply 0, so that all our price and XVA notation refers to the new deal at time 0.

The bank and its client are default prone with zero recovery and no collateralization (no variation or initial margins). We denote by J and J_1 the survival indicators of the bank and its client at time 1 (both being assumed alive at time 0), with default probability of the bank $\mathbb{Q}(J = 0) = \gamma$ and no joint default for simplicity, i.e $\mathbb{Q}(J = J_1 = 0) = 0$. The bank wants to charge to its client an add-on (or obtain from its client a rebate, depending on whether the bank is seller or buyer), denoted by CA, accounting for its expected counterparty default losses and funding expenditures, as well as a KVA risk premium.

The all-inclusive XVA add-on to the entry price for the deal, which we call funds transfer price (FTP), follows as

$$\mathrm{FTP} = \underbrace{\mathrm{CA}}_{\text{Expected costs}} + \underbrace{\mathrm{KVA}}_{\text{Risk premium}}. \tag{27}$$

4.1 Cash Flows

The counterparty risk related cash flows affecting the bank before its default are its counterparty default losses

$$\mathscr{C}^{\circ} = (1 - J_1)\rho^{+}$$

and its funding expenditures $\mathscr{F}^{\circ}$. Accounting for the to-be-determined add-on CA (the KVA paid by the client at time 0 is immediately transferred by the bank management to the shareholders), the bank needs to borrow $(\mathrm{MtM} - \mathrm{CA})^{+}$ unsecured or invest $(\mathrm{MtM} - \mathrm{CA})^{-}$ risk-free, depending on the sign of $(\mathrm{MtM} - \mathrm{CA})$, in order to pay $(\mathrm{MtM} - \mathrm{CA})$ to the client. In accordance with Assumption 2.5, unsecured borrowing is "fairly" priced as its valuation (cf. Definition 2.3) $\gamma \times$ the amount borrowed by the bank, so that the funding expenditures of the bank amount to

$$\mathscr{F}^{\circ} = \gamma(\mathrm{MtM} - \mathrm{CA})^{+} \tag{28}$$

(deterministically in this one-period setup). We assume further that a fully collateralized back-to-back market hedge is set up by the bank in the form of a deal with a third party, with no entrance cost and a payoff to the bank of $-(\rho - \mathrm{MtM}) = -he$ at time 1, irrespective of the default status of the bank and the third party at time 1.

Collecting all cash flows, the result of the bank over the year is (cf. the proof of Lemma 3.2 in [10] for a detailed calculation)

$$ba = -(1 - J_1)\rho^{+} - \gamma(\mathrm{MtM} - \mathrm{CA})^{+} + (1 - J)\big(\rho^{-} + (\mathrm{MtM} - \mathrm{CA})^{+}\big) + \mathrm{FTP}. \tag{29}$$

Introducing further

$$\mathscr{C}^{\bullet} = (1 - J)\rho^{-},\ \mathscr{F}^{\bullet} = (1 - J)(\mathrm{MtM} - \mathrm{CA})^{+}, \tag{30}$$

we thus have

$$\begin{aligned} ba &= \underbrace{\mathrm{FTP} - \overbrace{(\mathscr{C}^{\circ} + \mathscr{F}^{\circ})}^{ca}}_{sh} + \underbrace{\mathscr{C}^{\bullet} + \mathscr{F}^{\bullet}}_{bh=cl} \\ &= -\underbrace{\big(\mathscr{C}^{\circ} - \mathscr{C}^{\bullet} - \mathrm{FTP}\big)}_{co} - \underbrace{\mathscr{F}^{\circ} - \mathscr{F}^{\bullet}}_{fu}, \end{aligned} \tag{31}$$

where the identification of the different terms follows from their financial interpretation.

4.2 Static XVA Formulas

In a static one-period setup, there are no unilateral versus first-to-default issues and no bank accounts involved (as no rebalancing of the trading strategies at intermediate time points is necessary), hence no $\mathrm{CVA}^{\mathrm{CL}}$ issue either (cf. Sect. 3.3). Moreover we assume no collateralization, hence there is nothing related with initial margin. As a consequence, contraliabilities in this context reduce to the DVA and the FDA. Accordingly, the FTP and wealth transfer formulas (8) and (9) reduce to

$$\begin{aligned} &\mathrm{FTP} = \mathrm{CA} + \mathrm{KVA} = \mathrm{CVA} - \mathrm{DVA} + \mathrm{CL} + \mathrm{KVA}, \\ &\mathrm{SH} = \mathrm{KVA}, \mathrm{BH} = \mathrm{CL}, \mathrm{CO} = -\mathrm{CL} - \mathrm{KVA}, \end{aligned} \tag{32}$$

where

$$\mathrm{CA} = \underbrace{\mathbb{E}\mathscr{C}^{\circ}}_{\mathrm{CVA}} + \underbrace{\mathbb{E}\mathscr{F}^{\circ}}_{\mathrm{FVA}}. \tag{33}$$

Hence, using also (28),

$$\mathrm{CVA} = \mathbb{E}\big[(1 - J_1)\rho^{+}\big],\ \mathrm{FVA} = \frac{\gamma}{1+\gamma}(\mathrm{MtM} - \mathrm{CVA})^{+}, \tag{34}$$

and

$$\mathrm{CL} = \underbrace{\mathbb{E}[(1 - J)\rho^{-}]}_{\mathrm{DVA}} + \underbrace{\mathbb{E}[(1 - J)(\mathrm{MtM} - \mathrm{CA})^{+}]}_{\mathrm{FDA} = \gamma(\mathrm{MtM} - \mathrm{CA})^{+} = \mathrm{FVA}}. \tag{35}$$

As for the KVA, it is meant to remunerate the shareholders at some hurdle rate h (e.g. 10%) for the risk on their capital, i.e.

$$\mathrm{KVA} = h\mathrm{E}. \tag{36}$$

Moreover, as the bank shareholders are in effect CVA and FVA traders in this setup (where market risk is hedged out and there is no MVA), we may add that it would be natural to size E by some risk measure (such as value at risk or, better, expected shortfall, at some "sufficiently high" level, e.g. 97.5%) of the trading loss(-and-profit) of the shareholders given as (cf. (31))

$$L = \mathscr{C}^{\circ} + \mathscr{F}^{\circ} - \mathrm{CA} = -sh + \mathrm{KVA} \tag{37}$$

(i.e. the pure trading loss not accounting for the KVA risk premium). Hence, in the static setup, our hurdle rate h is nothing but the return on equity (ROE).

Note moreover that

$$CA - CL = CVA + FVA - (DVA + FDA) = CVA - DVA, \tag{38}$$

as FVA = FDA.

5 Derivative Management: From Hedging to XVA Compression

The global financial crisis of 2008–2009 emphasized the incompleteness of counterparty risk. XVAs represent the ensuing switch of paradigm in derivative management, from hedging to balance sheet optimization. In this section we illustrate this evolution by a discussion of two potential applications of the XVA metrics in the optimization mode, beyond the basic use of computing them and charging them into prices for some target hurdle rate $h > 0$ that would be set by the management of the bank.

Of course, for any potential application of the XVA metrics to be practical, one needs efficient XVA calculators in the first place: XVAs involve heavy computations at the portfolio level, which yet need sufficient accuracy so that trade incremental numbers are not in the numerical noise of the machinery. In practice, banks mostly rely on exposure-based XVA computational approaches, based on time 0 XVAs reformulated as integrals of market expected exposures against relevant CDS or funding curves. This is somehow enhanced by the regulation, which requires banks to compute their mark-to-future cubes[2] already for the determination of their credit limits (through potential future exposures, i.e. maximum expected credit exposures over specified periods of time calculated at some quantile levels). Exposure-based approaches are also convenient for computing the XVA sensitivities that are required for XVA hedging purposes[3] (see [29, 33], or [11]). The most advanced (but also quite demanding) implementations are nested Monte Carlo strategies optimized with GPUs (see [1, 6]).

5.1 Capital/Collateral Optimisation of Inter-Dealer Trades

In this section, we consider a bilateral derivative market with banks and clients. Each bank is supposed to have a CDS curve which is used as its funding curve. Each bank also uses a certain hurdle rate for passing cost of capital to clients. We assume that clients hold a fixed portfolio of derivative trades with one or more of the banks and are indifferent to trades between dealers. We want to identify capital/collateral

[2]Prices of spanning instruments at future time points of different scenarios, from which expected exposure profiles are easily deduced.

[3]Hedging of spread risks, as jump-to-default risk can hardly be hedged.

optimisation inter-dealer trades that would be mutually beneficial[4] to the shareholders of two dealer counterparties and find a way to achieve a "Pareto optimal equilibrium", in the sense that there exists no additional mutually beneficial inter-dealer trade.

Considering a tentative inter-dealer trade, with contractually promised cash flows ρ, between a tentative "seller bank" a (meant to deliver the cash flows ρ) and a tentative "buyer bank" b (meant to receive ρ), we denote the corresponding FTPs as (cf. (23)):

$$\begin{aligned}\mathrm{FTP}_a(\rho) &= \Delta\mathrm{UCVA}_a(\rho) + \Delta\mathrm{FVA}_a(\rho) + \Delta\mathrm{MVA}_a(\rho) + \Delta\mathrm{KVA}_a(\rho)\\ \mathrm{FTP}_b(-\rho) &= \Delta\mathrm{UCVA}_b(-\rho) + \Delta\mathrm{FVA}_b(-\rho) + \Delta\mathrm{MVA}_b(-\rho) + \Delta\mathrm{KVA}_b(-\rho).\end{aligned}$$

Note that all incremental XVA terms are entity specific as they depend on the endowment of each dealer, i.e. of their current portfolio. The FVA and the MVA also depend on the entity specific funding spread. Finally, the KVA depends additionally on the target hurdle rates set by bank managements.

Consistent with Assumption 2.9 and Theorem 2.12, bank a would be happy to sell the contract at any price $\geq \mathrm{MtM} + \mathrm{FTP}_a(\rho)$, whereas bank b would be happy to buy it at any price $\leq \mathrm{MtM} - \mathrm{FTP}_b(-\rho)$. Hence the sales of the contract ρ from a to b may be a win-win provided

$$\mathrm{FTP}_a(\rho) + \mathrm{FTP}_b(-\rho) < 0. \tag{39}$$

However, allocated shareholder capital sets a constraint on trading. As a consequence of the new trade, shareholder capital at risk (cf. Sect. 2.5) of each bank $i = a, b$ changes by the amount $\Delta\mathrm{SCR}_i$. Hence the sales can only occur if

$$\mathrm{SCR}_i + \Delta\mathrm{SCR}_i \leq \mathrm{SHC}_i, \ \text{ for } i = a, b, \tag{40}$$

where SHC_i is the shareholder capital (at risk or uninvested) for bank i.

In conclusion:

Proposition 5.1 *The transaction whereby bank a agrees on delivering the additional future cash-flows ρ to bank b (contractually promised cash-flows, ignoring counterparty risk and its capital and funding consequences) can be a win-win for both parties if and only if (39) holds, subject to the constraint (40). In this case, if the transaction occurs at the intermediate price*

$$\begin{aligned}&\mathrm{MtM} + \mathrm{FTP}_a(\rho) - \frac{1}{2}\big(\mathrm{FTP}_a(\rho) + \mathrm{FTP}_b(-\rho)\big)\\ &\qquad = \mathrm{MtM} - \mathrm{FTP}_b(-\rho) + \frac{1}{2}\big(\mathrm{FTP}_a(\rho) + \mathrm{FTP}_b(-\rho)\big)\end{aligned}$$

paid by bank b to bank a, then the shareholders of both banks a and b mutually benefit of a positive net wealth transfer equal to

[4] In the sense of Assumption 2.9.

$$\Delta w = -\frac{1}{2}\big(\mathrm{FTP}_a(\rho) + \mathrm{FTP}_b(-\rho)\big) > 0. \tag{41}$$

5.2 Optimal Liquidation of the CCP Portfolio of a Defaulted Clearing Member

Another potential application of the XVA metrics is for dealing with default or distress resolutions. Specifically, we consider the problem of the liquidation of the CCP derivative portfolio of a defaulted clearing member, dubbed "defaulted portfolio" for brevity henceforth. Here CCP stands for a central counterparty (also called clearing house, see [5, 12, 27] for references). A CCP nets the contracts of each of its clearing members (typically broker arms of major banks) with all the other members and collects variation and initial margins in the same spirit as for bilateral trades, but at the netted portfolio level for each member. In addition, a CCP deals with extreme and systemic risk through an additional layer of protection, called default fund, contributed by and pooled among the clearing members. We denote by μ_i the (nonnegative) default fund refill allocation weights, which determine how much each surviving clearing member must contribute to the refill of the default fund in case the latter has been eroded by the default of a given member, e.g. μ_i proportional to the current default fund contributions (DFC) of the surviving members. We denote by $L^\star$ the negative of the sum between the margins and the default fund contribution of the defaulted member. We denote by MtM the mark-to-market[5] of the defaulted portfolio at the liquidation time where it is reallocated between the surviving members.

In a first stage, we assume that the defaulted portfolio is proposed as an indivisible package to the surviving members, which are just left with the freedom of proposing a price for global novation, i.e. replacing the defaulted member in all its future contractual obligations related to the defaulted portfolio.

One may then consider the following reallocation and pricing scheme. The CCP computes the FTPs of each surviving member corresponding to every other member receiving the portfolio of the defaulted member, i.e. their respective incremental XVA costs in each of these alternative scenarios.

The scenario giving rise to the lowest aggregated FTP, say member 1 receiver of the defaulted portfolio, is implemented. Accordingly, member 1 recovers the portfolio of the defaulted member and an MtM amount of cash from the CCP, whereas each surviving member (1 itself included) receives its corresponding FTP from the CCP, so that everybody is indifferent to the reallocation in MtM and XVA terms. The ensuing overall liquidation loss (or negative of the gain, if negative) is $L = L^\star + \mathrm{MtM} + \mathrm{FTP}_1$, where FTP_1 denotes the aggregated FTPs of the surviving members corresponding to this reallocation of the portfolio to member 1. In addition:

- If $L > 0$, meaning erosion by L of the default fund, then each surviving member (1 included) pays $L \times \mu_i$ to the CCP as refill to the default fund;

[5] Counted, sign-wise, as a debt to the other clearing members.

- If $L \leq 0$, then the CCP uses the residual $(-L)$ for, prioritarily, reimbursing any non-consumed DFC and IM (in this order) of the defaulted member to its liquidator and, if there remains a surplus after that, distributing it to the surviving clearing members proportional to their μ_i.

Let us now assume that the defaulted portfolio would instead be rewired with a surviving member corresponding to an aggregated FTP greater than FTP_1, say member 2, along with the required amounts of cash making each survivor indifferent to the reallocation in MtM and XVA terms. Then one must replace L by $L' = L^\star + \mathrm{MtM} + \mathrm{FTP}_2$ in the bullet points above, where FTP_2 is the aggregated FTP of all clearing members when member 2 receives the defaulted portfolio. As $\mathrm{FTP}_2 > \mathrm{FTP}_1$ by definition, therefore $L' > L$, implying that everybody would end up worse off than in the first alternative.

Last, we assume that, instead of being reallocated as an indivisible package, the defaulted portfolio is divided by the CCP into sub-packages reallocated one just after the other to possibly different clearing members. Iterating the above procedure, we denote by MtM^l the mark-to-markets and by FTP^l the recursively cheapest aggregated FTPs of the successive lots l (cheapest over consideration of the different scenarios regarding which survivor receives the lot l). The overall liquidation loss (or negative of the gain, if negative) becomes

$$L'' = L^\star + \sum_l (\mathrm{MtM}^l + \mathrm{FTP}^l) = L^\star + \mathrm{MtM} + \sum_l \mathrm{FTP}^l \leq L,$$

because of the multiple embedded optimizations (so that, in particular, $\sum_l \mathrm{FTP}^l \leq \mathrm{FTP}_1$). Hence, everybody would now end up better off than in the first alternative.

5.3 XVA Compression Cycles

In the context of Sect. 5.1, a solution to reach Pareto optimal equilibrium with no additional mutually beneficial inter-dealer trades would be to run iterative bidding cycles to discover trades that could be profitable to two banks. By running virtual bidding cycles, we can also answer other questions such as how to find a Pareto optimal reallocation of the portfolio of one given bank in case this portfolio is being liquidated (cf. Sect. 5.2). Another related application could be how to perform a partial liquidation in case a bank is in regulatory administration and the objective is to restore compliance with regulatory capital requirements.

Such systematic, market-wide bidding cycles would be very useful to optimise bilateral derivative markets by releasing costs for funding and capital into the dividend stream of participating broker dealers. A practical implementation problem, however, is that bilateral portfolios are held confidentially by each dealer and cannot be disclosed as an open bidding cycle would require: In order for a bidding cycle to be realistic and implementable in real life situations, it has to be designed in such a way that trade data at each bank is handled securely and not revealed to the other participants except when a mutually beneficial trade is identified and there is consensus on both sides to discover it.

Likewise, in the default resolution setup of Sect. 5.2, one should be careful that, in the course of the process, the CCP does not disclose to members any information, direct or indirect (i.e. via XVAs or incremental XVAs) relative to the portfolio of the other members.

A possible solution to this portfolio confidentiality problem could be as follows. Suppose that an XVA calculator is present within the firewall of each bank. This XVA calculator would be separate from the internal XVA calculator but hopefully will not deviate too much from it. Internal calculators, to avoid model risk, would be seeded with precisely the same calibrated models and guaranteed to produce identical results when loaded with identical portfolios under identical conditions and market data.

Suppose also that these calculators can share confidentially information with each other corresponding to each trade or set of trades which would be a candidate for novation. In this case, the existence of win-win trades would be discovered by the calculators without trade information itself being revealed.

Once the existence of a win-win trade such as (39)–(40) is detected by the XVA calculator, the two parties would then have to take the initiative to communicate with each other to exploit the opportunity if they so choose. Namely, the hypothetical seller would know which trade (or set of trades) would be worthwhile selling and to which peer. The hypothetical seller will be notified and it will be up to her to decide whether to start a conversation with the peer. In particular, before proceeding the seller has to verify that the calculator result on entry prices is acceptably accurate. If the seller then opens a communication channel with the peer, she will disclose the nature of the trades in question and ask the peer to verify a price on his own internal systems. If both internal systems agree that there is a trade opportunity, then the trade takes place.

In the future such largely automated XVA compression cycles could favorably replace the current XVA compression procedures that monopolize hundreds of quants twelve hours in a row in major tier 1 banks.[6]

Appendix: Connections with the Modigliani–Miller Theory

The Modigliani–Miller celebrated invariance result is in fact not one but several related propositions, developed in a series of papers going back to the seminal [39] paper. These propositions are different facets of the broad statement that the funding and capital structure policies of a firm are irrelevant to the profitability of its investment decisions. See e.g. [14, 36, 42] for various discussions and surveys. We

[6]Source: David Bachelier, panel discussion Capital & margin optimisation, Quantminds International 2018 conference, Lisbon, 16 May 2018.

emphasize that we do not need or use such result (or any negative form of it) in our paper, but there are interesting connections to it, which we develop in this section.

Modigliani–Miller Irrelevance, No Arbitrage, and Completeness

Modigliani–Miller (MM) irrelevance, as we put it for brevity hereafter, was initially understood by its authors as a pure arbitrage result. They even saw this understanding as their main contribution with respect to various precedents, notably [43]'s law of conservation of investment value (see Sect. 2.3). So, quoting the footnote page 271 of [39]:

> See, for example, Williams [21, esp. pp. 72–73]; David Durand [3]; and W. A. Morton [15]. None of these writers describe in any detail the mechanism which is supposed to keep the average cost of capital constant under changes in capital structure. They seem, however, to be visualizing the equilibrating mechanism in terms of switches by investors between stocks and bonds as the yields of each get out of line with their 'riskiness.' This is an argument quite different from the pure arbitrage mechanism underlying our proof, and the difference is crucial.

But, thirty years later, judging by the footnote page 99 in [36], the view of Miller on their result had evolved:

> "For other, and in some respects, more general proofs of our capital structure proposition, see among others, Stiglitz (1974) for a general equilibrium proof showing that individual wealth and consumption opportunities are unaffected by capital structures; See Hirshleifer (1965) and (1966) for a state preference, complete-markets proof; Duffie and Shafer (1986) for extensions to some cases of incomplete markets"

Non-arbitrage and completeness are intersecting but non-inclusive notions. Hence, implicitly, in Miller's own view, MM invariance does not hold in general in incomplete markets (even assuming no arbitrage opportunities). As a matter of fact, we can read page 197 of [28]:

> "When there are derivative securities and markets are incomplete the financial decisions of the firm have generally real effects"

and page 9 of [25]:

> "As to the effect of financial policy on shareholders, we point out that, generically, shareholders find the span of incomplete markets a binding constraint. This yields the obvious conclusion that shareholders are not indifferent to the financial policy of the firm if it can change the span of markets (which is typically the case in incomplete markets). We provide a trivial example of the impact of financial innovation by the firm. DeMarzo (1986) has gone beyond this and such earlier work as Stiglitz (1974), however, in showing that shareholders are indifferent to the trading of existing securities by firms. Anything the firm can do by trading securities, agents can undo by trading securities on their own account. Indeed, any change of security trading strategy by the firm can be accomodated within a new equilibrium that preserves consumption allocations. Hellwig (1981) distinguishes situations in which this is not the case, such as limited short sales."

Regarding MM irrelevance or not in incomplete markets (including some of the references that appear in the above quotations and other less closely related ones): [14, 31, 32, 38] deal with the impact of the default riskiness of the firm; [13, 37] discuss the special case of banks, notably from the angle of the bias introduced by government repayment guarantees for bank demand deposits; [22] tests empirically MM irrelevance for banks, concluding to MM offsets of the order of half what they should be if MM irrelevance would fully hold.

The XVA Case

A bit like with limited short sales in [32], a (seemingly overlooked) situation where shareholders may "find the span of incomplete markets a binding constraint" is when market completion or, at least, the kind of completion that would be required for MM invariance to hold, is legally forbidden. This may seem a narrow situation but it is precisely the XVA case, which is also at the crossing between market incompleteness and the presence of derivatives pointed out as the MM 'non irrelevance case' in [28]. The contra-assets and contra-liabilities that emerge endogenously from the impact of counterparty risk on the derivative portfolio of a bank (cf. Definition 2.7) cannot be "undone" by shareholders, because jump-to-default risk cannot be replicated by a bank: This is practically impossible in the case of contra-assets, for lack of available or sufficiently liquid hedging instruments (such as CDS contracts with rapidly varying notional on corporate names that would be required for replicating CVA exposures at client defaults); It is even more problematic in the case of contra-liabilities, because a bank cannot sell CDS protection on itself (this is forbidden by law) and it has a limited ability in buying back its own debt (as, despite the few somehow provocative statements in [37], a bank is an intrinsically leveraged entity).

As a consequence, MM irrelevance is expected to break down in the XVA setup. In fact, as seen in the main body in the paper, cost of funding and cost of capital are material to banks and need be reflected in entry prices for ensuring shareholder indifference to the trades.

More precisely, the XVA setup is a case where a firm's valuation is invariant to funding strategies and, still, investment decisions are not. The point here is a bit subtle. Saying that the value of a company is independent of financing strategies does not imply that investment decisions do not depend on financing strategies. There two numbers we can look at: the value of the equity E and the sum of the value of equity and debt, $E + D$. Equity holders will naturally seek to optimize E and will accept an investment opportunity if ΔE is positive. Williams' law implies that equity plus debt, $E + D$, stays invariant under a certain financial transaction. But this does not imply in general that shareholders are indifferent to the transaction: Shareholders are indifferent if $\Delta E = 0$, not if $\Delta(E + D) = 0$. To go from Williams' wealth conservation law to MM irrelevance, we have to assume complete markets or, at least, the availability of certain trades to shareholders. Namely, assuming shareholders can and do change financing strategy, then, even if we start with $\Delta E < 0$ for a given transaction (but

$\Delta(\mathrm{E}+\mathrm{D})=0$), we may conclude that equity shareholders are actually indifferent as there exists a change in financing strategy for which $\Delta\mathrm{E}=0$. However, in the XVA case, the bank cannot freely buy back its own debt, so such a change is not possible and only Williams' wealth conservation law remains, whereas MM irrelevance breaks down: See Sect. 4 for illustration in a pedagogical static setup.

References

1. Abbas-Turki, L., Diallo, B., Crépey, S.: XVA principles, nested Monte Carlo strategies, and GPU optimizations. Int. J. Theor. Appl. Financ. **21**, 1850030 (2018)
2. Albanese, C., Andersen, L.: Accounting for OTC derivatives: funding adjustments and the re-hypothecation option. Working paper (2014). https://papers.ssrn.com/sol3/papers.cfm?abstract_id=2482955
3. Albanese, C., Andersen, L.: FVA: what's wrong, and how to fix it. Risk Mag. 54–56 (2015)
4. Albanese, C., Andersen, L., Iabichino, S.: FVA: accounting and risk management. Risk Mag. 64–68 (2015). Long preprint version available as https://doi.org/10.2139/ssrn.2517301
5. Albanese, C., Armenti, Y., Crépey, S.: XVA Metrics for CCP Optimisation. Statistics & Risk Modeling. (preprint on https://math.maths.univ-evry.fr/crepey) (2020)
6. Albanese, C., Bellaj, T., Gimonet, G., Pietronero, G.: Coherent global market simulations and securitization measures for counterparty credit risk. Quant. Financ. **11**(1), 1–20 (2011)
7. Albanese, C., Brigo, D., Oertel, F.: Restructuring counterparty credit risk. Int. J. Theor. Appl. Financ. **16**(2), 1350010 (29 p) (2013)
8. Albanese, C., Caenazzo, S., Crépey, S.: Credit, funding, margin, and capital valuation adjustments for bilateral portfolios. Probab. Uncertain. Quant. Risk **2**(7), 26 (2017)
9. Albanese, C., Crépey, S.: Capital valuation adjustment and funding valuation adjustment. Working paper (2019). https://math.maths.univ-evry.fr/crepey. (First, very preliminary version: arXiv:1603.03012 and ssrn.2745909, March 2016)
10. Albanese, C., Crépey, S., Hoskinson, R., Saadeddine, B.: XVA analysis from the balance sheet. Working Paper (2019). https://math.maths.univ-evry.fr/crepey
11. Antonov, A., Issakov, S., McClelland, A., Mechkov, S.: Pathwise XVA Greeks for early-exercise products. Risk Magazine (January) (2018)
12. Armenti, Y., Crépey, S.: Central clearing valuation adjustment. SIAM J. Financ. Math. **8**(1), 274–313 (2017)
13. Balling, M.: Modigliani–Miller, Basel 3 and CRD 4. SUERF Policy Notes **2**(1), 1–8 (2015)
14. Baron, D.: Default risk and the Modigliani–Miller theorem: a synthesis. Am. Econ. Rev. **66**(1), 204–212 (1976)
15. Basel Committee on Banking Supervision: Consultative document: fundamental review of the trading book: a revised market risk framework (2013). http://www.bis.org/publ/bcbs265.pdf
16. Basel Committee on Banking Supervision: Consultative document: application of own credit risk adjustments to derivatives (2012)
17. Bichuch, M., Capponi, A., Sturm, S.: Arbitrage-free XVA. Math. Financ. **28**(2), 582–620 (2018)
18. Brigo, D., Pallavicini, A.: Nonlinear consistent valuation of CCP cleared or CSA bilateral trades with initial margins under credit, funding and wrong-way risks. J. Financ. Eng. **1**, 1–60 (2014)
19. Burgard, C., Kjaer, M.: In the balance. Risk Magazine, pp. 72–75 (2011)
20. Burgard, C., Kjaer, M.: Funding costs, funding strategies. Risk Magazine, pp. 82–87. Preprint version (2013). https://ssrn.com/abstract=2027195
21. Burgard, C., Kjaer, M.: Derivatives funding, netting and accounting. Risk Magazine, March 100–104. Preprint version (2017). https://ssrn.com/abstract=2534011

22. Cline, W.: Testing the Modigliani–Miller theorem of capital structure irrelevance for banks. Peterson Institute for International Economics Working Paper Series (2015)
23. Committee of European Insurance and Occupational Pensions Supervisors. QIS5 technical specifications (2010). https://eiopa.europa.eu/Publications/QIS/QIS5-technical_specifications_20100706.pdf
24. Crépey, S., Sabbagh, W., Song, S.: When capital is a funding source: the XVA anticipated BSDEs. SIAM J. Financ Math. Forthcoming (2020)
25. Duffie, D., Sharer, W.: Equilibrium and the role of the firm in incomplete market. Stanford University, Working Paper No. 915 (1986). https://www.gsb.stanford.edu/faculty-research/working-papers/equilibrium-role-firm-incomplete-markets
26. Eisele, K.-T., Artzner, P.: Multiperiod insurance supervision: top-down models. Eur. Actuar. J. **1**(1), 107–130 (2011)
27. European Parliament: Regulation (EU) no 648/2012 of the European parliament and of the council of 4 July 2012 on OTC derivatives, central counterparties and trade repositories (2012). http://eur-lex.europa.eu/LexUriServ/LexUriServ.do?uri=OJ:L:2012:201:0001:0059:EN:PDF
28. Gottardi, P.: An analysis of the conditions for the validity of Modigliani–Miller Theorem with incomplete markets. Econ. Theory **5**, 191–207 (1995)
29. Green, A., Kenyon, C.: Efficient XVA management: pricing, hedging, and allocation using trade-level regression and global conditioning (2014). arXiv:1412.5332v2
30. Green, A., Kenyon, C., Dennis, C.: KVA: capital valuation adjustment by replication. Risk Magazine, pp. 82–87 (2014). Preprint version KVA: capital valuation adjustment available at https://doi.org/10.2139/ssrn.2400324
31. Hagen, K.: Default risk, homemade leverage, and the Modigliani–Miller theorem: note. Am. Econ. Rev. **66**(1), 199–203 (1976)
32. Hellwig, M.: Bankruptcy, limited liability, and the Modigliani–Miller theorem. Am. Econ. Rev. **71**(1), 155–170 (1981)
33. Huge, B., Savine, A.: LSM Reloaded—differentiate xVA on your iPad Mini (2017). https://doi.org/10.2139/ssrn.2966155
34. International Financial Reporting Standards: IFRS 13 fair value measurement (2012). http://www.iasplus.com/en/news/2011/May/news6909
35. International Financial Reporting Standards: IFRS 4 insurance contracts exposure draft (2013)
36. Miller, M.: The Modigliani–Miller propositions after thirty years. J. Econ. Perspect. **2**(4), 99–120 (1988)
37. Miller, M.: Do the M & M propositions apply to banks? J. Bank. Financ. **19**, 483–489 (1995)
38. Milne, F.: Choice over asset economies: default risk and corporate leverage. J. Financ. Econ. **2**(2), 165–185 (1975)
39. Modigliani, F., Miller, M.: The cost of capital, corporation finance and the theory of investment. Econ. Rev. **48**, 261–297 (1958)
40. Salzmann, R., Wüthrich, M.: Cost-of-capital margin for a general insurance liability runoff. ASTIN Bull. **40**(2), 415–451 (2010)
41. Swiss Federal Office of Private Insurance: Technical document on the Swiss solvency test (2006). https://www.finma.ch/FinmaArchiv/bpv/download/e/SST_techDok_061002_E_wo_Li_20070118.pdf
42. Villamil, A.: The Modigliani–Miller theorem. In: The New Palgrave Dictionary of Economics (2008). http://www.econ.uiuc.edu/~avillami/course-files/PalgraveRev_ModiglianiMiller_Villamil.pdf
43. Williams, J.B.: The Theory of Investment Value. Harvard University Press, Cambridge (1938)
44. Wüthrich, M., Merz, M.: Financial Modeling, Actuarial Valuation and Solvency in Insurance. Springer, Berlin (2013)

CPSIA information can be obtained
at www.ICGtesting.com
Printed in the USA
LVHW062317250320
651244LV00001B/9

9 789811 515750